FOOD SAFETY

Researching the Hazard in Hazardous Foods

ADVANCES IN HOSPITALITY AND TOURISM BOOK SERIES

BOOKS IN THE SERIES:

Food Safety: Researching the Hazard in Hazardous Foods
Editors: Barbara Almanza, PhD, RD, and Richard Ghiselli, PhD

Strategic Winery Tourism and Management: Building Competitive Winery Tourism and Winery Management Strategy
Editor: Kyuho Lee, PhD

Sustainability, Social Responsibility and Innovations in the Hospitality Industry
Editor: H. G. Parsa, PhD
Consulting Editor: Vivaja "Vi" Narapareddy, PhD
Associate Editors: SooCheong (Shawn) Jang, PhD, Marival Segarra-Oña, PhD, and Rachel J. C. Chen, PhD, CHE

Managing Sustainability in the Hospitality and Tourism Industry: Paradigms and Directions for the Future
Editor: Vinnie Jauhari, PhD

Management Science in Hospitality and Tourism: Theory, Practice, and Applications
Editors: Muzaffer Uysal, PhD, Zvi Schwartz, PhD, and Ercan Sirakaya-Turk, PhD

FOOD SAFETY

Researching the Hazard in Hazardous Foods

Edited by

Barbara Almanza, PhD, RD, and Richard Ghiselli, PhD

Apple Academic Press

TORONTO NEW JERSEY

Apple Academic Press Inc.
3333 Mistwell Crescent
Oakville, ON L6L 0A2
Canada

Apple Academic Press Inc.
9 Spinnaker Way
Waretown, NJ 08758
USA

Printed in the United States of America on acid-free paper

International Standard Book Number-13: 978-1-926895-70-3 (Hardcover)

Library of Congress Control Number: 2014930948

Library and Archives Canada Cataloguing in Publication

Food safety: researching the hazard in hazardous foods /edited by Barbara Almanza, PhD, RD, and Richard Ghiselli, PhD.

Includes bibliographical references and index.
ISBN 978-1-926895-70-3 (bound)

1. Food industry and trade--Safety measures--Research--United States. 2. Food industry and trade--Safety regulations--United States. 3. Hospitality industry--Safety measures--Research--United States. 4. Food adulteration and inspection--United States. 5. Food contamination--Research--United States. I. Almanza, Barbara A., author, editor of compilation II. Ghiselli, Richard, author, editor of compilation

TX531.R48 2014 363.19'20973 C2014-900308-0

Apple Academic Press also publishes its books in a variety of electronic formats. Some content that appears in print may not be available in electronic format. For information about Apple Academic Press products, visit our website at **www.appleacademicpress.com** and the CRC Press website at **www.crcpress.com**

DEDICATION

We would like to dedicate this book to the advancement of research in the hospitality and tourism industries. Food safety and sanitation are critical concerns to these industries. This effort reflects those concerns. Operators cannot risk the safety or health of their customers and must design systems and processes to minimize the risk.

We would also like to dedicate this collection to the MS and PhD students who will be generating future research studies. Hopefully they can use these contributions as a springboard.

Lastly, we would like to recognize the encouragement and support that we receive from our institutions and families. They allow us to engage in activities that we consider quite enjoyable. Our special thanks to Gil, Jacob, Matthew, Corina, and Carol, Carl, Eric, Olivia, and Brian.

— Barbara Almanza, PhD, RD, and Richard Ghiselli, PhD

CONTENTS

ABOUT THE EDITORS

Barbara Almanza, PhD, RD

Barbara Almanza, PhD, RD, has become nationally recognized for her research on health issues related to the hospitality industry, particularly in the area of restaurant food safety. With 25 years of teaching about food safety and more than 50 research, text, technical, conference, and trade publications in food safety, she has contributed to a better understanding of the foodservice inspection process, how that information is shared, and its impact on consumers. Because of ongoing concerns regarding the safety of the global food supply, her research has been beneficial to the hospitality industry and consumers through its value in advocating better food handling practices and minimizing the risks associated with hazards in food.

Richard Ghiselli, PhD

Richard Ghiselli, PhD, is a professor and head of the School of Hospitality and Tourism Management at Purdue University. Previously he was the Director of the School of Hotel and Restaurant Administration at Oklahoma State University. Dr. Ghiselli has 20 years of teaching and research in the hospitality industry. In addition to his experience in education, he has ten years of industry experience as a club manager, food service director, and general manager. Has also spent time as a chef and has a degree from the Culinary Institute of America. His research in food safety and the environmental impact of the foodservice industry has included a strong focus on hospitality industry applications and use of research information to improve operation.

LIST OF CONTRIBUTORS

Barbara A. Almanza
Professor, Purdue University, West Lafayette, IN 47907, United States

Carl Behnke
Assistant Professor, Purdue University, West Lafayette, IN 47907, United States

Huey Chern Boo
Associate Professor, Universiti Putra Malaysia, Serdang, Selangor, Darul Ehsan, 43400, Malaysia

Wei Leong Chan
Lecturer, YTL-International College of Hotel Management, Kuala Lumpur, Malaysia

Jin-Kyung Choi
Assistant Professor, Woosong University, Daejeon, Korea

Robin B. DiPietro
Associate Professor, University of South Carolina, Columbia, SC 29208, United States

Jeff Fisher
Purdue University, West Lafayette, IN 47907, United States

Richard Ghiselli
Professor, Purdue University, West Lafayette, IN 47907, United States

A. Scott Gilliam
Director, Food Protection Program, Indiana State Department of Health, Indianapolis, IN 46204, United States

Brenda H. Halbrook
Director, Office of Food Safety, USDA Food and Nutrition Service, Washington, D.C.

Sheryl F. Kline
Professor, University of Delaware, Newark, DE 19716, United States

Jim Mann
Executive Director, Handwashing For Life® Institute, Libertyville, IL 60048, United States

Jack "Jay" Neal
Assistant Professor, University of Houston, Houston, TX 77004, United States

Haeik Park
Purdue University, West Lafayette, IN 47907, United States

Kevin R. Roberts
Associate Professor, Kansas State University, Manhattan, KS 66506, United States

Kevin Sauer
Associate Professor, Kansas State University, Manhattan, KS 66506, United States

Sarah Slette
Enteric Epidemiologist, Indiana State Department of Health, Indianapolis, IN 46204, United States

Jeannie Sneed
Professor, Kansas State University, Manhattan, KS 66506, United States

Soobin Seo
Assistant Professor, The Ohio State University, Columbus, OH 43210, United States

Dennis Stearns
Professor from Practice, Seattle University School of Law, of Counsel, Marler Clark, LLP, PC, Seattle, WA 98122, United States

Mary Stiker
Food Defense Coordinator, Indiana State Department of Health, Indianapolis, IN 46204, United States

Sandra Sydnor
Assistant Professor, Purdue University, West Lafayette, IN 47907, United States

Lionel Thomas
Assistant Professor, North Carolina Central University, Durham, NC 27707, United States

Ameet Tyrewala
Professor, Algonquin College, Ottawa, Ontario, Canada

Kelly Way
Associate Professor, University of Arkansas, Fayetteville, AR 72701, United States

PREFACE

After much litigation of foodborne illness lawsuits for 20 years, food has become a "contact sport," much like football or, for the rest of the world, soccer. Whether it is dining at home or out, the thought of how the food might well poison you is never far from my mind. Years ago, I asked my long suffering spouse why we seemed never to be invited over to friends' homes for dinner after hosting more than a few overcooked meals at our home. She simply said, "you make them nervous about food."

When *Escherichia coli* O157:H7 crashed into the food industry's awareness during the Jack-in-the-Box outbreak of 1992/1993 (no one had paid attention to the McDonald's *E. coli* outbreak a decade earlier), hamburgers began to take on an ominous air at the Marler household as opposed to the "Happy Meal." After I saw Brianne Kiner in the hospital shortly after she had come out of a coma—she was still on dialysis, had endured surgery to have her large intestine removed and had suffered multiple seizures—never was a burger served at our home. My kids were not allowed to eat them while their friends did. As you can well imagine, there were more than a few odd discussions with parents about meal preparation for sleepovers. But, to see a child's life forever changed, or ended, because they consumed a hamburger does change your perspective on what is considered safe—especially for your kids.

In the decades since, more food was either taken off the shopping list or if consumed it was with an unnatural gulp. *E. coli* found its way into sprouts, juice, lettuce, spinach and even cookie dough. *Salmonella* stayed on chicken (it's allowed to be there by odd USDA/FSIS decisions). *Salmonella* slipped into peanut butter, mangoes, tomatoes (or was that peppers?) and even pot pies. *Listeria* continued to be a pest in deli meats and cheeses, and expanded its deadly toll to cantaloupe. After 20 years of this buffet, it is easy to see why food began to look less like something to be enjoyed, and more something to be wary of.

The decades have not been without food safety successes. During the height of the yet another summer *E. coli* outbreak linked to hamburger in 2002, I penned an opinion piece for the *Denver Post* entitled, "Put me out of business." I banked on the animus that lawyers have—slightly below used car salesmen, yet comfortably above members of Congress—to convince the beef industry that I was making too much money off its failure to get ahold of the deadly pathogen. The beef and restaurant industries responded (likely more to the fact that *E. coli* was listed as an adulterant by the USDA/FSIS and to requirements for increased cook temperatures), and my firm's *E. coli* income linked to hamburger dropped from 90% to nearly zero today. That's a success by anyone's measure.

There remain challenges to a safer table as this book clearly lays out. With 48,000,000 fellow citizens sickened each year, 125,000 hospitalized and 3,000 deaths linked to food consumption, and with an increasing population of those facing some form of compromised immune system, the farm-to-fork continuum continues to be confronted with persistent and emerging risks.

This book is a smorgasbord of the challenges facing the production of food on an ever-increasing global scale. However, this book also sets forth practical solutions to many of what seems like intractable problems facing the hospitality industry.

It is true that foodborne illness has been with us from the beginning and will continue to remain a challenge for an ever-increasing population. But, what this book makes clear is there are people and institutions ignoring that reality to try to prevent a next Brianne, and to make food not something to be feared, but savored.

— William D. Marler, Esq.
Marler Clark, LLP PS
The Food Safety Law Firm, Seattle, Washington

LIST OF ABBREVIATIONS

AMC	Active Managerial Control
AMS	Agricultural Marketing Service
ANSI	American National Standards Institute
APHA	American Public Health Association
CBRN	Chemical, Biological, Radiological and Nuclear
CCFS	Certified in Comprehensive Food Safety
CDC	Centers for Disease Control and Prevention
CERC	Crisis and Emergency Risk Communication
CFP	Conference for Food Protection
CFS	Certified Food Scientist
CFSAN	Center for Food Safety and Applied Nutrition
CFSAN	The Center for Food Safety and Applied Nutrition
CP-FS	Certified Professional-Food Safety
DHHS	The Department of Health and Human Services
DVFA	Danish Veterinary and Food Administration
EPA	The Environmental Protection Agency
EPIA	Egg Products Inspection Act
FDA	The Food and Drug Administration
FIFO	First In, First Out
FMIA	Federal Meat Inspection Act
FOOD	Foodborne Outbreak Online Database
FSA	Farm Service Administration
FSIS	The Food Safety and Inspection Service
FSMA	Food Safety Modernization Act
GAPs	Good Agricultural Practices
GRAS	Generally Recognized As Safe
HACCP	Hazard Analysis and Critical Control Point
HAI	Healthcare Acquired Infections
HUS	Hemolytic Uremic Syndrome
HW/EH	Handwashes/Employee Hour
IAFP	International Association for Food Protection
IFT	Institute of Food Technologists
ISDH	Indiana State Department of Health
MOW	Meals on Wheels
MRSA	Methicillin Resistant *Staphylococcus aureus*

NACMCF	National Advisory Committee on Microbiological Criteria for Food
NCFPD	National Center for Food Protection and Defense
NEHA	National Environmental Health Association
NFSMI	National Food Service Management Institute
NMFS	The National Marine Fisheries Service
NOAA	National Oceanic and Atmospheric Administration
NRA	National Restaurant Association
NRAEF	National Restaurant Association Educational Foundation
NRW	North Rhine-Westphalia
OECD	Organization for Economic Cooperation and Development
PAH	Polycyclic Aromatic Hydrocarbons
PHF	Potentially Hazardous Foods
PIC	Person-In-Charge
QSCV	Quality, Service, Cleanliness and Value
QSR	Quick Service Restaurants
REHS/RS	Registered Environmental Health Specialist/Registered Sanitarian
RFID	Radio Frequency Identification
RTE	Ready-To-Eat
SCCT	Situational Crisis Communication Strategy
SNS	Social Network Sites
SSOPs	Sanitation Standard Operating Procedures
STEC	Shiga Toxin *Escherichia coli*
TFE	Temporary Foodservice Establishments
USDA	The United States Department of Agriculture
USDC	U.S. Department of Commerce
VSP	Vessel Sanitation Program
WHO	World Health Organization
WOM	Word-of-Mouth

PART 1

BACKGROUND INFORMATION

CHAPTER 1

IMPORTANCE OF FOOD SAFETY IN RESTAURANTS

RICHARD GHISELLI, PhD

Professor, Purdue University

CONTENTS

1.1 THE RESTAURANT INDUSTRY

There are an estimated 980,000 restaurants in the U.S. What is more, total restaurant and foodservice sales are expected to approach $660.5 billion in 2013 (NRA, December 11, 2012). This level of sales represents an increase of 3.8% over 2012 – slightly below the average annual restaurant growth rate of 3.92% from 2005 through 2012. Figure 1 shows restaurant-industry sales from 2005 through 2012, and then forecasts them out to 2020 using a modest rate of 3.0% annually. At this level they will reach $800 billion by the end of the second decade – more than double the sales at the start of the century.

Restaurants in the traditional sense are just one facet of the industry. The restaurant industry encompasses all meals and snacks prepared away from home, including all takeout meals and beverages (NRA Forecast, 2013). Accordingly the restaurant-industry includes eating places, bars and taverns, managed services, lodging places, retail, vending, recreation and mobile services, noncommercial restaurant services, and military restaurant services.

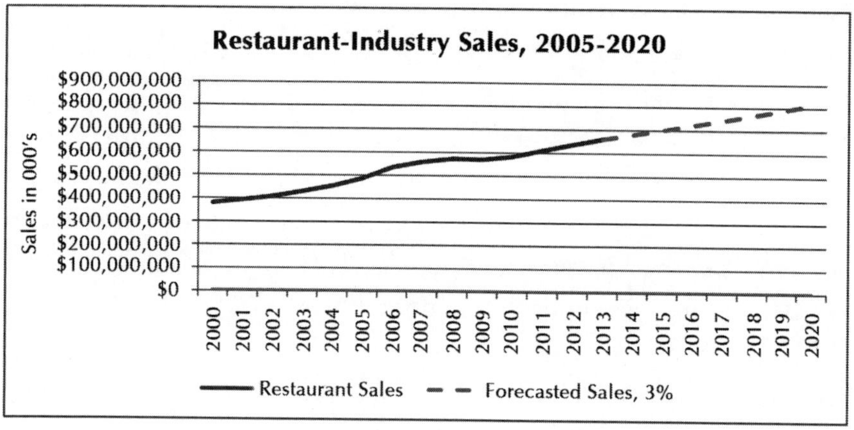

FIGURE 1 Restaurant Industry Sales Forecast to 2020 assuming a growth rate of 3.0% (after 2013).

An increase in the population growth has fueled some of the growth. Sales have also grown as the proportion of the food dollar spent on food away-from-home has increased. Specifically, 26.2% of total food expenditures were away-from-home in 1960; by 2009, the amount was 48.6% (USDA Economic Research Service Food CPI and Expenditures). Figure 2 shows US per capita food expenditures for food away-from-home from 1980–2009.

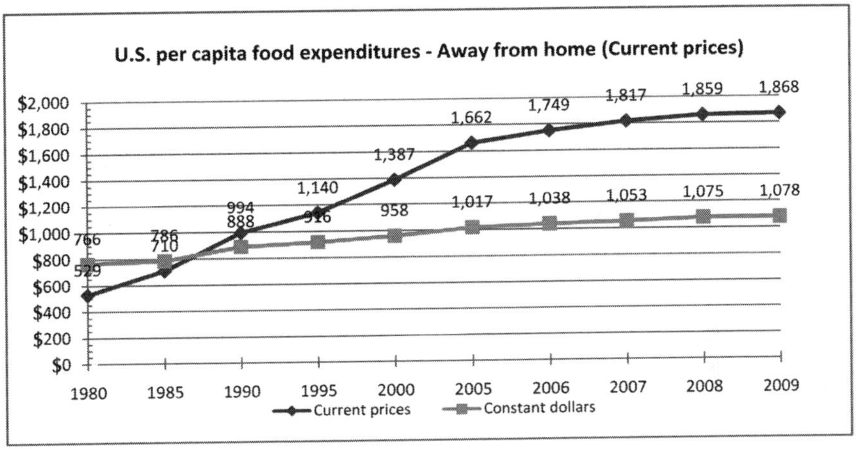

FIGURE 2 U.S. per capita food expenditures away from home in current and constant dollars.

As might be expected there is some variation in the number of meals eaten at restaurants by age. For example, those between the 25 and 34 years old consume, on average, 250 meals at restaurants annually whereas those less than 18 and more than 65 consume about 160 (*see* Figure 3).

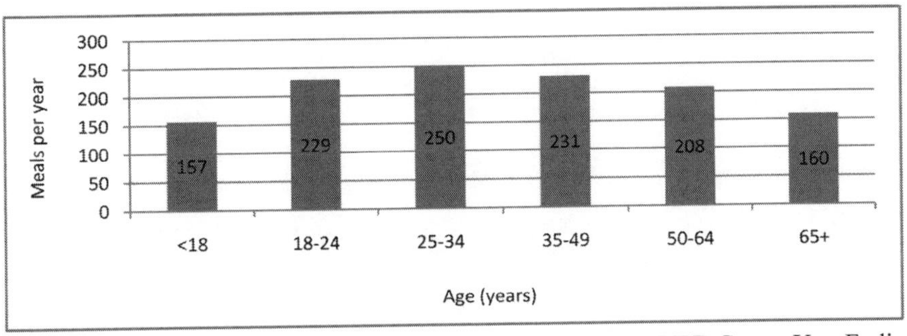

FIGURE 3 Annual restaurant meals per person by age. Source: NPD Group, Year Ending May 2009.

1.2 WHY SANITATION IS IMPORTANT

The size and scope of the industry are compelling reasons why sanitation and safe food-handling practices are important. But there are several other reasons as well. First of all sanitation is a legal requirement. Health inspectors are supposed to inspect restaurants on a regular basis to make sure local regulations are being followed.

Another reason is that sanitation may help prevent a food poisoning outbreak. Most of the food-related illnesses that are caused by restaurants are the result of unsanitary food handling practices. A third is that maintaining safe and sanitary conditions can help preserve food quality. Storing, preparing or serving food in unsanitary conditions could adversely affect the food quality and taste. Also, sanitation protects the brand name. Having just one case of food poisoning can destroy the business, so training employees to follow safe food handling practices is a necessity. Keeping things clean and sanitary will gain the trust of customers and employees.

As indicated the size and scope of the industry are daunting reasons. The National Restaurant Association (2011) estimates that nearly half of all U.S. adults have worked in the restaurant industry. Currently there are 12.9 million employees in the restaurant-industry workforce. Furthermore, job opportunities will continue to be available as the industry expects to add 1.3 million positions in the next 10 years. The job outlook for selected food service occupations is shown in Table 1.

Given the sheer number of employees and the transitory nature of the business, systems and processes that foster safe food-handling practices are essential. A further consideration is the size of most restaurants; 93% of eating-and-drinking place businesses have fewer than 50 employees (2011), and many are independents. These factors can affect the amount and quality of training provided to food service employees.

TABLE 1 Job Outlook for Food Service Occupations from 2010 through 2020.

Occupation	Job Outlook, 2010–2020*
Bartenders	9%
Chefs and head cooks	–1%
Cooks	8%

TABLE 1 *(Continued)*

Occupation	Job Outlook, 2010–2020*
Food and beverage serving and related workers	12%
Food preparation workers	10%
Waiters and waitresses	9%
Food Service Managers	–3%

*The projected rate of change in employment for the 10-year timeframe between 2010 and 2020. The projected rate of change in employment for all occupations between 2010 and 2020 is 14.3 percent. **Bureau of** Labor Statistics, U.S. Department of Labor, Occupational Outlook Handbook, 2012–2013 Edition, Food Service Managers, on the Internet at http://www.bls.gov/ooh/management/food-service-managers.htm (visited October 26, 2012).

1.3 CONSUMER PERCEPTIONS

There can be little doubt that consumers are concerned with the wholesomeness of the food they purchase and eat. This presents unique challenges for restaurants and other food-preparation facilities since consumers may not be privy to the food-handling practices. As a result they may associate safe and wholesome with cleanliness. Quality might be considered in the same way. In fact Quality, Service, Cleanliness and Value (QSCV) are the four pillars upon which McDonald's built their success.

Cleanliness appears to be fundamental. In a study of consumers 50 years and older, cleanliness was the most important reason when evaluating dining options: the cleanliness of kitchens in restaurants, the cleanliness of dining areas and tables, and the cleanliness of rest rooms in restaurants were the highest rated factors (Ghiselli, Li, and Almanza, 2012). The next most important criteria were the taste and quality of the food and the quality of service.

In another study dirty dishware, restrooms and odor were the top three reasons adults 18 years and older would never return to a restaurant (Cintas Corporation, 2010). Table 2 lists the reasons identified for not returning to a particular restaurant.

TABLE 2 Reasons for not returning to a particular restaurant.

Reason	% of Respondents
Unclean Dishware	86%
Unpleasant Odors	85%
Dirty Restrooms	75%
Unclean Tables	74%
Poor Service	74%
Dirty Floors	68%
Staff Appearance	65%

Cintas Corporation, 2010.

This survey of more than 1,000 also discovered that 86% of U.S. adults equate the cleanliness of a restaurant's restroom with the cleanliness of its kitchen.

How do customers define cleanliness? In a study of consumer perceptions seven factors were found to explain restaurant cleanliness (Yoo, and Seung Ah, 2012). They are listed in Table 3 with the items that delineate them. Understanding consumer perceptions will assist food service operators in their efforts to present safe and wholesome environs.

TABLE 3 Factors that Define Customer Perceptions of Restaurant Cleanliness.

Factor	Dimensions
Restaurant Interior Appearance	Tablecloths
	Windows or windowsills
	Open kitchen
	Floor and carpet
	Food contact surface
Server's Appearance	Uniform
	Accessories
	Hair style
	Hands and nails
Restroom Personal Hygiene	No soap
	No hot water
	No paper towels or drying device

TABLE 3 *(Continued)*

Factor	Dimensions
Restroom Appearance	Odor
	Floors
	Trash
Server's Behavior	Smoking
	Coughing and sneezing
	Bare hand contact
Food Condition	Freshness
	Temperature
	Presentation
Signage	Employee hand washing signage
	Restaurant inspection score posted
	Vermin

Yoo, Seung Ah. (2012). Customer Perceptions of Restaurant Cleanliness: A Cross Cultural Study. Master of Science Thesis. Virginia Polytechnic Institute and State University. http://scholar.lib.vt.edu/theses/available/etd-07232012–152526/unrestricted/Yoo_SA_T_2012.pdf

1.4 FOODBORNE ILLNESS

The Centers for Disease Control and Prevention define foodborne illness as a (public) health problem resulting from consuming contaminated foods or beverages. There are a number of disease-causing microbes or pathogens that can contaminate foods. Further, poisonous chemicals or other harmful substances can cause foodborne diseases if they are present in food. Foodborne illness is sometimes called "foodborne disease," "foodborne infection," or "food poisoning."

FoodSafety.gov, an official U.S. government web site managed by the U.S. Department of Health and Human Services, defines a food illness outbreak in the following manner:

> *When two or more people get the same illness from the same contaminated food or drink, the event is called a foodborne outbreak.*

Public health officials investigate outbreaks to control them, so more people do not get sick in the outbreak, and to learn how to prevent similar outbreaks from happening in the future.

Foodborne illness is a major health concern in the U.S. The Centers for Disease Control and Prevention estimates that 1 in 6 Americans gets sick from foodborne diseases each year; this amounts to 48 million people (CDC). Further, the CDC estimates that 128,000 are hospitalized, and 3,000 die of food borne illness annually (Bottemiller, January 3, 2012).

While estimates have varied over the years, a recent calculation indicates that the annual cost of foodborne illness is $77 billion (Bottemiller, January 3, 2012). This includes economic estimates for medical costs, a measure for pain, suffering, and functional disability, and illness-related death. It does not include costs associated with reduced consumer confidence, recall losses, or litigation, nor does it included the cost to public health agencies that respond to illnesses and outbreaks.

In order to assist food control authorities at various levels of government, the U.S. Food and Drug Administration publishes the *Food Code.* The *Food Code* is a "scientifically sound technical and legal" model upon which to develop food safety rules to be consistent with national policy. The most recent version is the 2009 Edition. The *Food Code* lists the government's recommendations for foodservice regulations. The FDA recommends its adoption – but each state decides whether to adopt the *Food Cod*e (or some modified form thereof). Most food regulations affecting restaurant and foodservice operations are written at the state level.

Other government agencies involved in regulating food and food safety include:

1. The United States Department of Agriculture (USDA). Food Safety is among the USDA's mission areas – to ensure that the nation's commercial supply of meat, poultry, and egg products is safe, wholesome, and properly labeled, and packaged. It is responsible for inspection and quality grading of meat, meat products, poultry, dairy products, egg and egg products, and fruit and vegetables shipped across state lines.
2. The Food Safety and Inspection Service (FSIS) is the public health agency in the U.S. Department of Agriculture responsible for ensuring that the nation's commercial supply of meat, poultry, and egg products is safe, wholesome, and correctly labeled and packaged.

3. The Food and Drug Administration (FDA). The FDA is responsible for protecting the public health by assuring that foods are safe, wholesome, sanitary and properly labeled. It issues the *FDA Food Code* jointly with the USDA and Centers for Disease Control and Prevention. The FDA Food Code represents the FDA's advice for a uniform system.
4. The Center for Food Safety and Applied Nutrition (CFSAN) works to assure that the food supply is safe, sanitary, wholesome, and honestly labeled (FDA). The Center's primary responsibilities as they relate to FOOD include:
   - the safety of substances added to food, for example, food additives (including ionizing radiation) and color additives;
   - the safety of foods and ingredients developed through biotechnology;
   - seafood and juice Hazard Analysis and Critical Control Point (HACCP) regulations;
   - regulatory and research programs to address health risks associated with foodborne, chemical, and biological contaminants;
   - regulations and activities dealing with the proper labeling of foods (e.g., ingredients, nutrition health claims);
   - regulations and policy governing the safety of dietary supplements, infant formulas, and medical foods;
   - food industry postmarket surveillance and compliance;
   - industry outreach and consumer education;
   - cooperative programs with state, local, and tribal governments; and
   - international food standard and safety harmonization efforts.
5. The Centers for Disease Control and Prevention (CDC). The following are among the responsibilities of the CDC as they pertain to ensuring safe and wholesome food:
   - Investigate outbreaks of foodborne illness; study the causes and control of disease;
   - provide educational services in the field of sanitation; and
   - manage the Vessel Sanitation Program – an inspection program for cruise ships.

6. The Environmental Protection Agency (EPA). The EPA sets air and water standards, regulates the use of pesticides (including sanitizers), and regulates the handling of waste.
7. The National Marine Fisheries Service (NMFS). The NMFS is an arm of the Department of Commerce, and operates a voluntary inspection program for fish processing operations.

A number of microorganisms or agents can lead to illness or infection *via* food systems; these include bacteria, viruses, parasites, mold, and chemical agents. The leading causes of most reported foodborne illnesses are bacteria and viruses (FoodSafety.gov). According to the CDC, 51% of the foodborne illness outbreaks in the years 2006–2008 were caused by Norovirus (CDC, http://www.cdc.gov/norovirus/index.html). Bacteria were the cause in 38% of the foodborne illness outbreaks. Chemicals were next at 6%. Parasites and Other were the causes in the remaining 5%.

Microbial food poisonings or foodborne illnesses mainly fall into one of two categories. The first is food infection, where the microorganism itself grows inside the body and is the source of symptoms. The second is food intoxication, where a chemical or natural toxin (often produced as a by-product of bacteria – known as an exotoxin – present in the food) causes the symptoms or illness. Most bacterial food poisonings are actually food infections.

The Centers for Disease Control has indicated there are eight pathogens that account for the vast majority of illnesses, hospitalizations and deaths (EHE Newsletter, October 18, 2012). They are identified in Table 4. Also included are foods that are likely sources of contamination and some of the practices that will reduce the risk of pathogenic development. Of particular concern in foodservice are the following: *Salmonella*, *Campylobacter jejuni, Clostridium perfringens*, *Staphylococcus aureus*, *Escherichia coli*, *Listeria monocytogenes*, and *Norovirus* (FoodSarety.gov).

TABLE 4 Major Bacterial / Viral Concerns in Food Preparation and Delivery Systems.

Bacteria or toxin:							
Salmonella	***Campylobacter***	***Clostridium perfringens***	***Staphylococ-cus aureus***	***Toxoplasma gondii***	***Escherichia coli (E. coli)***	***Listeria monocyto-genes***	***Norovirus***
Illness:							
Salmonellosis	Campylobac – teriosis	Clostridium Perfringens gastroenteritis	Staphylococcus gastroenteritis	Toxoplasmosis	Diarrheagenic E. coli	Listeriosis	Norovirus gastroen-teritis
Infection/intoxication:							
Infection	Infection	Toxin-medi-ated infection	Intoxication	Infection	Infection or intoxication	Infection	Infection
Food Products or Source of Bacteria:							
Meat products Poultry and eggs Produces	Raw Poultry	Meat products Leftovers	Deli meats Egg Products	Meat products	Ground beef Contaminated produce	Raw meat Ready-to-eat food	Shellfish Ready-to-eat-foods

TABLE 4 *(Continued)*

Control measures:							
Thorough cooking Avoid cross contamination Good personal hygiene	Thorough cooking Good personal hygiene Avoid cross contamination	Thorough cooking Reheat foods to 165°F Hold cooked foods > 140°F	Proper cooling Keep foods below 40°F or above 140°F Good personal hygiene	Cook food to safe temperatures Avoid cross contamination	Thorough cooking and reheating Proper refrigeration Employee hygiene	Control time and temperature Observe processors directions Thorough cooking	Good personal hygiene Avoid barehand contact with ready-to-eat foods Use approved suppliers
Other:							
Bacteria killed in a few seconds at 165°F for poultry Eggs should be cooked to 155°F		One of the most common causes of foodborne illness in the U.S.	"Staph" Toxin is heat resistant		Heat leftover red meats to 165°F Heat ground meats to 155°F		Norwalk virus

Developed from information provided by CDC, FSIS, Gisslen, and Educational Foundation.

1.5 ARE RESTAURANTS TO BLAME?

Oftentimes restaurants are blamed for serving less than wholesome food. In particular, they have been identified more often than other food preparation sites as the source of food which has led to (foodborne) illness; grocery stores and homes are other frequently mentioned sites (Fein, Jordan-Lin, and Levy, 1995). More recently, respondents were twice as likely to say it is very common to get sick from food prepared at restaurants compared with food prepared at home (Fein et al., 2011) (*see* Table 5). Despite these accounts researchers have suggested that most foodborne illnesses are caused by foods prepared at home – not food from restaurants (Fein, Jordan-Lin, and Levy, 1995). One reason for the discrepancy may be that individuals are less likely to report making themselves sick; another may be they are less likely to associate an illness that results from contaminated food at home with the food.

TABLE 5 Risk Perception of Food Prepared at Home and at Restaurants by Year.

				% of Respondents	
	1993	**1998**	**2001**	**2006**	**2010**
Risk perceptions					
How common to get sick from foods prepared at home (% very common)	13.0	17.6	17.8	14.6	14.5
How common to get sick from food prepared at restaurants (% very common)	29.6	37.1	35.5	28.8	29.1
Seriousness of germs in food (% serious)	38.3	56.6	53.4	31.1	35.7

Fein, Sara B.; Lando, Amy M.; Levy, Alan S.; Teisl, Mario F.; Noblet, Caroline T. (2011, Sept.). Trends in U.S. Consumers' Safe Handling and Consumption of Food and Their Risk Perceptions, 1988 through 2010. Journal of Food Protection, Volume 74, Number 9, pp. 1513–1523 (11).

The CDC has developed the *Foodborne Outbreak Online Database* (FOOD) "*to make Foodborne Disease Outbreak Surveillance System data more available to the public and stakeholders.*" The database is an annual listing of foodborne disease outbreaks in the United States. It contains national information about single-state and multistate outbreaks including total illnesses, hospitalizations, deaths, reported food vehicle, pathogen species and serotype, etiologic status, and location of consumption (Centers for Disease Control and Prevention). In 2010, restaurants were identified more often than any other venue – including private homes. Close to 25% of the illnesses, however, were not associated with any particular location (Other). Table 6 indicates the frequency of foodborne outbreaks by location for 2010.

TABLE 6 Foodborne Outbreaks in the U.S. by Location of Consumption, 2010.

Location of Consumption	n	%
Banquet Facility	51	6.1%
Camp	5	0.6%
Caterer	16	1.9%
Church	19	2.3%
Fair/Festival	5	0.6%
Grocery Store	4	0.5%
Hospital/Nursing Home	8	1.0%
Picnic	8	1.0%
Prison	11	1.3%
Private Home	133	16.0%
Restaurant	299	36.0%
School	17	2.0%
Workplace	49	5.9%
Other	206	24.8%
Total	831	

Centers for Disease Control and Prevention (CDC). Foodborne Outbreak Online Database. Atlanta, Georgia: U.S. Department of Health and Human Services. Available from URL: http://wwwn.cdc.gov/foodborneoutbreaks. Accessed 11/10/2012.

Outbreaks are the major way we learn what foods are causing illness and how to prevent it. One of the difficulties in attributing illness to food sources is that outbreaks account for a small proportion of the total number of foodborne illnesses. Another is that more than half of foods reported are complex in that foods involved can be grouped into more than one commodity – for example, a casserole. Also, many outbreak investigations don't implicate a single food because of delays in reporting, or the outbreak is small. Further, not all pathogens cause foodborne illness outbreaks – for example, more than 60 million people in the U.S. may have *Toxoplasma gondii*. Table 7 shows the frequency of disease outbreaks by "source attribution."

TABLE 7 Foodborne disease outbreaks by type of food vehicle implicated, 1998–2008 (N = 13,405 outbreaks).

	%
Simple foods	24
Complex Foods	34
Unknown food	42

Source: Dana Cole, DVM, PhD. 2012, *Estimating the Sources of Foodborne Illness in the United States.* PowerPoint presentation to Enteric Disease Epidemiology Branch, Division of Foodborne, Waterborne, and Environmental Diseases. January 31, 2012.

The most frequently implicated (contaminated) foods include meat, poultry, seafood, leafy greens, dairy products and fruits-nuts. Other foods that have been involved include dairy products, vine vegetables, and eggs. The following chart from the CDC shows the relative source of illness in 1,565 single food commodity outbreaks during 2003–2008 (Figure 4).

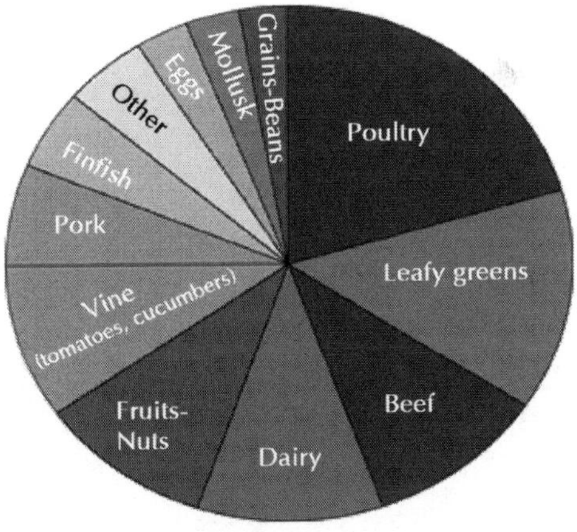

FIGURE 4 Causes of illness in 1,565 single food commodity outbreaks, 2003–2008. Source: Centers for Disease Control and Prevention. *Estimates of Foodborne Illness in the United States*. http://www.cdc.gov/foodborneburden/cdc-and-food-safety.html

The most commonly reported risky food consumption practices are eating raw eggs followed by eating raw fish and undercooked hamburgers. Consumption of raw fish has greatly increased since 1993; in fact by 2010 eating raw fish was more common than eating undercooked hamburger (Fein et al., 2011) (Table 8).

TABLE 8 Percentage of Respondents that Indicated Risky Food Consumption Behavior by Year.

				% of respondents		
	1988	**1993**	**1998**	**2001**	**2006**	**2010**
Risky consumption						
Eat undercooked hamburger	28.7	24.1	16.1	16.3	19.0	19.2
Eat raw oysters	16.3	10.0	12.1	11.2	12.3	
Eat raw clams	8.0	5.2	6.2	5.6	5.4	
Eat raw fish	8.4	9.4	15.2	17.9	25.6	
Eat raw eggs	53.6	39.1	41.6	37.8	41.3	
Eat steak tartar	5.4	3.6	3.4	5.0	5.9	

Fein, Sara B.; Lando, Amy M.; Levy, Alan S.; Teisl, Mario F.; Noblet, Caroline T. (September 2011). Trends in U.S. Consumers' Safe Handling and Consumption of Food and Their Risk Perceptions, 1988 through 2010. Journal of Food Protection, Volume 74, Number 9, pp. 1513–1523 (11).

1.6 RECENT FOOD SAFETY OUTBREAKS

The Hunter Public Relations' 9th Annual Survey identified the cantaloupe listeria outbreak as the top food story of 2011. Recall, cantaloupe from Jensen Farms, Colorado was identified as the source of *Listeria monocytogenes* infections (listeriosis) in August 2011. In total 33 deaths from outbreak-associated cases of listeriosis were reported to the Centers for Disease Control and Prevention. There were more than 140 persons infected in 28 states. Apparently workers had walked through pools of water where listeria was present and tracked the bacteria into the packing facility.

Later that year eating at a Mexican-style fast food restaurant chain was associated with some illnesses. A total of 68 individuals from 10 states were infected with a strain of *Salmonella* Enteriditis. An epidemiologic study comparing foods eaten by 48 ill and 103 well persons indicated that ill persons (62%) were significantly more likely than well persons (17%) to report eating at the chain in the week before illness. Among ill persons eating at the chain, 90% reported eating lettuce, 94% reported eating ground beef, 77% reported eating cheese, and 35% reported eating tomatoes. The epidemic curve was consistent with those observed in past produce-related outbreaks. The investigation indicated that contamination likely occurred before the product reached the restaurants.

There have been a number of outbreaks in 2012. One was an outbreak of *Salmonella* associated with peanut butter. Thirty-five persons in 19 states were infected; 63% were children under 10 years old. FDA testing found *Salmonella* bacteria in raw peanuts from a peanut processing facility. As a result a recall was conducted to include raw and roasted shelled and in-shell peanuts processed in the facility (and the products using them).

Another multistate outbreak in 2012 involved prepackaged leafy greens. A total of 33 persons were infected with Shiga toxin-producing *Escherichia coli* 0157:H7. Traceback investigations identified State Garden of Chelsea, Massachusetts as a common producer, but a source of contamination was not identified. Wegmans, a major regional supermarket chain, voluntarily recalled its 5-ounce and 11-ounce clam shell packages of Organic Spinach and Spring Mix blend, a product produced by State Garden.

1.7 HACCP

In recent years public health officials have placed greater emphasis on reducing and/or eliminating potentially unsafe practices and conditions that may endanger consumers – especially in away-from-home situations. The misgivings that many people have about food prepared away from home should point out the need to food service operators to vigorously embrace safe food handling practices and to actively market their efforts. One approach has been HACCP (Hazard Analysis and Critical Control Point).

HACCP involves incorporating procedures that attempt to minimize the opportunities for contamination and the growth of pathogenic microorganisms. The approach is preventative in nature rather than corrective, and its purpose is to reduce the number of critical violations (and situations) that cause foodborne illness. Also, HACCP provides a record of safety and sanitation on an on-going basis. The seven basic steps of the HACCP system include:

1. performing a hazards assessment;
2. identifying critical control points;
3. establishing critical limits;
4. establishing monitoring requirements;
5. establishing corrective actions;
6. establishing effective record keeping systems; and
7. establishing a verification system.

The following procedures are usually performed to accomplish this:

1. recipes are reviewed to identify potentially dangerous foods – flow diagrams are generally used to trace potentially dangerous foods through an operation.
2. preparation procedures are reviewed to identify potential sources of contamination in the process.
3. preparation is observed to verify that procedures are being followed.
4. testing is done to determine that preparation procedures are safe – this includes taking temperature measurements.
5. preparation is monitored during service
   i. to observe hot holding practices.
   ii. to observe *a la minute* preparation practices.
6. Critical Control Points are identified.
7. Implementation and monitoring.

1.8 SETTING HACCP PRIORITIES

HACCP is a valuable "tool" in efforts to ensure food safety, but setting it up takes time, money, and knowledge about the specific foodservice system. Inexperienced users of HACCP may be puzzled at the sizable number of areas where HACCP controls could be set up. Ideally, a HACCP system defines only CRITICAL points in the safe handling of food. Because of this, foodservice operators should set priorities in handling the most potentially hazardous foods (*see* Figure 4).

Some criteria for setting HACCP priorities are food related, some are foodservice-related, some are labor-related and some are customer-related.

HACCP priorities are not based on quality concerns, but instead on safety concerns – although these two may be linked. In addition, HACCP priorities should be set for foods where no other "system controls" are in place. For example, if the standard is to color code ground beef as it is received, stored, and thawed, the use of ground beef that is out of date is not very likely making this less of a priority.

Analysis of HACCP should be done for each individual operation. Priorities may differ based on an operation's product types, equipment, employees, and service style. For instance, handling and cooking raw meat products is clearly not a concern where only precooked, portion controlled products, which require simple reheating prior to service, are purchased.

1.8.1 CRITERIA FOR SETTING HACCP PRIORITIES

1. Potentially hazardous foods with the least amount of preservative treatment are higher on the priority list than those receiving no treatment. Preservative treatments include temperature control, additives and packaging. Examples include: fresh ground beef (higher priority) vs. frozen ground beef, round steak (higher priority) vs. hot dogs, unpackaged beef roasts (higher priority) vs. cryovac packaged beef. This criterion is food-related.
2. Foods that are more likely to be contaminated at the source are higher on the priority list; examples include poultry (*Salmonella*), ground meats (*E. coli*), and eggs (*Salmonella*). This criterion is also food-related.
3. Foods which are handled more during their preparation are higher on the priority list; examples include stuffed meats and breaded products. This criterion is labor-related.
4. Foods which "naturally" have a shorter recommended shelf life because of their chemical composition are also higher on the priority list. Examples include ground meats (particularly pork) and fish. This criterion is food-related.
5. Foods which have any holding period prior to service, particularly where they will be transported outside the foodservice are higher on the priority list because temperature control must be more closely monitored. Particularly important in this category are foods

that are being catered for a special event outside the foodservice as well as foods, which are picked up by customers for carry-out service. This criterion is foodservice-related.

6. Foods that are being eaten by more vulnerable groups of people may require stricter standards in HACCP, although foods should be handled in all foodservices so that they are safe to eat. Examples of foodservices where more vulnerable groups of people are served include hospitals, nursing homes and other health care foodservices, schools, and daycares. This criterion is customer-related.

1.9 IDENTIFYING CRITICAL CONTROL POINTS IN FOOD SYSTEMS

Critical Control points can be defined as instances where a "… loss of control may result in (an) unacceptable health risk" (FDA, 1995). In essence they are the points in a system where intervention could prevent or reduce the probability that a food will become contaminated or adulterated and, as a result, decrease the chance of foodborne illness. In foodservice settings these include the food handling practices that, if controlled, could reduce the possibility that consumers will receive potentially harmful food.

Studies have found that keeping potentially hazardous foods at room temperature for prolonged periods of time has contributed most often to foodborne illness. Other practices that have frequently led to or been implicated in outbreaks of foodborne illness include: inadequate cooking; improper holding of hot food; improper cooling of food; an extended length of time between preparation and consumption; inadequate reheating of food previously cooked and cooled; improper cleaning of equipment; cross contamination; and infected food handlers. By preventing or controlling these situations, the probability that food will become contaminated or adulterated should be reduced. HACCP, in essence, involves monitoring these critical points or practices.

While there are controls that will be applicable to all food service systems, critical control points are to a certain extent system specific. That is, operational demands will dictate which criteria are applicable. For example, operations that prepare all food items to order, and do not prepare or serve any foods that must be held or stored will have different concerns than one that holds food in a steam table for long periods of time, or one that precooks and then chills food for service at a later time or date. In general food systems can be classified into the following four types (Spears, 2012):

1. Conventional Food Service
   a. foods are purchased in various stages of preparation.
   b. production, distribution, and service are completed on site.
   c. meals are usually served without extensive holding.
2. Commissary Food Service – centralized production area with remote satellites.
3. Ready Prepared Food Service
   a. food items are produced and held chilled or frozen until heated for service (cook-chill or cook-freeze).
   b. Two stages – initial cooking and final heating before service.
4. Assembly Serve Food Service.
   a. minimal cooking concept using completely processed foods.
   b. only storage, assembly, heating and service are required.

1.10 CRITICAL LIMITS

Critical limits are measures or criteria in food production and delivery systems that must be reached or satisfied in order to reduce the possibility of serving contaminated food. Unsafe food handling practices that have led to outbreaks of foodborne disease help define these limits in that they reveal the faults or oversights that might allow harmful microorganisms to live and grow. The preventative measures or preparation criteria that address these lapses will minimize the risks to consumers.

The limits themselves are often defined by regulatory agencies. One of the agencies prescribing limits is the Food and Drug Administration, US Department of Health and Human Services, Public Health Service; this agency's recommendations are detailed in the *Food Code.* Other agencies charged with protecting the public include state and local health departments. In Indiana, for example, the Indiana State Department of Health (ISDH) – through the local health departments – monitors food service operations; critical limits as defined by ISDH are enumerated in the manual *Food Service Sanitation Requirements*. For example, Indiana requires that:

> *Potentially hazardous foods that have been cooked and then refrigerated shall be reheated rapidly to 165°F, or higher throughout before being served or before being placed in a hot food storage facility. (ISDH, Sec. 5(f)).*

Another source of critical limits is the National Restaurant Association. The NRA offers a National Food Safety Certification Program for food service personnel, SERVSAFE[â]; food safety control measures are detailed in the certification Coursebook, *Applied Foodservice Sanitation*.

To help ensure safe food throughout, critical limits should be established for the flow of foods throughout the system. The flow of food in a *Conventional Food Service System* (Figure 5) usually includes the following steps:

FIGURE 5 Flow of Foods in a Conventional Food Service System.

1. In a number of situations food is cooked and then cooled and stored for later use. When that is the case there are additional concerns. The flow of food in a *Ready-Prepared Food Service System* would be more like that shown in Figure 6.

FIGURE 6 Flow of Foods in a Ready-Prepared Food Service System.

2. To help ensure safe food throughout, critical limits should be identified in the flow of food. Table 9 is listed with some of the critical concerns and/or limits at each phase in the flow of food in a food system.

TABLE 9 Critical Concerns and/or Limits in the Flow of Food through a Food Service System.

Flow of Food	*Critical Concerns and/or Limits*
Purchasing	Suppliers/vendors should be approved and reputable
Receiving	Frozen products should be frozen upon arrival Refrigerated products should be 40°F or below on arrival Damaged products should be inspected for possible contamination Refrigerated and frozen products should be put away within 30 min of delivery Refrigerated and frozen products should be dated and rotated Use by dates should be current
Storage	All food should be labeled and dated Proper rotation practices – First In, First Out (FIFO) Frozen food products should be kept frozen Refrigerated food products should be kept at 40°F or below Dry storage areas should be clean and dry, and the temperature between 50°F and 70°F
Preparation	Hands should be kept clean Working surfaces and utensils should be clean and sanitized Acceptable methods of thawing frozen food products should be used Fruits and vegetables should be washed thoroughly before preparation Foods should not sit out at room temperature for extended periods of time
Cooking	Minimum internal cooking temperatures should be established
Holding	Holding temperatures and times for hot and cold foods should be identified Food should be protected from contamination – e.g., Sneeze guards
Cooling	Cooling processes should be established Foods should be cooled down to 40°F within 6 h (from 135°F)
Reheating	Potentially hazardous foods that have been cooked and then refrigerated should be reheated rapidly to 165°F, or higher
Service	Proper hand washing techniques and good personal hygiene should be practiced Bare-hand contact with ready-to-eat foods should be minimized Proper food temperatures should be maintained Service ware should be handled properly

3. Other practices and processes that could affect the wholesomeness of the food should also be monitored. One of those "other prac-

tices" is dishwashing. The limits depend upon the type of machine involved; see Table 10.

TABLE 10 Critical Concerns and/or Limits with Dishwashing.

Type of machine	*Critical Concerns and/or Limits*
Manual Dishwashing	1. The wash cycle should be at least 110°F 2. Final sanitizing hot water should be at least 171°F
High –Temperature	Final sanitizing rinse should be at least 165°F (stationary rack, single temperature machine) Final sanitizing rinse should be at least 180°F
Chemical-Sanitizing	Temperature guidelines should be followed

Source: SERVSAFE[ª] Coursebook, 5th ed. (2008). National Restaurant Association.

These limits and processes should be incorporated into the operating procedures. Moreover, to the extent that the menu defines the operational processes, reviewing the recipes is an important step. When designing a HAACP system the recipes should be reviewed and ingredients traced from storage to service. This is facilitated with flow charts or diagrams that follow potentially dangerous foods through the system to the customer. The following chart maps a relatively simple menu item – a BLT – from storage to service. Steps where there may be some cause for concern can be identified. In this case a critical control point might be where the mayonnaise is held in a cold table (Figure 7).

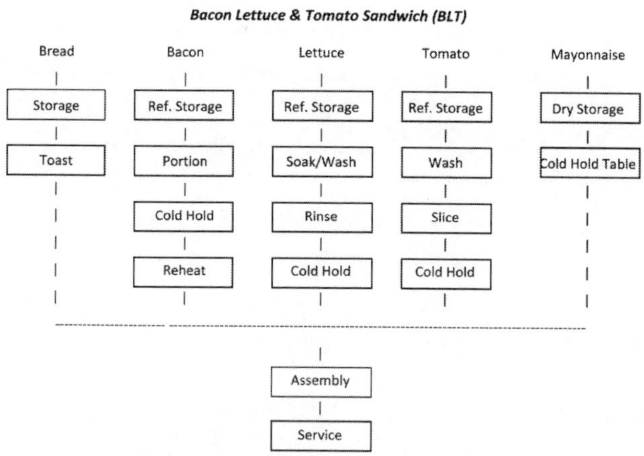

FIGURE 7 Flow of Ingredients for a BLT.

The same approach holds for more complex menu items. After reviewing the flow the next step is to write recipes that identify the steps or points in the process where intervention can reduce the risk of serving contaminated food. Consider the recipe for *New England Clam Chowder*. Suppose the procedure in which the chowder is usually handled in the operation consists of the stages given in Figure 8.

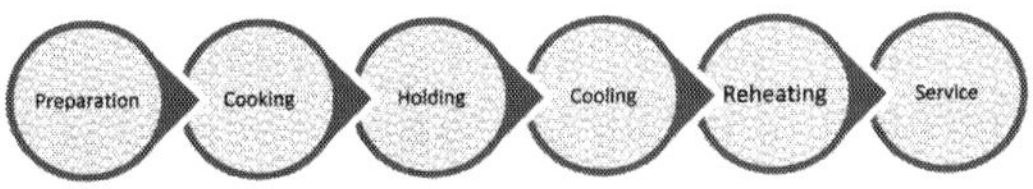

FIGURE 8. "HAACPized" Recipe for New England Clam Chowder.

New England Clam Chowder

Ingredients	**Amount**
Canned minced clams	2 qt.
Water	1 ½ qt
Salt pork, fine dice	10 oz.
Onions, small dice	1 lb.
Flour	4 oz.
Potatoes, small dice	2 lb.
Milk, hot	2 ½ qt.
Heavy Cream	1 cup
Salt, White pepper	to taste

Preparation

1. Drain the clams.
2. Dice potatoes.
3. Dice onions.
4. Mince salt pork.

Cooking

5. Combine the clam juice and the water in a saucepan. Bring to a boil.
6. Remove from the heat and keep the liquid hot (140°F) for step 10.
7. In a heavy saucepan or stock pot, render the salt pork over medium heat.
8. Add the onions and cook slowly until they are soft, but do not brown.

9. Add the flour and stir to make a roux. Cook the roux slowly for 3–4 min, but do not let it brown.

10. Using a wire whip, slowly stir the clam liquid and water into the roux. Bring to a boil, stirring constantly to make sure the liquid is smooth.

CCP

11. Add the potatoes. Simmer until tender. (Temperature must be at least 140°F.)

12. Stir in the clams and the hot milk and cream. Heat gently but do not boil.

13. Season to taste with salt and white pepper.

14. Finish cooking to a product temperature of 165°F or higher for at least 15 sec.

Holding

15. Hold chowder at 140°F or higher for service. Do not mix new with old.

Cooling

16. Cool in shallow pans with a product depth not to exceed 2 inches. Cool from 140°F to 70°F within 2 h and from 70°F to 40°F within an additional 4 h – for a total cooling time of 6 h.

17. Store in refrigerated unit at a product temperature of 40°F or lower. Cover.

Reheating

18. Reheat chowder to a product temperature of 165°F or higher for at least 15 sec within 2 h.

Standard Operating Procedures: Measure all internal product temperatures with a cleaned and sanitized thermometer. Properly wash hands and exposed parts of arms before handling food, after handling raw food, and after any interruption that might contaminate hands. Wash, rinse, and sanitize all equipment and utensils before and after use. Return all ingredients to refrigerated storage if preparation is interrupted.

From Gisslen, *Professional Cooking* (6th ed.), p. 240, and Educational Foundation, *A Practical Approach to HACCP Coursebook*, pp. 78–79.

In this case, the soup is prepared, cooled, and reheated at service. At each stage there are steps and processes that need to be identified and controlled – and in the recipe the Critical Control Points are identified, and desired food-handling practices incorporated (*see* Figure 8).

1.11 HACCP CHECKLISTS/LOGS

In conjunction with a review of the recipes, checklists/logs that monitor the Critical Control Points in the flow of food throughout the operation should be developed; other practices that are associated with or could affect food safety could also be included. The checklists should be designed after reviewing standard operating procedures as detailed in the company's operation manual, and after discussing practices and procedures with both managerial and nonmanagerial employees.

On a daily basis the operation should be reviewed by management at regular intervals – conceivably every 4 h. Since one of the objectives is to develop a practicable and workable plan, the scheme should be reviewed and modified after a trial period. Station checklists should be designed to reduce the time requirements on management; also, they should be designed to include monitoring activities that can be performed by hourly employees in the normal course of their duties – with the notion that eventually these activities will be incorporated into their job descriptions. Station checklists can be kept on clipboards in the areas or stations where the activities are performed. In order to assign accountability and resolve any potential questions, both managers and hourly employees should be required to initial their records and actions on the checklists. The following are among the records that should be maintained:

1. Receiving Log
2. Refrigeration Log
3. Food Safety Checklist
4. Cooking and Reheating Temperature Log
5. Cooling Temperature Log
6. Damaged or Discarded Product Log.

The checklists that are developed will be specific to the operation. Figure 9 is an example of a management checklist that monitors certain aspects of a foodservice operation that a manager could complete on a walk-through. It focuses mostly on temperature, but includes other items that encourage food safety and/or might be included on a health inspector's inventory.

FOOD SAFETY CHECKLIST

Date ____________________ Observer ____________________

Directions: Use this checklist daily to determine areas in your operations requiring corrective action. Record corrective action taken and keep completed records in a notebook for future reference.

Mgmt-1. Receiving & Storage

Cold Storage:	7:30 AM	11:30 AM	3:30 PM	7:30PM
Temperature of walk-in freezer	____°F	____°F	____°F	____°F
Temperature of walk-in refrigerator	____°F	____°F	____°F	____°F
Temperature of shake base freezer	____°F	____°F	____°F	____°F
Raw meats stored separately or on separate shelves or on lower shelves (Circle)	Y / N	Y / N	Y / N	

Mgmt-2. Back-of-House Preparation Area

	7:30 AM	11:30 AM	3:30 PM	7:30PM
Temperature of Chili	____°F	____°F	____°F	____°F
Temperature of Soup, identify ________	____°F	____°F	____°F	____°F
Temperature of water in steam table	____°F	____°F	____°F	____°F
Cutting board for potentially hazardous food clean (Circle)	Y / N	Y / N	Y / N	Y / N
Effective sanitizing solution available	Y / N	Y / N	Y / N	Y / N
Soap available at hand washing sink	Y / N	Y / N	Y / N	Y / N

Mgmt-3. Grill Area

	7:30 AM	11:30 AM	3:30 PM	7:30PM
Temp. inside service refrigerator under cold table	____°F	____°F	____°F	____°F
Temp. of mayonnaise on cold table	____°F	____°F	____°F	____°F
Cold table above refrigerator covered (Circle)	Y / N	Y / N	Y / N	Y / N
Sandwich boards cleaned and sanitized	Y / N	Y / N	Y / N	Y / N

Notes: ______________________________

FIGURE 9 Sample Food Safety Checklist.

Both managerial and station checklists will more than likely include monitoring temperatures. Although they may not be CCPs, indirect measures should also be included because they could influence the safety of the food. For example, monitoring/inspecting the cleanliness of equipment could affect food safety. Because of these considerations, the final checklists may include monitoring some situations and/or items that may not be critical in and of themselves, such as:

- the cleanliness of cutting boards and equipment;
- the availability of effective sanitizing solution;
- the availability of soap and paper towels at hand washing stations;
- the placement of raw meat products relative to other foods.

1.12 MONITORING – COMPLIANCE/CONFORMITY

Implementing HACCP requires a bit of time, effort, and paperwork. In most cases the information collected will be for a specific point in time – and corrective action can be taken. As a next step the information can be compiled; this enables management to monitor the overall system. For example suppose there are six cold holding units in a restaurant (as indicated in Table 11). At week's end (or other appropriate period) the number of times a unit is out of conformity with prescribed standards can be calculated. Refrigerators would be considered out of compliance if the temperature was above 40°F, and freezers if the temperature was more than 10°F above freezing or 10°F.

TABLE 11 Frequency That Cold Holding Units Were Out of Compliance With Standards.*

	Areas					
	Storage/ receiving	**Griddle area**	**Sandwich prep**	**Drive up ____**	**Cold food prep**	**Fountain area**
Walk-in refrigerator	26.1%					
Walk-in freezer	2.2%					
Service refrigerator	9.0%	8.8%	20.0%			
Service ref. under cold table		14.7%	23.5%			
Pie/salad refrigerator			46.6%			
Service freezer		98.1%		35.6%		
Milk compartment			0%			

* Notes:
Refrigerators are considered out of compliance if the temperature is above 40°F.
Freezers if the temperature is more than 10°F.

Viewed in this way problem areas become apparent. In this case, the service refrigerator in the griddle area and the service refrigerator in the drive up were out-of-compliance less than 10% of the time, whereas the walk-in refrigerator and the service refrigerator under the cold table in the drive up were out-of-compliance a quarter of the time. In addition to refrigerator temperatures, food products often held in cold tables above refrigerated units – like mayonnaise – should be monitored. The frequency that these products are not in accordance with the standards can be tallied, and an "out-of-compliance" table prepared.

Besides cold foods, management should also be concerned about the number of times food products are being held at temperatures that might be considered too hot. While wholesomeness may not be a problem at temperatures greater than 175°F, quality may be an issue since products deteriorate more quickly. Consumer safety may also be a consideration.

1.12.1 TIME AND COST CONSIDERATIONS

In a study of the time and cost to implement and use HACCP checklists in a quick-service restaurant, researchers found that it would have cost $6,697 per store (Almanza and Ghiselli, 1998). As a percentage of system-wide sales, this would have represented less than 1% (0.72%). Based on the average, the length of time it took to complete the checklists was close to 2 h of management time in an operation that was open 24 h per day. At first this may be cause for concern – but if more than one purpose could be addressed in the process, it might be more palatable. That is, going through the checklist would be more (cost) effective if two or more ends could be accomplished at the same time. A logical secondary purpose would be to inspect the quality of the food; seemingly this is a natural accompaniment since management is checking each station anyway.

1.13 FINAL THOUGHTS

The nature of the service industry – in particular the commercial foodservice sector – does not always allow systems and processes to be implemented in the way they were designed. Commonly resources, especially labor, must be shifted to where there is the greatest or most immediate need. In many cases management must support other employees; often times they must fill-in for them. Because of this situation, management may be limited in the amount of time and energy it can devote to any one task. Nonetheless, management cannot risk the safety or health of consumers. The proper balance between competing commitments and responsibilities must be found. Spending time monitoring the safety of the food is essential in any business where customers are a top priority.

Recognizing that the overall purpose of a HACCP system is to monitor and correct critical situations or processes that may directly or immediately result in an unacceptable health risk, there are nevertheless many circumstances and/or practices that could (indirectly) affect the safety and wholesomeness of food products. For example, improper cleaning of equipment has been implicated in some outbreaks of foodborne illness. Since management may not be able to observe firsthand every employees' actions, merely looking at a piece of equipment may not be enough to determine whether it was cleaned and sanitized or just wiped off. That is, "checking off" an item on a checklist that asks whether the equipment has been cleaned and sanitized may not be enough. Suppose, however, that employees were required to keep effective sanitizing solution in their areas. While this measure is no guarantee that it will be used, and management still may not be able to determine whether or not the equipment was cleaned and sanitized, the situation has been framed so that employees may be more likely to use the sanitizing solution – and with it the likelihood of serving safe and wholesome food. Since management cannot be in all CRITICAL places at all CRITICAL times, the question becomes whether these "tangential" practices or measures can be operative cues for employees and suitable checkpoints for a HACCP system?

1.14 THINGS TO COME

The NPD Group, a market research firm, found that concern levels about food safety for 2012 were not remarkably different from those in previous years. In particular approximately 60 percent of U.S. consumers indicated they were somewhat or slightly concerned about the safety of the American food supply, 25 percent were extremely or very concerned and 15 percent were not concerned at all. (NPD Group, Oct. 2012).

"How concerned are you about the safety of the U.S. food supply?"
% of adults who are...

	2010 Average	**2011 Average**	**Jan-August 2012 Average**
Extremely concerned	10	9	9
Very concerned	16	16	16
Somewhat concerned	29	28	29
Slightly concerned	31	31	30
Not at all concerned	14	15	15

Source: The NPD Group/Food Safety Monitor

Nevertheless, there are difficulties. Among the food safety issues receiving quite a bit of attention at the current time are food safety following disasters and the increasing demand for, and supply of, imported foods. Additionally, the CDC has identified the following food safety challenges:

- Changes in our food production and supply;
- Changes in the environment leading to food contamination;
- A rising number of multistate outbreaks;
- New and emerging germs, toxins, and antibiotic resistance;
- New and different contaminated foods, such as prepackaged raw cookie dough, bagged spinach, and peanut butter, causing illness.

(http://www.cdc.gov/foodsafety/cdc-and-food-safety.html)

1.14.1 FOOD-AWAY-FROM-HOME

The move away from in-home food preparation continues and there has been a proliferation of vendors selling food including food trucks and mobile vendors. These new approaches and other creative ways to entice consumers – like take-out food stations in parking garages – will also present new challenges to food safety. Some are suggesting that home cooking will still be around, but largely as a hobby in the affluent countries.

1.14.2 FOOD SAFETY MODERNIZATION ACT (FSMA)

In January 2011 President Obama signed into law the Food Safety Modernization Act (FSMA) According to the U.S. Food and Drug Administration

(FDA) this legislation is the "the most sweeping reform of our food safety laws in more than 70 years." The goal is to ensure the food supply is safe by "shifting the focus from responding to contamination to preventing it." This will require food facilities to identify and evaluate the hazards in their systems and processes, and implement and monitor effective measures to prevent contamination (see http://www.fda.gov/Food/FoodSafety/FSMA/ucm247559.htm#Scope). This approach defines HACCP. HACCP is a prevention-based food safety system that assesses the inherent risks to a food product or process, determines the necessary steps that will control the risks, sets up a system to monitor those steps, reacts to any potential deviations from the standards that might occur, and documents these efforts. Perhaps the greatest advantage to the use of HACCP is that it is prevention based; to a certain extent it puts control into the hands of the foodservice operator. Because of this, health departments may start asking to view HACCP documents as a part of an operation's sanitation inspection.

1.14.3 EDUCATION IS ESSENTIAL

The World Health Organization has been sensitive of the need for greater food safety worldwide. "*Every day people all over the world get sick from the food they eat. This sickness is called foodborne disease and is caused by dangerous microorganisms and/or toxic chemicals. Most foodborne disease is preventable with proper food handling.*"

To address this need they have identified the five keys to Safer Food (WHO, 2006). The five keys are:

1. Keep Clean;
2. Separate raw and cooked;
3. Cook thoroughly;
4. Keep food at safe temperatures; and
5. Use safe water and raw materials.

For each key core – information is presented, safe food handling practices identified, and training suggestions provided. For example, the core information for "Keep Clean" includes the following:

- Wash your hands before handling food and often during food preparation;
- Wash your hands after going to the toilet;

- Wash and sanitize all surfaces and equipment used for food preparation;
- Protect kitchen areas and food from insects, pests and other animals.

1.14.4 TECHNOLOGY TO THE RESCUE

The availability of reliable, rapid and accepted test systems to detect the presence or absence of contamination would be a boon to food service. Speed, selectivity, and sensitivity are among the most important features. "It should be able to detect pathogens in very low concentrations of the samples and must be suitable for real-time monitoring as well." It also needs to be cost effective (Velusamy et al., 2010). There are a number of possibilities on the horizon.

1.15 ADDITIONAL RESOURCES

The website www.FoodSafety.gov is the "gateway" to government food safety information.

Government:

- CDC Food Safety Web Site
- FDA (U.S. Food and Drug Administration)

Extension:

- Purdue Extension Food Safety website http://www3.ag.purdue.edu/extension/Pages/FoodSafety.aspx

KEYWORDS

- **critical limits**
- **Food Code**
- **foodborne illness**
- **HACCP**

REFERENCES

800-CDC-INFO (800–232–4636) TTY: (888) 232–6348 – Contact CDC–INFO. http://www.cdc.gov/foodsafety/cdc-and-food-safety.html Retrieved Oct. 19, 2012.

Ali, A.A.; Spencer, N.J. (1996). Hazard Analysis and Critical Control Point Evaluation of School Food Programs in Bahrain. Journal of Food Protection, 59(3), 282–286.

Almanza, B.A.; Ghiselli, R.F. (1998) Implementation and Cost of HACCP in Grill Type Operations. Journal of College and University Foodservice, 10(2), 107–124.

Bottemiller, H.(January 3, 2012). Annual Foodborne Illnesses Cost $77 Billion, Study Finds. Food Safety News. Retrieved Oct. 19, 2012 from http://www.foodsafetynews.com/2012/01/foodborne-illness-costs-77-billion-annually study-finds/

Bryan, F.L. (1990). Hazard Analysis Critical Control Point (HACCP) Systems for Retail Food and Restaurant Operations. Journal of Food Protection, 53(11), 978–983.

Bureau of Labor Statistics, U.S. Department of Labor, Occupational Outlook Handbook, 2012–13 Edition, Food Service Managers, on the Internet at http://www.bls.gov/ooh/management/food-service-managers.htm (visited October 26, 2012).

Centers for Disease Control and Prevention (CDC). Foodborne Outbreak Online Database. Atlanta, Georgia: U.S. Department of Health and Human Services, Center for Disease Control and Prevention. Available from URL: http://wwwn.cdc.gov/foodborneoutbreaks. Accessed MM/DD/YYYY.

Centers for Disease Control and Prevention. http://www.cdc.gov/foodsafety/facts.html#what Centers for Disease Control and Prevention, 1600 Clifton Rd. Atlanta, GA 30333, USA

Cintas Corporation. (Nov. 23, 2010). Dirty Dishware, Restrooms and Odor Are Top Three Reasons Why U.S. Adults Would Never Return to a Restaurant. Retrieved Oct. 16, 2012. http://www.cintas.com/company/news_media/press_releases/Top-Three-Reasons-Adults-Would-Never-Return-to-a-Restaurant.aspx

EHE Newsletter, October 18, 2012. Food Poisoning and Food Safety Volume 12, Number 43. Retrieved from https://www.eheandme.com/news_articles/549646588 on Nov. 8, 2012.

Fein, S.A.; Jordan-Lin, C.T.; Levy, A.S. (1995). Foodborne Illness: Perceptions, Experience, and Preventive Behaviors in the United States. Journal of Food Protection, 58 (12), 1405–1411.

Fein, Sara B.; Lando, Amy M.; Levy, Alan S.; Teisl, Mario F.; Noblet, Caroline, T. (September 2011). Trends in U.S. Consumers' Safe Handling and Consumption of Food and Their Risk Perceptions, 1988 through 2010. Journal of Food Protection, Volume 74, Number 9, pp. 1513–1523 (11).

Food and Drug Administration. (1993). Food Code. Washington, D.C.: U.S. Department of Health and Human Services, Public Health Service, Food and Drug Administration.

Food and Drug Administration. (1995). Food Code. Washington, DC: US Department of Health and Human Services, Public Health Service.

Food and Drug Administration. (April 2006). Managing Food Safety: A Manual for the Voluntary Use of HACCP Principles for Operators of Food Service and Retail Establishments. OMB Control No. 0910–0578. http://www.fda.gov/Food/FoodSafety/RetailFoodProtection/ManagingFoodSafetyHACCPPrinciples/Operators/default.htm

Food Safety and Inspection Service. (December, 1990). FSIS Facts Bacteria That Cause Foodborne Illness (FSIS-40). Washington, DC: United States Department of Agriculture.

FoodSafety.gov. Causes of Food Poisoning. Retrieved on Oct. 19, 2012 from http://www.foodsafety.gov/poisoning/causes/index.html

FoodSafety.gov. Retrieved October 18, 2012 from http://www.foodsafety.gov/poisoning/responds/

Ghiselli, Li.; Almanza. (2013). Foodservice design: Assessing the importance of physical features to older consumers. Journal of Foodservice Business Research, 17(4).

Gisslen, W. (2011). Professional Cooking (6th ed.). New York: John Wiley and Sons.

Hunter Public Relations. (Uploaded December 8–2011). Hunter Public Relations 2011 Annual Survey of the Top Food Stories. Retrieved on Oct. 15, 2012 from http://www.youtube.com/watch?v=dcEzjvLGrgw.

Indiana State Department of Health. (1990). Food Service Sanitation Requirements (IN 410-IAC 7–15.1). Indianapolis, IN: Indiana State Department of Health.

King, P. (1992). Implementing A HACCP Program. Food Management, 54 (12), 54+.

Lorenzini, B. (1995). Here Comes HACCP. Restaurants and Institutions, 105 (1), 119.

National Restaurant Association Educational Foundation. (2008). Servsafe Coursebook (5th ed). Chicago: National Restaurant Association Solutions.

NPD Group. (2012, October). U.S. Consumers Are Only Somewhat or Slightly Concerned About the Safety of U.S. Food Supply Despite Frequent Food Safety Outbreaks. Retrieved on December 29, 2012 from https://www.npd.com/wps/portal/npd/us/news/press-releases/us-consumers-areonly somewhat-or-slightly concerned-about-the-safety-of-us-food-supply despite-frequent-food-safety-outbreaks-reports-npd/.

NRA. (December 11, 2012) Restaurant industry will grow, outpace national job growth in 2013 despite sustained challenges. Posted by Annika Stensson on December 11, 2012–8:26 AM. Retrieved from http://www.restaurant.org/nra_news_blog/2012/12/restaurant-industry-will-grow-outpace-national-job-growth-in-2013-despite-sustained-challenges.cfm on December 13, 2012.

Todd, E.C.D. (1996). Worldwide Surveillance of Foodborne Disease: the Need to Improve. Journal of Food Protection, 59(1), 82–92.

U.S. Census Bureau. Number of Firms, Number of Establishments, Employment, and Annual Payroll by Enterprise Employment Size for the United States, All Industries: 2010. Release Date: 10/2012. http://www.census.gov/econ/susb/ Accessed: 11/12/2012.

USDA Economic Research Service Food CPI and Expenditures http://www.ers.usda.gov/Briefing/CPIFoodAndExpenditures/Data/Expenditures_tables/table1.htm

Velusamy, V.; Arshak, K.; Korostynska, O.; Oliwa, K.; Adley, C. (2010). An overview of foodborne pathogen detection: In the perspective of biosensors. Biotechnology Advances, 28, 232–254.

World Health Organization. (2006). Five Keys to Safer Food Manual. Department of Food Safety, Zoonoses and Foodborne Diseases. WHO Press, Geneva, Switzerland.

CHAPTER 2

HISTORY OF FOOD SAFETY AND FOOD SAFETY REGULATION IN THE UNITED STATES

BARBARA ALMANZA, PhD, RD
Professor, Purdue University

JEFF FISHER, MS
Purdue University

CONTENTS

2.1 EARLY LEGISLATION

Although discussed for 25 years and proposed in hundreds of prior bills (FDA, 2009a), the history of food safety regulation in the United States finally started with a book by Upton Sinclair titled *The Jungle* that was published in 1906. Although it was a fictional account of the lives of its characters, it accurately and graphically depicted the unsanitary conditions in the Chicago meat packing industry. Sinclair's book states "There would be meat stored in great piles in rooms; and the water from leaky roofs would drip over it, and thousands of rats would race about on it. It was too dark in these storage places to see well, but a man could run his hand over these piles of meat and sweep off handfuls of the dried dung of rats. These rats were nuisances, and the packers would put poisoned bread out for them, they would die, and then rats, bread, and meat would go into the hoppers together."

In contrast to today's slow moving political environment, public outrage from this book was so strong that Congress drafted and passed the first food protection law, the Pure Food and Drug Act, and it was signed into law by Theodore Roosevelt in the same year that the book was published. This was closely followed in the same year by the Meat Inspection Act of 1906, which made inspection of meat processing facilities mandatory.

Although this law was heralded as a landmark in consumer protection in the United States, two problems existed with this first law. One was that the burden of proof for lack of safety was on the government (not the food manufacturer). The second was that the penalties for not following this regulation were relatively weak.

To enforce the law, Harvey Wiley as the head of the U.S. Department of Agriculture was charged with testing the safety of the foods that were in the marketplace, a more than daunting task. To test products, he used what became referred to as a "Poison Squad" which consisted of a group of prisoners who were willing to consume the test products in exchange for shorter prison sentences. If the prisoners became sick, the products were deemed to be unsafe. Although clearly not an effective method by today's standards and certainly not a method that would be approved by an Institutional Review Board on the Use of Human Subjects as Research, it did represent a first effort in the United States to provide food safety regulation and predates most countries' efforts to regulate food safety.

A second major food safety law was passed in 1938. It was called the Food, Drug, and Cosmetic Act and it also gained momentum as a result of public outrage over another event. That event was related to a pharmaceutical product that was developed by the S.E. Massengill Company. A chemist for S.E. Massengill by the name of Harold Watkins was given the assignment to development a liquid form of an antibiotic called sulfanilamide so that young children could be given this medicine. Unfortunately, he used di-ethylene glycol (a major component of antifreeze that happens to be toxic) as the solvent to make the antibiotic into a liquid. He added caramel, amaranth, and raspberry extract to make it taste good for young children. Watkins did sample the product, but his tolerance was likely greater as an adult and he only tasted small amounts. Because the burden of proof was on the government according to the first food protection law (the Pure Food and Drug Act of 1906) the company was able to go ahead and offer the product for sale without having it undergo further testing. Before anyone knew that the product was deadly, physicians started prescribing "Elixir Sulfanilamide" in 1937 and more than 100 deaths occurred, many of which were children (FDA, 2006a).

Because of this event, in a matter of a few months the second major food safety law in the U.S. was drafted, passed, and signed into law by Franklin Delano Roosevelt as the Food, Drug, and Cosmetic Act of 1938. As result of the deaths related to their product, the S.E. Massengill Co. was fined for misbranding (penalties were relatively weak from the first Food and Drug Law), and Harold Watkins was fired by S. E. Massengill (he later committed suicide).

One of the most important changes in the 1938 Food, Drug, and Cosmetic Act was that it put the burden of proof for lack of safety on the manufacturer. This was significant as manufacturers were subsequently required to conduct extensive safety testing. Today, safety testing of new food additives or ingredients includes seven types of toxicity tests: metabolism testing (an evaluation of how the ingredient or food additive changes as it is being metabolized); acute testing (for a single lethal dose); subacute testing (over a relatively short time period); chronic testing (over the lifetime of the animals being tested); carcinogenicity testing (for cancer); mutagenicity testing (to determine mutation effects on the offspring of the animals fed the additive); and teratogenicity testing (to see if there are birth defects in the offspring of the animals fed the additive). The results of these tests must then be submitted to the government for approval prior to being able to bring these products into the marketplace.

Numerous laws related to safety of foods and other products were passed following this regulation. Most notably among them was the 1958 Delaney Clause to the Food, Drug, and Cosmetic Act. Senator Delaney from Wisconsin proposed this amendment, which stated that no additive or ingredient could be put into food products in any amount that was shown to cause any type of cancer in any animal. Although this appears to be a sound idea, since this time science has progressed to the point that it is no longer enforceable. For example, extremely small trace amounts of chemicals may now be detected and very minute cancer risks for extremely rare types of cancer may be calculated. The U.S. government as a result has decided to use what is called a "de minimus" approach to enforcing this anticancer legislation so that minimal acceptable levels of risk and/or maximum additive levels are now used to help determine acceptability for safe food products.

Tremendous improvements in public health occurred with all of these regulations. During the early 1900s, contaminated food, water, and milk caused foodborne infections such as tuberculosis, botulism, typhoid fever, and scarlet fever (CDC, 1999). Even before vaccines and antibiotics were available, the new food protection laws created changes in hand washing, sanitation, pasteurization, refrigeration, and better animal care and processing that resulted in a safer food supply. In 1900, for example (prior to the Pure Food and Drug Act), approximately 100 cases of typhoid fever occurred per 100,000 people in the United States. Following the implementation of the first food protection law (the Pure Food and Drug Act of 1906), this number dropped to 33.8 by 1920 and dropped even lower to 1.7 in 1950 following the passage of the second food protection law (the Food, Drug, and Cosmetic Act of 1938) (CDC, 1999). Improvements in food safety have continued. In the 1940s approximately 300–400 cases of trichinosis were diagnosed every year in the U.S. with 10–20 deaths. By contrast, from 1991 through 1996, only 38 cases were reported and only three deaths occurred (CDC, 1999).

2.1.1 THE HISTORY OF FOOD INSPECTIONS BY THE FOOD AND DRUG ADMINISTRATION AND THE STATE HEALTH DEPARTMENTS

The Food and Drug Administration (FDA), in its role of protecting product safety has evolved extensively in the more than 100 years since it began (FDA, 2006a). Early descriptions of U.S. Food and Drug Inspectors state that the agency started with no more than 45 inspectors (FDA, 1913) and they covered more than 3 million square miles of the country with a population of approximately 90 million (roughly a ratio of one inspector per 2 million people). Today, the FDA in partnership with state health departments or its agencies conducts 43,000 establishment inspections in the country annually and the FDA conducts more than 900 inspections abroad. In addition, they review approximately 12 million imported products, conduct 110,000 import field examinations, and analyze about 46,000 product samples every year. This represents oversight of 128,000 U.S. establishments, including 62,000 food companies, 37,100 medical device firms, 18,900 human pharmaceutical companies, and 11,800 companies that manufacture animal medications or medicated feeds (FDA, 2006a).

The history of food establishment or restaurant regulation started in 1934 with the "Restaurant Sanitation Regulations" proposed by the U.S. Public Health Service in cooperation with the Conference of State and Territorial Health Officers and the National Restaurant Code Authority (FDA, 2009b). An Ordinance Regulating Eating and Drinking Establishments was recommended and revised several times during the 1930s and 1940s. Eventually, regulations for vending of foods and beverages were also recommended (1957) and retail food stores (grocery stores) were also added.

In some states, as regulations were passed, three separate food related codes were used for restaurants, grocery stores, and vending machines so that enforcement of these sometimes overlapping but diverse regulatory areas became more challenging. Most importantly, federal recommendations were updated periodically, but could not keep pace with the rapid growth of businesses that offered food products for sale. Sale of ready to eat foods expanded into areas that previously may not have even required inspection such as gas stations, shopping malls, drug stores, airports, street corners, and convenience stores. As food sales expanded, restaurants sold food products for home consumption (similar to grocery stores) and grocery stores sold food for immediate consumption (similar to restaurants). Vending machines could be located almost anywhere with relatively little oversight. Consistency in food safety regulation was challenging so that a significant revision in food establishment food safety regulation became necessary by the early 1990s.

To assist state governments with the need to comprehensively regulate the sale of food products, the U.S. Public Health Service and Food and Drug Administration integrated recommendations for all establishments selling food and proposed a thorough and completely revised set of guidelines in 1993 titled the Food Code. This federal Food Code was intended to serve as a set of recommendations based on the latest scientific evidence and expert judgment for states' use in developing their state regulations. The Food Code has been subsequently widely adopted by states and other jurisdictions in the United States (such as Washington, D.C. and Puerto Rico).

Although the federal Food Code was originally designed to be revised every two years (with revisions in 1993, 1995, 1997, 1999, and 2001), individual states found it difficult to keep up and pass changes from the revised Food Code through their state legislature every two years. Since 2001, the FDA Food Code has been revised every four years (2005, 2009) with supplements provided in the two-year intervals between the revisions (2003, 2007).

2.2 NEWER ISSUES AND EVOLVING LEGISLATION

2.2.1 FOODBORNE ILLNESS OUTBREAKS AND THEIR LEGISLATIVE IMPACT

The Centers for Disease Control and Prevention have been tracking national foodborne illnesses data since at least 1996 (CDC, 2012), and as early as 1952 for some illnesses reported as infectious hepatitis, botulism, and gastro-enteritis (U.S. National Office of Vital Statistics, 1952). Methods for obtaining data through surveillance have improved with better tracking and analysis providing more accurate data, particularly in the estimates provided in 2011. Although the U.S. estimates may sound high, most countries do not even track their illness data.

Knowledge has improved regarding which microorganisms cause foodborne illnesses in food establishments. Twenty-five years ago when food safety was taught to restaurant management students, emphasis was placed on the big five bacteria of *Staphylococcus aureus*, *Salmonella*, *Clostridium perfringens*, *Clostridium botulinum*, *Campylobacter jejuni* and the parasite *Trichinella spiralis*. Viruses were barely even mentioned. Today's estimates from CDC (2012) now include 31 known pathogens (including viruses, and parasites).

Limitations still exist for today's CDC estimates including the fact that the proportion of illnesses caused by a pathogen obtained from a source other than food are not well estimated. In addition, differences in health-care seeking behavior for different demographic groups mean that CDC estimates only provide information for the country as a whole, not for individual or different demographic groups within the United States. Even the sample used in the analyzes represents only a fraction of the actual cases so that it may not present as accurate an estimate as would be possible with a larger sample. Given these limitations, CDC estimates that 1,000 foodborne illness disease outbreaks occur annually in the U.S., resulting in 48 million illnesses, 128,000 hospitalizations, and 3,000 deaths (CDC, 2012).

In spite of improvements in the regulatory systems, the negative medical, emotional, as well as financial consequences of foodborne illness outbreaks are still tremendous. In fact, the Centers for Disease Control and Prevention estimates are alarming in that one in six Americans are thought to become sick every year because of what are considered largely preventable problems in providing safe food (FDA, 2013b). As a result, food safety inspection regulations continue to evolve. Although decreases in some infections have occurred (*Campylobacter, Listeria, STEC 0157, Shigella*, and *Yersinia*), overall incidence of some important pathogens, such as, *Salmonella* remain unchanged (although illnesses from some forms of *Salmonella* have increased and other forms have decreased) and others such as *Vibrio* infection have actually increased (CDC, 2013).

Foodborne illness data are also important in that they lead to very specific changes in regulations as health officials and food establishments learn from these unfortunate events. This is essential in the dynamic field of food safety as newer pathogens (or perhaps unrecognized) emerge, or newer or alternative processing systems are developed, or as a result of the expansion of the food industry in both its size and in its diversity. Although the original task of providing safe food for 90 million Americans with 45 inspectors in the early 1900s was daunting, the task appears to be even more difficult today. This is in part because today's large-scale production facilities (farms and processing plants) are larger. If food safety problems occur, the number of people affected is huge and the impact is widely distributed geographically. In addition, food imports and exports have expanded so that even global food issues are now bigger and more relevant within the United States. As reported in 2011 (FDA, in press), 20% of fresh vegetables, 50% of fresh fruit, and 80% of seafood are imported into the United States. Margaret Hamburg, FDA Commissioner states that, "Today we recognize that to successfully protect U.S. public health, we must think, act, and engage globally. Our interests must be broader than simply those within our own borders." (FDA, in press). Finally, concerns have grown regarding both intentional as well as unintentional contamination issues related to food safety.

There are many examples of changes in food safety guidelines as knowledge has grown about the causes of foodborne illness. For example, early restaurant regulations for raw eggs allowed them to be stored at room temperature because the eggshells were believed to protect the eggs inside from contamination. Then, research (and outbreaks of foodborne illness) showed that asymptomatic chickens infected with *Salmonella enteritidis* could lay eggs with the bacteria already inside the egg shell (Guard-Petter, 2001). As a result, regulations started requiring eggs to be treated as potentially hazardous foods, which must be refrigerated.

One of the most famous examples of a foodborne illness outbreak was the *E. coli 0157:H7* outbreak that occurred in some Jack in the Box restaurants in late 1992 into 1993. This devastating outbreak also caused one of the most significant changes in regulation regarding ground beef preparation. At the time the federal guideline (that most states used) for the cooking endpoint for ground beef was 145°F (63°C). The state regulation in Washington where the outbreak started however, was higher at 155°F (68°C) (Benedict, 2011). Unfortunately, Jack in the Box used the lower federal guideline rather than the state regulation, which they were required to use in Washington. When contaminated meat products were shipped to their restaurants from their supplier, the lower cooking endpoint was not adequate to destroy the *E. coli 0157:H7* and the extremely disastrous outbreak occurred in which almost 600 people were estimated to become sick and four children died (Marler Clark, 1993). Of the 600 people who became sick, at least 40 people developed hemolytic uremic syndrome as a result of eating the contaminated product, a disease associated with a lifetime of medical consequences (Marler Clark, 1993).

As a result of this *E. coli 0157:H7* devastating outbreak, the federal guideline was raised to 155°F (68°C) and other states moved forward to revise their regulations as well. For their actions in this outbreak, Jack in the Box was reported to lose $160 million in the 18 months post outbreak and later paid $50 million in individual and class action settlements (Marler Clark, 1993).

Foodborne illness outbreaks have led to other regulatory changes. In particular, additional foods have been placed on the potentially hazardous foods list, also known as temperature controlled for safety foods. The foods that originally appeared on the list were the obvious foods of beef, pork, chicken, turkey, fish and dairy products. In the last 25 years however, more recent additions to this include foods that would have surprised the first health inspectors in the early 1900s. They include: cut melons, garlic in oil mixtures, cooked vegetables, baked potatoes, sliced tomatoes, and most recently, cut leafy greens. It is important that food establishments are made aware of these changes because managers or owners may have been born during a time when at least some of these foods were not considered potentially hazardous and were left unrefrigerated.

The foodborne illness outbreak associated with cut leafy greens became nationally famous in 2006 (FDA, 2006b) following news coverage that reported *E. coli 0157:H7* was associated with fresh spinach from a farm in California. In this outbreak, 199 illnesses occurred in 26 states, which resulted in at least three deaths (FDA, 2006b). Concern was also expanded to other leafy greens, including lettuce. Recommendations were then made to store fresh produce for both quality, and most importantly, for safety reasons in a clean refrigerator at a temperature of 41°F (5°C) or below.

Similarly, cut melons have been associated with *Salmonella* in several outbreaks dating back to at least 1990 (FDA, 2001). Garlic in oil mixtures (Colorado State University Extension, 1998) and baked potatoes (Penn

State Extension, 2013) have been associated with *Clostridium botulinum*, tomatoes with *Salmonella* (Hedberg, et al., 1999), and cooked vegetables with various bacteria depending on what other foods are combined with them. Cut tomatoes may be surprising to some people because of their acid pH, however, the acidity of tomatoes is variable. In recent years, low acid tomatoes have become increasingly popular.

Another example of a change in food safety regulation occurred as a result of drinking water contamination with a protozoan parasite, *Cryptosporidium*, in the Milwaukee municipal water supply (CDC, 2011). In this outbreak, 403,000 people became ill. This led to the adoption of a federal mandate for all water utilities to test for *Cryptosporidium* once a month.

A final example of a wide scale foodborne illness outbreak with regulatory implications occurred with peanut butter produced by a company in Blakely, Georgia in the fall of 2008 into the winter of 2009. The Peanut Corporation of America became nationally famous in this outbreak, in which 714 cases of *Salmonella typhimurium* occurred in 46 states as a result of contaminated peanut butter from their production facility (FDA, 2011a). Because peanut butter was sold as an ingredient for other products, this outbreak created one of the largest recalls of food products in the United States with 2,800 products identified as containing peanut butter that were produced by this plant.

Up to this point in time, recall ability of the FDA was very limited (in the earlier *E. coli 0157:H7* foodborne illness outbreak with Jack in the Box for example, the eventual recall was only able to obtain 20% of the implicated product according to Marler Clark in 1993). In the Peanut Corporation of America outbreak, this limited recall ability became a significant factor in being able to control the outbreak. Because peanuts are used as a food ingredient for everything from candies, cookies and snack crackers, to peanut butter and even entrees, the amount of contaminated food was enormous beyond anyone's expectations. The management of this outbreak would have been greatly facilitated if the FDA had the ability to immediately issue a mandatory recall for the contaminated products when the contamination was first discovered. This and other related outbreak issues were addressed with the Food Safety Modernization Act.

2.2.2 EXPANDED NEEDS AND THE FOOD SAFETY MODERNIZATION ACT OF 2011

On January 4, 2011, President Obama signed into law the FDA Food Safety Modernization Act. The purpose of the Act, according to FDA (2013b) is to ensure the safety of the U.S. food supply by shifting the focus of federal regulators to preventing contamination rather than responding to it. It is considered by the FDA (2013c) to be the most sweeping reform to food safety laws in more than 70 years.

There are five major elements to the new law (FDA, 2013c). They are preventive controls (throughout the entire food chain), inspection and compliance (including how often FDA should inspect food producers), imported food safety (so that for the first time importers must verify that their foreign suppliers have adequate preventive controls and also that FDA will able to accredit qualified third party auditors to certify that foreign food facilities are complying with U.S. food safety standards), response (which will include mandatory recall authority for the first time), and enhanced partnerships (with all food safety agencies including federal, state, local, and territorial, tribal, and foreign) to achieve health goals.

Work on implementing some of these elements have been prioritized to take place first, such as the mandatory recall authority, others are expected to require additional time to issue the regulations and guidance documents after input from all stakeholders is obtained. Although many of the measures were intended to be implemented within two years of enacting the law, more time has been needed to complete the final rules. Additional funding (not yet available) will also be required to fully implement all of these elements.

Based on experiences from prior events such as the historic Peanut Corporation of America and Jack in the Box foodborne illness outbreaks where a more rapid and vigorous response would have minimized the size of the outbreaks, mandatory recall is expected to be an essential tool in protecting food safety. In addition, FDA will have expanded administrative detention for suspect products that are potentially in violation of the law, suspension of registration of a facility if it determines that the food produced in that facility poses a reasonable probability for serious health consequences or death so that food may not be distributed from that facility, and enhanced product tracing abilities for both domestic and imported foods. Finally, FDA will be able to issue rules to establish record-keeping requirements for facilities that manufacture, process, pack or hold high-risk foods.

The U.S. receives imported food from more than 150 countries. Because many of these countries do not have as strict a set of standards as the U.S., the Food Safety Modernization Act will give authority for the first time to help ensure that imported products meet U.S. food safety standards (FDA, 2013c). More specifically, for the first time, importers will be tasked with the responsibility of verifying that their foreign suppliers have adequate controls in place to assure that their food products are safe. In addition, because of the size, number, and geographic location of these foreign production facilities, the FDA is expected to set up a qualification program and certification for third party inspectors to ensure that foreign facilities comply with U.S. food safety standards. FDA is also expected to establish a voluntary program for importers that would allow for an expedited review and entry of their foods (if the facility is also certified). Finally, similar to the authority for detaining suspect foods from U.S. facilities, the FDA will also be able to deny entry to foods from a foreign facility if the FDA is denied access by the facility or country in which the facility is located.

In regards to inspection and compliance, the Food Safety Modernization Act emphasizes that preventive standards to minimize food safety will only be effective if food producers and processors comply with them (FDS, 2013c). Oversight for inspection and compliance has therefore been strongly incorporated into the Act. New tools with this Act include: a mandated inspection frequency based on risk that will occur more frequently (for example, all high-risk domestic facilities will be inspected within five years of enactment and no less than every three years thereafter, and with regards to imported food the FDA is expected within one year to inspect at least 600 foreign facilities and double the number inspected every year over the next five years); more FDA access to company records in regards to industry food safety plans and documentation records for those plans; and the establishment of an accredited laboratory food testing program for suspect foods.

2.3 FOOD DEFENSE

Food defense is a newer issue that is also being addressed through the Food Safety Modernization Act of 2011, although it was not the first regulation to cover the issue of food safety in food and animal feed facilities. Earlier regulation was set up in the Public Health Security and Bioterrorism Preparedness and Response Act of 2002. Food defense is a focused effort to protect the food supply against intentional contamination (FDA, 2013c). Intentional contamination may occur as a result of sabotage, counterfeiting, terrorism, or other illegal, intentionally harmful acts. It includes all types of potential contaminants such as biological, chemical, or radiological hazards that have been placed in foods or the production environment. This represents a significant change from the focus of traditional systems where the emphasis is placed on unintentional or perhaps accidental food safety problems, which might occur from bacterial or viral contamination.

The Food Safety Modernization Act requires FDA to issue regulations for food defense (FDA, 2013c). In this regard, the FDA will issue regulations specifying effective and science-based mitigation strategies to protect the food supply chain from intentional contamination at vulnerable points. Prior to this law, food facilities were not required to implement mitigation strategies to protect foods against intentional contamination.

In fact, however, foodborne illnesses related to intentional contamination have been documented since at least the 1980s. One of the most commonly cited cases occurred with the Rajneesh cult in 1984 in Oregon. This cult was seeking to influence voting and the subsequent local election results by making people sick. In this outbreak, 751 people were made ill and 45 were hospitalized after some cult members placed *Salmonella* in salad bars the day before the election. Classified as the first known bioterrorist event in the U.S. by the CDC, two cult members served time in prison for their actions (CDC, 2011).

Part of the process of protecting food against intentional contamination includes registration of approximately 42,000 domestic and foreign food facilities (FDA, 2006c) as required by the Public Health Security and Bioterrorism Preparedness and Response Act of 2002. This registration requirement does not apply to "retail food establishments" (grocery stores and other similar food establishments that sell food directly to the consumer), restaurants, nonprofit food establishments, or certain other facilities such as farms or fishing vessels. The purpose of registration was to provide FDA with information about human food and animal feed facilities so that they quickly detect and respond to actual or potential threats to the U.S. food supply.

Numerous tools have been developed by the FDA to assist food establishments with protecting their food. They include the Food Defense Plan Builder, an interactive software program that may be downloaded at no charge from the FDA website (FDA 2013d). FDA designed this program to assist food facilities with developing voluntary, personalized food defense plans using company information to set up broad mitigation strategies, a vulnerability assessment, focused mitigation strategies, and an action plan that includes emergency contacts and supporting documents. The value of a well-established food defense plan is that it provides a written document that can reduce the food establishment's vulnerability from intentional contamination and help in protecting their food products.

2.4 STATE REGULATORY CHANGES

State regulation and its enforcement at the state or local level is the important final line of defense in food safety. Most states have adopted some version of the guidelines from the FDA Food Code to use as their state regulations. The version of the FDA Food Code used in state regulations depends on the frequency of legislative changes within that state. Because of its variability from one state to another, the history of state regulation is difficult to summarize. Three of the most current issues for states in their regulatory efforts are related to consistency in health inspector education and training requirements, certification of food handlers or managers, and decreases or limitations in fiscal resources needed for food safety enforcement.

Standardization of education and training requirements for health inspectors has been discussed for several years, but still varies widely from state to state and even from one local jurisdiction to another within a state. Some jurisdictions only require a high school diploma and on the job training. Other jurisdictions require additional education (such as a college degree), examinations, or credentialing.

Mandatory certification of food handlers or managers has also been a change in recent years in many state's regulations. As of 2011 (FDA, 2011b), 49 out of 50 states had adopted some version of the FDA Food Code (which is offered as the most current guidelines based on scientific evidence) which states that "...upon request the person in charge shall demonstrate to the regulatory authority knowledge of foodborne disease prevention, application of the hazard analysis critical control point principles, and the requirements of this code. The person in charge shall demonstrate this knowledge by compliance with the code, by being a certified food protection manager who has shown proficiency of required information through passing a test that is part of an accredited program, or by responding correctly to the inspector's questions as they relate to specific food operation" (FDA, 2009b). Most states have interpreted this to require evidence in the form of a certification from a source that has been approved by the Conference for Food Protection. As of this time, mandatory certification of restaurant managers (and/or food handlers in at least two states) is required in Alabama, Alaska, California, Connecticut, Florida, Georgia, Idaho, Illinois, Indiana, Louisiana, Maine, Massachusetts, Michigan, Minnesota, Mississippi, New Jersey, New York, North Carolina, Ohio, Oregon, Pennsylvania, Rhode Island, South Dakota, Texas, Utah, Washington, and Wisconsin (National Restaurant Association, 2013).

The number of states with mandatory certification has been steadily increasing since the first state (which was Washington) started requiring certification of food handlers in the 1950s (Almanza, and Nesmith, 2004). Since 2004 when only 17 states required certification (Almanza, and Nesmith, 2004), the number of states requiring certification has increased by about one state per year. The 27 states currently requiring certification now outnumber the states that do not require certification. Certification is also expanding in those states where certification is not mandatory because some local jurisdictions may require certification.

Where certification is mandatory, regulations vary state to state. Certification regulations include who must be certified (managers or food handlers), whether the certified individual must be on site at all times, recertification, approved examinations, training requirements, exemptions, allowed times for coming into compliance for new establishments and for turnover, fees, instructor requirements, and the way in certification is checked (Almanza, and Nesmith, 2004).

In general, most states require certification of a single responsible individual (generally an owner, manager, or supervisor) per site (although this individual may be off site for brief time periods if a knowledgeable person in charge has been designated to be responsible during these time periods)

(Almanza, and Nesmith, 2004). Recertification is generally required every three or five years (depending on the state) and most states use a Conference for Food Protection approved examination (although some states use their own examination) (Almanza, and Nesmith, 2004). Other certification regulations tend to vary considerably from state to state.

Finally, as with many government agencies, a major regulatory concern for many state and local health departments or agencies is a diminishing budget. In fact, one study found that in a two-year period from January 2008 to December 2009, local health departments lost 15% of their entire workforce (UPI, 2010). In Indiana, two to three counties do not have a designated local health inspection officer for food establishments because of recent cost control measures. Robert Pestronk, the executive director of the Nation Association of County and City Health Officials, has stated that "Unfortunately, a lack of federal, state, and local budget resources is straining an already fragile public health system" and has suggested that there are serious gaps in the ability to respond to health crises (UPI, 2010).

2.5 SUMMARY

The U.S. has a significant and dynamic history of food safety regulation dating back over a hundred years. Guidelines and regulations have evolved as new scientific evidence or foodborne illness outbreak experiences have expanded our knowledge about the causes of foodborne illness. In spite of this regulatory history however, foodborne illness in the U.S. is still a major health concern. From a global standpoint it is also a concern. Although an estimate of 46 million annual cases of foodborne illness in the U.S. sounds high, most other countries are not tracking their illnesses.

Estimates of foodborne illness are helpful in setting goals for illness reduction and for changes in regulatory guidelines. They also underscore the importance of new legislation such as the Food Safety Modernization Act. Even though the U.S. is actively engaged through federal, state, and local agencies in minimizing foodborne illness, more effort is needed as outlined in the Food Safety Modernization Act (FDA, 2011a). It is hoped that the five elements of this Act will move the U.S. to a new and stronger era of food protection. They include: preventive controls, inspection and compliance, imported food safety, response (including mandatory recall authority), and enhanced partnerships with all federal, state, and local food safety agencies.

The responsibility of protecting the safety of the food supply in the U.S. was an enormous task in 1906 with the first food protection law. Today, the task is even more difficult, in spite of broader and more comprehensive laws and a larger and more expanded inspection network. Regulatory challenges are amplified because of growth in the food industry. Regulations now must cover a huge number of food establishments operating in diverse locations that are receiving food from both domestic and global sources. Today's challenges include standardizing health inspector training, strengthening requirements for food establishments (such as manager or food handler certification), and budgetary issues.

In fact, foodborne illnesses still occur today in spite of such strong efforts to prevent them from occurring. Unfortunately, the consequences of food safety problems are still enormous, medically, financially, as well as emotionally for everyone involved. Brianne Kiner, one of the children who became ill and went into a coma for 42 days after eating *E. coli* contaminated ground beef in a Jack in the Box restaurant in 1993, clearly demonstrates the devastating impact on an individual. Facing a lifetime of serious medical concerns, she was awarded the largest personal-injury award in Washington state history of $15.6 million (Benedict, 2011; Marler Clark, 1993). Although Brianne Kiner's experience manifested the most serious consequences of a foodborne illness, she is unfortunately only one of 48 million people that become sick every year from foodborne illness in the United States.

Looking into the future one hundred years from today, the laws and inspection system will undoubtedly have evolved into an even more complex and comprehensive system. Hopefully, they will keep pace with the consumers' ever changing lifestyle which pushes the restaurant and foods industries to offer new kinds of products in more venues using different methods of processing, preparation, and service. Globally, the food industry will also expand. The hope is that with the dynamic system of food protection in the U.S., the regulatory changes will impact food safety so that foodborne illnesses will become a much more unusual, and perhaps even a rare occurrence.

KEYWORDS

- ***Campylobacter jejuni***
- ***Clostridium botulinum***
- ***Clostridium perfringens***
- **de minimus**
- **Elixir Sulfanilamide**
- ***Salmonella***
- ***Staphylococcus aureus***
- ***Trichinell aspiralis***

REFERENCES

Almanza, B. A. and Nesmith, M.S. (2004). Food safety certification regulations in the United States. *Journal of Environmental Health, 66*(9), 10–13.

Benedict, J. (2011). Poisoned: The true story of the deadly *E. coli* outbreak that changed the way Americans eat. Buena Vista, VA: Inspire Books.

CDC. (1999). Achievements in public health, 1900–1999: Safer and healthier foods. Atlanta, GA: United States Centers for Disease Control and Prevention. Retrieved from http://www.cdc.gov/mmwr/preview/mmwrhtml/mm4840a1.htm

CDC. (2011). History of the EIS, Epidemic Intelligence Service. Atlanta, GA: United States Centers for Disease Control and Prevention. Retrieved from http://www.cdc.gov/EIS/History/

CDC. (2012). Questions and answers: 2011 estimates. Atlanta, GA: United States Centers for Disease Control and Prevention. Retrieved from http://www.cdc.gov/foodborneburden/questions-and-answers.html

CDC. (2013). Trends in foodborne illness in the United States, 2012. Atlanta, GA: United States Centers for Disease Control and Prevention. Retrieved from http://www.cdc.gov/features/dsfoodnet2012/

Colorado State University Extension. (1998). Oil infusions and the risk of botulism. *Food Safety Notebook, 9*(4). Retrieved from http://www.ext.colostate.edu/safefood/newsltr/v2n4s 08.html

FDA. (1913a). How to become a U.S. Food and Drug Inspector, Silver Spring, MD: United States Food and Drug Administration reprint from the Civil Service Magazine. Retrieved from http://www.fda.gov/AboutFDA/WhatWeDo/History/Overviews/ucm109718.htm

FDA. (2001). Retail food safety program information manual: Safe handling practices for melons. Silver Spring, MD: United States Food and Drug Administration. Retrieved from http://www.fda.gov/Food/GuidanceRegulation/RetailFoodProtection/IndustryandRegulatory AssistanceandTrainingResources/ucm217290.htm

FDA. (2006a). FDA law enforcement: Critical to public safety. Silver Spring, MD: United States Food and Drug Administration's FDA Consumer Magazine. Retrieved from http://www.fda.gov/AboutFDA/WhatWeDo/History/FOrgsHistory/ORA/ucm084102.htm

FDA. (2006b). FDA statement on foodborne *E. coli 0157:H7* outbreak in spinach. Silver Spring, MD: United States Food and Drug Administration. Retrieved from http://www.fda.gov/NewsEvents/Newsroom/PressAnnouncements/2006/ucm108761.htm

FDA. (2006c). Compliance and policy guide – Registration of food facilities under the Public Health Security and Bioterrorism Preparedness and Response Act of 2002. College Park, MD: Food and Drug Administration. Retrieved from http://www.fda.gov/Food/GuidanceRegulation/FoodFacilityRegistration/ucm081616.htm

FDA. (2009a). Centennial of FDA. Silver Spring, MD: United States Food and Drug Administration. Retrieved from http://www.fda.gov/AboutFDA/WhatWeDo/History/CentennialofFDA/default.htm

FDA. (2009b). *Food Code 2009.* College Park, MD: United States Department of Health and Human Services, Public Health Service, Food and Drug Administration. Retrieved from http://www.fda.gov/Food/FoodSafety/RetailFoodProtection/FoodCode/FoodCode2009/default.htm

FDA. (2011a). Food Safety Modernization Act: Biennial report to Congress on the Food Emergency Response Network. College Park, MD: United States Food and Drug Administration. Retrieved from http://www.fda.gov/Food/GuidanceRegulation/FSMA/ucm271966.htm

FDA. (2011b). Real progress in food code adoptions. College Park, MD: United States Food and Drug Administration. Retrieved from http://www.fda.gov/Food /GuidanceRegulation/RetailFoodProtection/FederalStateCooperativePrograms/ucm108156.htm

FDA. (2013b). Food Safety Modernization Act: Frequently asked questions. College Park, MD: United States Food and Drug Administration. Retrieved from http://www.fda.gov/Food/GuidanceRegulation/FSMA/ucm247559.htm

FDA. (2013c). Background on the FDA Food Safety Modernization Act (FSMA). College Park, MD: United States Food and Drug Administration. Retrieved from http://www.fda.gov/NewsEvents/PublicHealthFocus/ucm239907.htm

FDA. (2013d). Food defense tools and educational materials: Food Defense Plan Builder. College Park, MD: United States Food and Drug Administration. Retrieved from http://www.fda.gov./Food/FoodDefense/ToolsEducationalMaterials/ucm349888.htm

FDA. (in press). Global engagement. Silver Spring, MD: United States Food and Drug Administration. Retrieved from http://www.fda.gov/downloads/AboutFDA/ReportsManualsForms/Reports/ucm298578.pdf

Guard-Petter, J. (2001). The chicken, the egg and *Salmonella enteritidis*. *Environmental Microbiology, 7*: 421–430.

Hedberg, C.W.; Angulo, F.J.; White, K.E.; Langkop, C.W.; Schell, W.L.; Stobierski, M.G.; Schuchat, A.; Besser, J.M.; Dietrich, S.; Helsel, L.; Griffin, P.M.; McFarland, J.W.; Osterholm, M.T. (1999). Outbreaks of salmonellosis associated with eating uncooked tomatoes: Implications for public health. *Epidemiology and Infection, 122*: 385–393.

Marler-Clark. (1993). Jack-in-the-Box *E. coli* outbreak lawsuits – western states (1993). Seattle, WA: Marler-Clark, Attorneys at Law. Retrieved from http://www.marlerclark.com/case_news/view/jack-in-the-box-e-coli-outbreak-western-states

National Restaurant Association. (2013). State food safety regulatory requirements. Chicago, IL: National Restaurant Association. Retrieved from http://www.servsafe.con/regulatory/foodsafety

Penn State Extension. (2013). Baked potatoes and botulism. University Park, PA: The Pennsylvania State University. Retrieved from http://extension.psu.edu/food/safety/food-preservation/faq/baked-potatoes-and-botulism

Sinclair, U. (1906). *The Jungle.* Mineola, NY: Dover Publications.

U.S. National Office of Vital Statistics. (1952). Provisional statistics for specified noticeable diseases in the United States for week ended January 5, 1952. *Morbidity and Mortality Weekly Report, 1*(1), 1–8. Retrieved from: http://stacks.cdc.gov/view/cdc/291/

UPI. (2010). Funding threatens state health departments. *UPI Health News.* Retrieved from http://www.upi.com/Health_News/2010/12/15/Funding-threatens-state-health-depts/UPI-10991292395720/

CHAPTER 3

MICROORGANISMS IN FOODS: A HISTORICAL PERSPECTIVE

JAY NEAL, PhD

Assistant Professor, University of Houston

CONTENTS

Millions of years ago microbes were the first forms of life on the planet. Microbes inhabited the air, water and soil. As hunters and gathers, humans consumed microbes in and on their foods. During this time, humans did not preserve food nor did they have a method to do so. Gradually, they began to produce food and through fermentation, microbes became part of their diet. Records of fermentation of foods date back to 4000 B.C. In addition, other methods for food preservation emerged including sun drying, smoking, freezing and salting. The Romans were known for food preservation of meats other than beef and were known for packing prawns and other meats in snow, according to Seneca (Jay, Loessner, and Golden, 2005). It is estimated that wine and cheese production developed around 1000 B.C.

While there is little historical record of foodborne illness prior to the 19th century, religious dietary laws and food taboos may reflect awareness of life threatening illnesses from the consumption of foods containing disease-causing microorganisms, also known as pathogens (Erbgurth, 2004). An early example of a food taboo does exist from the 10th century when Emperor Leo VI of Byzantium (886–911) forbids the production of blood sausages (Erbgurth, 2004). In fact, the bacterium *Clostridium botulinum* was named because of its pathological association with sausages (Latin word for sausage is *botulus*) (Erbguth, 2004). Reports of ergot poisoning (caused by *Claviceps purpurea*), a fungus found on rye and other grains, caused thousands of deaths during the Middle Ages. Some have speculated that the Salem witchcraft crises may have been attributed to ergot poisoning (Caporael, 1976). Antonie van Leeuwenhoek was the first to observe microorganisms in the 1670s, which he described as "little animals" in rainwater using a primitive microscope (Montville and Matthews, 2005). Both Francesco Redi (1626–1697) and Lozzaro Spallanzani (1729–1799) worked to disprove the theory of spontaneous generation of life. Prior to their research, it was common belief that certain forms of life could arise from inanimate matter (i.e., maggots from meat) known as spontaneous generation. However, it was not until the 1860s that Louis Pasteur disproved spontaneous generation and the scientific community accepted that life can only come from other life (Montville and Matthews, 2005).

Between 1789 and 1793, Nicolas Appert discovered a food preservation process that would become the basis of food processing 200 years later (Jay et al., 2005; Montville and Matthews, 2005) He established a cannery in 1802 and quickly began exporting foods to other countries as well as rations for the French navy. The first person to understand and apply food microbiology to solving food quality and food safety issues was Louis Pasteur. His first research in the 1830s was concerned with souring of milk. Later, he applied heat to destroy undesirable microorganisms in beer and wine and this process was named after him; pasteurization. His study of fermentation led him to formulate the germ theory of infectious disease. He then began to study various specific diseases. At the same time, Robert Koch earned his medical degree and began studying anthrax and identified the bacteria as *Bacillus anthracis*. Koch established criteria for proving that a bacterium caused a disease. The criteria became known as Koch's postulates.

3.1 KOCH'S POSTULATES

1. The microorganism must be found in abundance in all organisms suffering from the disease, but should not be found in healthy organisms.
2. The microorganism must be isolated from a diseased organism and grown in pure culture.
3. The cultured microorganism should cause disease when introduced into a healthy organism.
4. The microorganism must be reisolated from the inoculated, diseased experimental host and identified as being identical to the original specific causative agent.

In addition to Koch's postulates, two other innovations from his lab have had a significant impact on microbiology. One of Koch's assistants, Robert Petri, invented the petri dish for plating samples, however, they used gelatin for media. Gelatin does not solidify in warm temperatures and another assistant in Koch's lab, Walter Hesse, complained to his wife Fanny about this dilemma. Mrs. Hesse was an established jam and jelly maker and used agar as a thickening agent and suggesting using agar as part of the plating method (Montville and Matthews, 2005). It is interesting to note that a cook significantly impacted food bacteriology forever linking these two endeavors. The traditional foodborne pathogens (*Salmonella*,

Staphylococcus aureus, *Bacillus cereus*, *Clostridium perfringens*) were identified in the early 20th century. While great progress was made during this time, some experts suggest that only one percent of all the bacteria in the biosphere can be detected by cultural methods in the lab suggesting that 99% of bacteria remain undiscovered (Montville and Matthews, 2005).

During the late 1950s the Pillsbury Corporation and the U.S. Army Natick Laboratory developed a risk-based approach to food safety that would later become known as Hazard Analysis and Critical Control Points (HACCP). With the development of HACCP for NASA and the space program, researchers moved away from endpoint testing of products to proactive risk assessment resulting in fewer people becoming sick. In spite of such improvements however, foodborne illness is still considered a serious public health threat today.

3.2 PATHOGENS

A foodborne pathogen is an organism that causes disease through food. While more than 250 diseases are known to be transmitted by food (Bryan, 1982), Scallon, et al. (2011) reported that each year, 31 major pathogens acquired in the United States caused 9.4 million episodes of foodborne illness, 55,961 hospitalizations, and 1351 deaths. Of these 31 pathogens, 21 are bacteria, five are parasites and five are viruses (*see* Table 1). The majority of illnesses are caused by norovirus, nontyphoidal *Salmonella*, *Clostridium perfringens*, and *Campylobacter* spp. These microorganisms may be naturally present on the food item (i.e., chicken), brought in from the field on produce boxes, or acquired through cross contamination during deliveries or through human contact. Knowing what microorganisms are causing illness and where they are coming from is important to know for allocating resources and prioritizing intervention strategies (Scallon et al., 2011). Unfortunately however; only a small percentage of illnesses are diagnosed and reported, so these data may not represent the whole picture. To diagnose an illness, first the ill person must seek medical care. Depending on the microorganism, the symptoms of the illness may not warrant a trip to the emergency room or a doctor's visit. Second, specimens had to have been submitted for lab testing. This may include samples from the suspected food item which may or may not still be available or a stool sample. See Koch's postulates above. Third, the lab must test and identify the causative agent. Unfortunately, this does not always occur and the lab may not be able to determine the source of the illness. The classic method for culturing bacteria was the plate count method. One of the limitations of this method is that it assumes that every cell forms one colony and that every colony originates from one colony (Montville and Matthews, 2005). Fortunately, better technology is now giving us better results. Nucleic acid (RNA and DNA)-based assays such as PCR, pulsed-field gel electrophoresis, ribotyping, plasmid typing, randomly amplified polymorphic DNA, and restriction fragment length polymorphism can differentiate and identify foodborne pathogens. Lastly, the illness will need to be reported to the public health authorities. This is a complex and time-consuming endeavor that may or may not identify the causative agent. Many outbreaks are not recognized due to their small size, long incubation periods or geographic dispersion (Dewaal, et al., 2006). The majority of illnesses are not reported because the symptoms are self-limiting, meaning that in a few days, the symptoms have either stopped or are manageable.

3.3 OUTBREAKS

The Centers for Disease Control and Prevention (CDC) define a foodborne disease outbreak as the occurrence of two or more similar illnesses resulting from the ingestion of a common food (CDC, 2013). Foodborne illness outbreaks are investigated by local or state health departments and the quality of these investigations varies depending on funding (Hoffman, et al., 2005; Dewaal et al., 2006). Once this information has been collected, the CDC collects data on foodborne disease outbreaks from all 50 states, the District of Columbia and Puerto Rico through the Foodborne Disease Outbreak Surveillance System (FoodNet). Data are collected for each outbreak, including the number of illnesses, hospitalizations, deaths, the etiological agent (when identified), the implicated food, and any other factors such as behaviors or food preparation methods (CDC, 2013). Multi-state foodborne outbreaks occurring in the last few years have been associated with the consumption of such food items as: spinach, romaine lettuce, basil, sprouts, celery, cantaloupe, tomatoes, peppers, ground beef, turkey, raw cookie dough, frozen pizzas, and frozen pot pies. It is not uncommon for outbreaks to be associated with multiple foods, including meat and nonmeat items and may occur as a result of cross contamination in which investigators are then not able to confirm a specific common food during an outbreak investigation. For example, several of the large multistate produce related outbreaks were caused by pathogens that live in the intestinal tracts of animals and have historically contaminated meat and poultry such as *Campylobacter* spp. and *E. coli* O157:H7. The five most commonly implicated single-food categories implicated in outbreaks are seafood, produce, poultry, beef and eggs. Therefore interventions directed at these commodities may reduce the risk of outbreaks. For example, the majority of egg outbreaks are associated with *Salmonella* Enteritidis; therefore interventions on the farm, through the distribution channel and in the kitchen should focus on that pathogen (Dewaal et al., 2006).

3.4 BACTERIA

Bacteria are single-celled microscopic organisms that can be found almost everywhere on the planet. The majority is not harmful and some may even be beneficial. However, some may be pathogenic. For the purpose of this chapter, we will focus only on the pathogenic types that contaminate food. Some bacteria cause infections and release endotoxins in the intestines of the infected person which later causes illness. An endotoxin is a toxin present in the bacteria's cell wall that is released when the bacteria dies. Illnesses caused by infection typically have a longer onset time of 12 to 48 h before the person starts showing symptoms such as vomiting or diarrhea. Foodborne intoxication is caused by exotoxins, which are created by bacteria during multiplication. They are highly toxic and most cannot be cooked out of the food product. Intoxications have a much shorter onset time ranging from 30 min to 12 h depending on how much of the toxin was consumed. In addition, some pathogens are spore formers (*Clostridium botulinum, Clostridium perfringens,* and *Bacillus cereus)*, which can survive adverse heat temperatures. By controlling the factors that contribute to the growth and survival of bacteria, managers can reduce the risk of foodborne illness outbreaks.

TABLE 1 Most common foodborne pathogens known to cause hospitalization and deaths in the United States.

Bacteria	Parasites	Viruses
Bacillus cereus	*Cryptosporidium* spp.	Astrovirus
Brucella spp.	*Cyclospora cayetanensis*	Hepatitis A virus
Campylobacter spp.	*Giardia intestinalis*	Norovirus
Clostridium botulinum	*Toxoplasma gondii*	Rotavirus
Clostridium perfringens	*Trichinella* spp.	Saporovirus
Escherichia coli O157:H7		
Shiga toxin *Escherichia coli* (STEC), non O157		
Enterotoxin *Escherichia coli*		
Diarrheagenic *Escherichia coli* non STEC and ETEC		
Listeria monocytogenes		
Mycobacterium bovis		
Salmonella spp., nontyphoidal		
S. enterica serotype Typhi		
Shigella spp.		
Staphylococcus aureus		
Streptococcus spp. group A, food-borne		
Vibrio cholera, toxigenic		
V. vulnificus		
V. parahaemolyticus		
Vibrio spp., other		
Yersinia enterocolitica		

3.5 INTRINSIC AND EXTRINSIC FACTORS

Bacteria in food need specific characteristics within the food itself to survive. These are known as *intrinsic factors*, which contribute to microbial growth and include things such as pH, moisture content (water activity), nutrient content or compounds added as preservatives. pH is a unit of measurement to rate the acidity or alkalinity of a food item and is measured on a scale of 1–14. Most bacteria prefer a pH range between 4.7 and 7, which is considered mildly acidic to neutral. However, bacteria can adapt to their environment and several pathogens such as *E. coli* O157:H7 and some *Salmonellas* can survive in slightly acidic conditions. One of the oldest food preservation methods is drying which is a direct result of removing water, a nutrient which microorganisms cannot grow without (Jay et al., 2005). The water requirement of microorganisms is described as water activity (a_w). Pure water has a a_w 1.00 and most fresh foods are above 0.99. Most spoilage bacteria cannot grow below a_w=0.91 however, some pathogens such as *Staphylococcus aureus* can grow as low as 0.86. In addition to water, in order to grow properly, microorganisms require a source of energy such as protein or carbohydrates, nitrogen, vitamins and minerals. Finally, certain naturally occurring substances as well as preservatives may be added to foods to prevent attack by microorganisms. Several plant species contain essential oils that have antimicrobial properties. These include: eugenol in cloves, allicin in garlic, cinnamon aldehyde and eugenol in cinnamon, allyl sothiocynate in mustard, eugenol and thymol in sage, and carvacrol (isothymol) and thymol in oregano (Jay et al., 2005).

Certain properties of the storage environment also influence the growth of bacteria and are known as *extrinsic factors*. These include temperature, relative humidity, presence and concentration of gases in the environment (such as oxygen), and presence of other microorganisms (competition). Unfortunately, microorganisms can grow over a wide range of temperatures. Customarily, microorganisms have been classified into three groups based on their temperature requirements for growth. The first group are *psychrotrophs* with an optimum range of 68°F–86°F (20°C–30°C). This temperature range is why certain spoilage bacteria such as *Pseudomonas* can grow at refrigerated temperatures. *Mesophiles* (the second group), include most pathogens, and grow optimally within a range of 86°F–113°F (30°C–45°C). Microorganisms that grow at hotter temperatures are classified in the third group and are referred to as *thermophiles*, however they are not known to grow in foods. The relative humidity of the storage environment is also critical. When the water activity of a food is low (to prevent microbial growth), it is important that the food be stored in dry conditions so that the food does not pick up moisture from the air to a point where microorganisms can grow. The presence of certain gases is also important. Some bacteria can multiply only in the presence of oxygen (*aerobes*) while others can only multiply when there is no oxygen (*anaerobe*). Some can multiply with or without the presence of oxygen (*facultative anaerobes*) depending on the environmental conditions. This is important in processing and packaging of foods. In addition, the cooking process removes oxygen from the food, which may lead to ideal conditions for anaerobes in the center of foods such as sauces, gravies or stews. Processing and packaging methods such as vacuum packaging and sous vide may prevent the growth of spoilage bacteria and some aerobic pathogens however; the removal of oxygen may create a suitable environment for anaerobic pathogens such as *Clostridium botulinum,* which is the organism that causes botulism. Other gases such as carbon dioxide (CO_2) are used frequently to control microorganisms in foods. Finally, another factor that contributes to the growth or prevention of bacterial multiplication is competition from other microorganisms. If there are different bacteria present on a food item such as *Pseudomonas,* which causes spoilage, and *Listeria monocytogenes,* which is a pathogen, they will be competing for the same nutrients, especially if the oxygen levels, temperature and pH are ideal.

3.6 BACTERIA GROWTH STAGES

Bacteria can be found in the environment, in the digestive tracts of domestic animals, wild animals and in and on people. When bacteria are introduced into a food item (usually through cross-contamination) they do not begin to grow immediately. There is an initial phase known as the *lag phase*, where the bacteria adjust to their new surroundings. During this phase, the bacteria are alive and can cause disease, depending on the number present, but they are not actively growing. Typically, bacteria are transferred in low amounts but need to grow to larger amounts in order to cause illness, but there are a few exceptions such as *Listeria monocytogenes,* which can cause illness with only a few cells. By manipulating the intrinsic and extrinsic factors (which might include using refrigeration or hot holding equipment), the lag phase can be increased which will prevent bacteria from growing to unsafe levels. Once bacteria adjust to the environmental conditions however, they will multiply using binary fission; one cell divides into two cells that divide into four cells that double into eight cells and so on. This is known as the *logarithmic phase* and the bacteria will continue to double as long as nutrients are available. Under ideal conditions most pathogens can double in number every 20 min. Some bacteria however, such as *Vibrio vulnificus*, which may be found in oysters, can double in as little as seven minutes. Eventually, the same number of bacteria produced by multiplication equals the same amount of bacteria that are dying. This is known as the *stationary phase.* Lastly, nutrients deplete or temperatures increase to unsafe levels and more bacteria are dying than are multiplying. This is known as the *death or decline phase.*

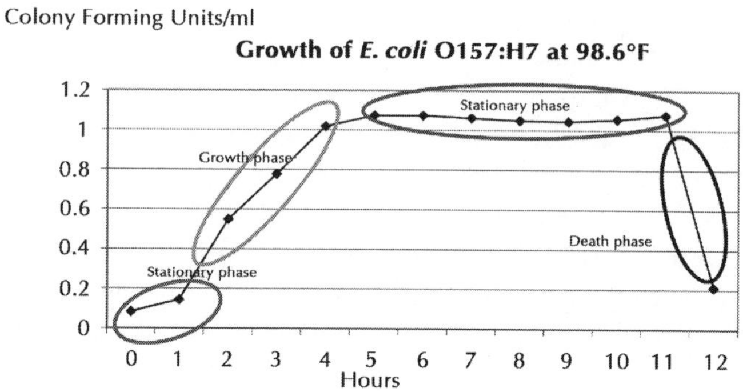

3.7 VIRUSES AND PARASITES

As mentioned above, viruses are the most common cause of foodborne illness outbreaks. Viruses consist of DNA or RNA surrounded by a protein shell; therefore, they are a more complex biochemical mechanism as compared to other living organisms. Viruses cannot reproduce on their own and require a living host. Some viruses remain outside of the cell and attach to specific receptor sites where they inject their genetic information (DNA or RNA) into the cell, which forces it to make genetic copies of the virus. Other viruses enter human cells through a watery pathway between proteins known as a protein channel. Once the virus has entered the cell, it infects the cell with its DNA or RNA, which causes the host cell to make genetic copies of the virus. These genetic copies remain within the host cell until the infected cell bursts releasing the copies, which then infect the surrounding healthy cells. Viruses can survive freezing and can remain on inanimate objects such as door handles, cutting boards and clothing for extended periods of time. Parasites are plants or animals that live off of other living plants or animals. They may be single celled or may be so large that they are visible to the naked eye (worms, for example). Certain parasites, such as *Trichinella spiralis,* which were once common in the U. S. have declined while others such as *Cryptosporidium* spp. and T*oxoplasma gondii* are being reported more frequently.

3.8 HURDLE TECHNOLOGY

It is important for food service managers to be aware of the intrinsic and extrinsic factors that contribute to bacterial growth because these factors can be used in combinations to protect and preserve food. Hurdle technology is a method, which deliberately incorporates multiple intrinsic and extrinsic factors in combinations to ensure that food is safe for consumption and perhaps extend the shelf life of the product. The most significant hurdles used in food preservation include temperature (high or low), water activity (a_w), redox potential (Eh), preservatives (nitrite, sorbate or sulfite), and competitive microorganisms such as lactic acid bacteria, which can reduce the pH as a result of its metabolism (Leistner, 2000). Each factor or approach may be considered a "hurdle" that the pathogen must overcome if it is to survive and grow within a food item. Hurdle technology may also combine multiple factors, which will reduce bacterial growth while still maintaining acceptable food quality. This is important because certain techniques such as reducing the water activity, if done by itself, might have a negative effect on food quality. For example, if the salt content of canned green beans is increased, the amount of water available within a food item will be reduced; however, the salt level might be so high that the taste is too salty for consumers' preferences. The goal of hurdle technology is to create a synergistic effect of hurdles that simultaneously hits different targets (cell membrane, DNA, enzyme systems, pH, a_w, Eh) and disturbs the stability (homeostasis) of the microorganism to a degree that the microorganism cannot overcome it.

3.9 SOURCES OF CONTAMINATION

For years, foodborne illness outbreaks were associated with eating at restaurants, potluck dinners and church picnics. Currently, there is a changing paradigm in which outbreaks are becoming associated with multistate outbreaks, as a result of a failure in the food safety network. Much of the contamination is associated with poor agricultural practices, which include not controlling animals (wild and domestic) around produce, contamination at processing facilities, and long distribution routes. It is imperative to realize that contamination can occur anywhere within the farm to fork continuum. Some outbreaks occur on the farm and food service managers need to ensure that they are getting foods delivered from reputable suppliers. At the same time, contamination may come from employees who come to work sick, may have sick family members at home or have poor personal hygiene.

3.9.1 FOODS ASSOCIATED WITH FOODBORNE ILLNESS OUTBREAKS

One of the first steps towards preventing foodborne illness outbreaks is to identify foods most commonly associated with outbreaks (Painter, et al., 2009). It is important to know which foods are most commonly associated with outbreaks in order to develop interventions both on the farm and within the kitchen. The five single-food categories most commonly implicated in outbreaks are: seafood, produce, poultry, beef and eggs (Dawaal et al., 2006). These food items may contain pathogens either in or on their surfaces, particularly eggs, poultry and beef. Linking particular foods to specific foodborne illness outbreaks allows for greater inspection of those foods and their production, processing or transport. For example, 40% of the seafood outbreaks are linked to improper holding temperatures, 40% of the produce outbreaks are related to poor personal hygiene, and an additional 30% of produce-linked outbreaks are associated with cross contamination. Inadequate cooking accounts for more than 40% of the beef related outbreaks (Dewaal et al., 2006). Produce may become contaminated from irrigation water contaminated with animal feces or at the packing shed through the washing process. All food items may become cross-contaminated by food service workers with poor personal hygiene. For example, if a produce linked outbreak was caused by a norovirus, this would indicate that the produce was either irrigated with contaminated water or was infected by human handling either at the packing shed or in the kitchen.

3.9.2 SEAFOOD

The majority of outbreaks associated with seafood including fin fish and crustaceans, are related to seafood being harvested from contaminated waters which may contain norovirus, Hepatitis A, *Vibrio vulnificus, Vibrio parahaemolyticus* as well as histamine, scombrotoxin, and ciguatoxin which comes from toxic algae. Several seafood commodities are inherently more risky due to several factors such as: the nature of the environment in which they live, their mode of feeding (filter feeding for molluscan shellfish), the season in which they are harvested, and how they are prepared and served (Iwamoto, et al., 2010). Mollusks, including oyster and clams are the most common seafood associated with foodborne illness, followed by finfish and lastly crustaceans.

Produce. Fresh fruits and vegetables are an important part of our diet, however, many of these are consumed raw and do not have a terminal kill step for pathogens such as cooking. The number of reported foodborne illness outbreaks associated with fresh produce as well as the variety of produce involved with outbreaks has increased in the past 30 years (Alkertruse and Swerdlow, 1996, Hedberg and Osterholm, 1994; Neal, Cabrera and Castillo, 2012; Sivapalasingam, et al., 2004). This can be attributed to changes in consumption patterns, changes in production and processing technologies, new sources of produce as well as the manifestation of pathogens such as *Salmonella* and *Escherichia coli* O157:H7 which have not been previously associated with raw produce (Burnett and Beuchat, 2001; Sivapalasingam et al., 2004; Hanning, Nutt and Ricke, 2009). The critical factors causing microbial contamination of produce include contaminated agricultural water used in irrigation or washing of equipment, biological soil amendments (composted manure), human health and hygiene issues, close proximity of animals (wild and domestic) to growing fields, and equipment contaminated by water or animal waste (Tarver, 2013). Bacteria can be introduced onto fruits and vegetables at any step from planting to consumption and once they are introduced, their colonization can have a tremendous effect on both the quality and safety of the product. The attachment and colonization of microorganisms on and in fresh produce have significant public health implications because sanitizers and decontamination treatments may not be able to remove or inactivate human pathogens within the food item (Beuchat, 2002; Frank, 2001). Bacteria attach to fruits and vegetables in pores, indentations and natural irregularities on the produce surface where there are protective binding sites as well as cut surfaces, puncture wounds, and cracks in the surface creating more opportunities for contamination (Sapers, 2001; Seo and Frank, 1999).

3.9.3 POULTRY AND EGGS

Campylobacter jejuni and nonTyphoidal *Salmonella* are two of the most common bacterial sources of foodborne illness. Both pathogens are found in poultry and can be introduced through multiple sources including feed, water, the environment, litter and bird-to-bird contamination. Chickens are coprophagic (consume feces) which is often the source of contamination leading to the colonization of entire flocks Once internalized, pathogens can colonize in the ceca and small intestines, but may also be found in the spleen and liver and reproductive track. Thus eggs can be contaminated both inside and outside of the shell. Conditions within the broiler house such as overcrowding also contribute to increased levels of contamination. Chickens can also become contaminated when placed in cages for transport to harvesting facilities. Chicken feces can adhere to feathers and may be difficult to remove and may spread pathogens throughout the harvesting facility. It is not uncommon for bacterial loads to be higher after harvesting (Heyndrickx, et al., 2002). Multiple methods for reducing the incidence of *Salmonella* and *Campylobacter jejuni* on poultry include: maintenance of salmonella-free breeding stock, fumigation of eggs for hatching, strict hygiene, cleaning and disinfecting in hatcheries and rearing houses between flocks, heat treatment of feed, clean clothing for workers, control of vermin, disinfection of water and the use of nipple drinking systems rather than open water, acidified feed, and competitive exclusion (Cory, et al., 2002). Competitive exclusion, also known as the Nurmi concept, is the process of introducing probiotic bacteria or mixed cecal contents of disease free adult birds to newly hatched chicks to prevent the colonization of pathogens within their digestive tracts. The use of organic acid washes pre and postharvest are common is processing facilities. Organic and free range chickens are becoming more popular because they do not use antibiotics or growth hormones and permit the chickens to have a more natural diet. While this may appear to be a healthy choice and a more humane treatment of the birds, unfortunately, these conditions contribute to higher counts of pathogens found on farms, within processing facilities, and on the birds postharvest.

3.9.4 BEEF

Every year, thousands of pounds of ground beef are recalled due to contamination by Shiga toxin-producing *Escherichia coli* O157:H7 and *Salmonella*. Sources of *E. coli* O157:H7 in beef include the hides, hooves and horns of animals entering beef harvesting facilities or the intestinal contents which may cross-contaminate carcasses during the harvest process

(Huffman, 2002). Over the past five years, the United States Department of Agriculture's Food Safety Inspection Services (FSIS) has reported approximately 60 food recalls for ground beef, ground beef products and beef trimmings that have tested positive for *E. coli* O157:H7, with the establishments reporting an estimated 21 million pounds of recovered product (USDA, 2011). The purpose of food recalls is to remove food from commerce when it is believed to be injurious or harmful for human consumption. This is important because the symptoms of STEC infections often include stomach cramps, diarrhea (which may develop into hemorrhagic colitis) and vomiting. Even more unfortunate is the fact that 5–10% of the individuals diagnosed with STEC infections develop hemolytic uremic syndrome (HUS), which may lead to renal failure (5%) and death (3–5%) (Mead and Griffin, 1998). The Centers for Disease Control and Prevention (CDC) reported that an average of 2,138 cases and 20 deaths in the U.S. are attributed to this pathogen annually (Scallan et al., 2011). Interventions such as water baths prior to harvesting, removal of hooves, horns and hides before evisceration, and hot lactic acid carcass washes all reduce the microbial load of the end product.

Restaurants are still an important source of foodborne illness outbreaks (approximately half) (Brown, 2013, Lynch, Painter, Woodruff and Braden, 2006). In the mid-1990s, the Food and Drug Administration (FDA) developed the National Food Retail Steering Committee, tasked with reducing the occurrence of the leading risk factors in foodborne illnesses in the food service industry, as identified by the Centers for Disease Control and Prevention (CDC). The leading risk factors were identified as (1) improper holding temperature, (2) contaminated equipment, (3) poor personal hygiene, (4) food from unsafe sources, (5) inadequate cooking. A few years later, a sixth risk factor was added, (6) other/chemicals. An example of improper holding temperature is when potentially hazardous foods (PHF), or temperature controlled for safety foods (TCS), are held below 135°F or above 41°F for more than four hours. An example of other/chemicals is not properly storing or identifying poisonous or toxic materials. Outbreak reporting is vital for proper foodborne illness surveillance and helps prevent future outbreaks. The greatest challenge with identifying the cause of outbreaks occurs because cross contamination can happen at almost any stage during food production (Allos, et al.,

2004). The most common situations associated with outbreaks in restaurants are infected food service employees, and employees handling food with improperly washed hands (Hedberg et al., 2006).

3.9.5 IMPROPER HOLDING TEMPERATURE

Two spore-forming bacteria, *Clostridium perfringens* and *Bacillus cereus* have traditionally been associated with foods linked to foodborne illness by inadequate hot holding. In addition, *Salmonella* and *Staphylococcus aureus* have also been identified as causative agents. This suggests that survival of vegetative cells and toxins as well as postcooking contamination are concerns in temperature control. Depending on the service style of a food service operation, some operations may want to hot hold certain foods such as soups for four to eight hours. Foods held at high temperatures for extended amounts of time however may have significant quality losses including color, texture, taste and nutrient content. Therefore, operations are recommended to limit hot holding times. Not only does this better maintain product quality, it also reduces the risk of foodborne illness. In addition, there is less chance of contamination of these hot held foods from the hands of infected workers, customers or contaminated equipment. Finally, when food is hot held at a constant temperature for an extended time, the surface temperature may be significantly lower than the interior of the food product. By covering the food item and stirring it often, the temperature can be evenly distributed throughout the food product which will also reduce the risk of postcooking contamination and ensure temperatures that are sufficient to prevent survival and growth of pathogens.

3.9.6 CONTAMINATED EQUIPMENT

Foodservice equipment has been implicated as a contributing source of pathogens via cross-contamination. One possible explanation is that *Listeria monocytogenes* can adhere and create biofilms on stainless steel and other materials commonly found in processing and retail operations (Beresford et al., 2001; Lin, 2006). Biofilms are defined as "a structured community of bacterial cells enclosed in a self-produced polymeric matrix adherent to an inert or living structure" (Costerton, Stewart, and Greenberg, 1999). In layman's terms, this means that bacteria can join together on surfaces and form a protective matrix around the group. Biofilms can survive and grow on stainless steel commonly found in food service operations and have been isolated on table tops, slicers, grinders, knives, and cutting boards. Slicing machines have been implicated in foodborne illnesses including typhoid fever, Hepatitis C and listeriosis (Bocket et al.; 2011; Howie, 1968; Neal, 2013; Vorst, Todd and Ryser, 2006). Commercial slicers and similar types of equipment can transfer *L. monocytogenes* onto deli meats depending on the microbial load found on the slicer blade (Lin et al., 2006). While environmental contamination and equipment have been implicated as contributing factors for the growth and survival of *L. monocytogenes* in both processing facilities and retail delis, the deli has proven to be more challenging to control. Ready to eat meat and poultry that are sliced in the foodservice operation are more likely (one estimate as high as five times) to cause listeriosis than are prepackaged products (Gombas et al., 2003; Endrikat et al., 2010). While collecting environmental samples, Sauders, et al. (2009) discovered *L. monocytogenes* in 60% of the retail establishments tested. Used kitchen towels in restaurants may also be a significant potential source of foodborne pathogens contributing to cross contamination events in the establishment. Towels used in restaurants may be contaminated with high levels of bacteria capable of transferring onto food preparation surfaces as well as food items prepared on these surfaces. Restaurants need to focus more attention on proper usage and storage of kitchen towels in the work place.

3.9.7 POOR PERSONAL HYGIENE

Employee behaviors have a significant impact on the safety of food served to the public. Pathogens can be easily spread from food workers' clothing, hands and arms, and hair. Food workers must wear clean clothes, aprons and hair restraints to prevent the spread of pathogens such as *Shigella, E. coli* O157:H7, *Salmonella,* and *S. aureus*. Employees must wash their hands properly in hand-washing sinks only, maintain trimmed fingernails, not wear jewelry, and not touch ready-to-eat foods with bare hands. Proper use of single use gloves also reduces the risk of spreading disease. Eating and drinking in designated areas and not allowing gum chewing or smoking will also reduce risks. Employees must be educated that they should not come to work if they have sores or boils on their arms or hands that have a persistent discharge. If an employee is sneezing, coughing or has a runny nose or discharge from the eyes, nose or mouth, they should be restricted from working with food. Should an employee have symptoms of vomiting, diarrhea, jaundice or sore throat with fever, they should report it to their manager and should not come to work. Employees must be excluded from coming to work if they have been diagnosed with norovirus, *Salmonella* Typhi, *E. coli* O157:H7, *Shigella,* or Hepatitis A because they are highly contagious and put the public's health at great risk.

3.9.8 FOOD FROM UNSAFE SOURCES

Multiple foodborne illness outbreaks have been traced back to food from unapproved sources, therefore, purchasing food from approved sources is critical. Unapproved sources include food prepared from a private home and unlicensed food manufacturers who have not been inspected or do not have the appropriate permits. Food establishments must be able to demonstrate that all food suppliers are approved and are permitted through the appropriate enforcement agency. Recently, farmers markets have become very popular and sell directly to restaurants, however, farmers selling directly to restaurants must be able to demonstrate that they are practicing Good Agricultural Practices (GAPs). Other regulations may also apply, depending on the jurisdiction. Shellstock tags are used to identify where oysters, clams and mussels were harvested and must be kept for 90 days.

3.9.9 INADEQUATE COOKING

Similar to improper holding, inadequate cooking can also lead to foodborne illness. As mentioned above, pathogens can be introduced to food items anywhere along the farm to fork continuum. Improper cooking may permit pathogens to survive in foods. Recommended cooking temperatures are established to target the most heat resistant pathogens (such as *Salmonella*), which will also kill more heat sensitive pathogens like *C. jejuni.*

3.10 IS OUR FOOD SAFE?

Given the amount of pathogens that can contaminate our food and the multiple routes and sources of potential contamination, one must ask if our food supply safe. It seems like we hear more and more about outbreaks and recalls every day. This is a complex issue that has many factors contributing to the safe supply of food however, based on the amount of food consumed every day in the U.S., the food supply is safe. This does not mean that there are no risks associated with eating. Foodborne illness caused by harmful bacteria and other pathogenic microorganisms in meat, poultry, seafood, dairy products and other foods is a significant public health problem in the U.S. There are risks associated with every daily decision we make, such as which route to drive to work. Similarly, there are risks associated with certain foods we eat. Certain foods have higher risks of contamination than others. For example, eating rare or medium rare ground beef may increase the risk of contracting *E. coli* O157:H7, eggs sunny side up may contain *Salmonella* or consuming oysters from unapproved waters may increase the risk of hepatitis A or *Vibrio vulnificus*.

Because foodborne illness can result in long-term health issues such as kidney dialysis, arthritis and may even cause death, much effort has been put into creating better surveillance and reporting of outbreaks. Outbreaks often have characteristic features in regards to incubation periods, duration of symptoms, the percentage of cases that experience specific signs and symptoms and the cause of the disease, also known as the etiology (Hedberg, et al., 2008). Several organizations track these trends for infections transmitted through foods. The Centers for Disease Control and Prevention, the FDA, USDA and ten state health departments work through a collaboration known as The Foodborne Diseases Active Surveillance Network (FoodNet) to collect data regarding foodborne illnesses. FoodNet personnel contact the clinical laboratories in Connecticut, Georgia, Maryland, Minnesota, New Mexico, Oregon, Tennessee, and selected counties in California, Colorado and New York to get reports of infections diagnosed in residents from these areas. The surveillance area within these ten states represents 47 million people or 15% of the U. S population. By soliciting reports of illness, it uses an active surveillance approach meaning that it does not wait until an outbreak occurs. It is constantly collecting information to track incidents and trends of infection from nine pathogens commonly associated with food: *Campylobacter, Cryptosporidium, Listeria, Salmonella,* Shiga-toxin producing *Escherichia coli* (STEC) O157 and nonO157, *Shigella, Vibrio* and *Yersinia.* Each year, the CDC posts a progress report on their website comparing the percentage change in that year compared to 2006–2008 (*see* Table 2).

TABLE 2 Changes in Microbial Incidence for 2012.

FOOD SAFETY PROGRESS REPORT FOR 2012

Disease Agents	Percentage change in 2012 compared with 2006–2008	2012 rate per 100,000 Population	2020 target rate per 100,000 Population	CDC estimates that...
Campylobacter	14% increase	14.30	8.5	For every *Campylobacter* case reported, there are 30 cases not diagnosed
Escherichia coli O157	No change	1.12	0.6	For every *E. coli* O157 case reported, there are 26 cases not diagnosed
Listeria	No change	0.25	0.2	For every *Listeria* case reported, there are 2 cases not diagnosed
Salmonella	No change	16.42	11.4	For every *Salmonella* case reported, there are 29 cases not diagnosed
Vibrio	43% increase	0.41	0.2	For every *Vibrio parahaemolyticus* case reported, there are 142 cases not diagnosed
Yersinia	No change	0.33	0.3	For every *Yersinia* case reported, there are 123 cases not diagnosed

For more information, see http://www.cdc.gov/foodnet/

Preliminary FoodNet 2012 Data

3.11 RECALLS

Food recalls are different from outbreaks. Recalls of foods contaminated or thought to be possibly contaminated with pathogens actually reduces the transmission of diseases (Dey, et al., 2013). Often recalls occur when manufacturers or distributors voluntarily recall products believed to be contaminated based either on their own findings or those of public health agencies, which may be local, state or federal. With the enactment of the Food Safety Modernization Act, the FDA has the authority to order an organization to recall a product that may contain pathogens or allergens.

Successful food recalls are often large collaborations between the regulatory agencies, distributors, processors and growers (Dey et al., 2013). Large food distributors, (Sysco Foods and U.S. Foods) use technology to track foods using bar codes to determine which potentially contaminated foods have been distributed, and which have been returned to the warehouse.

3.12 CONCLUSION

While foodborne illness is a significant risk to the public health affecting millions of people each year, it can be prevented. Food service managers and employees can reduce the risk of illness by being able to identify bacteria, viruses and parasites. Bacteria have specific intrinsic and extrinsic factors that contribute to growth. Hurdle technology can be implemented to reduce or eliminate pathogens from foods by controlling multiple intrinsic or extrinsic factors. The most foods most commonly associated with foodborne illness include seafood, produce, poultry, beef and eggs. Knowledge of where contamination can occur in the harvesting and processing of these foods can also reduce risks of illness. Lastly, behaviors and procedures of foodservice workers play a critical role in providing safe food because they are the last safety intervention that can be applied prior to serving the customer. Protocols such as purchasing food from safe sources, using thermometers to cook, cool and reheat food, and training employees in proper hygiene collectively can reduce the risk of foodborne illness and protect the public.

KEYWORDS

- ***Claviceps purpurea***
- ***decline phase***
- ***extrinsic factors***
- ***hurdle technology***
- ***intrinsic factors***
- ***lag phase***
- ***logarithmic phase***
- ***pathogen***
- ***stationary phase***

REFERENCES

Alkertruse, S. F.; Swerdlow, D. L. (1996). The changing epidemiology of foodborne disease. The American Journal of Medical Sciences, 311, 23–29.

Allos, B. M.; Moore, M. R.; Griffin, P. M.; Tauxe, R. V. (2004). Surveillance for sporadic foodborne disease in the 21st century: The FoodNet perspective. Clinical Infectious Diseases, 38(Supplement 3), S115-S120.

Beresford, M. R.; Shama, G.; Andrew, P. W. (2001). *Listeria monocytogenes* adheres to many materials found in food-processing environment. Journal of Applied Microbiology, 90, 1000–1005.

Beuchat, L. R. (2002). Ecological factors influencing survival and growth of human pathogens on raw fruits and vegetables. Microbes and Infection, 4, 413–423.

Bocket, L.; Chevaliez, S.; Talbodec, N.; Sobszek, A.; Pawlotsky, J. M.; Yazdanpanah, Y. (2011). Occupational transmission of Hepatitis C virus resulting from use of the same supermarket meat slicer. Clinical Microbiology and Infection, 17, 238–241.

Brown, L. G. (2013). EHS-Net restaurant good safety studies: what have we learned? Journal of Environmental Health, 75, 44–45.

Bryan, F. L. (1982). Diseases transmitted by foods: A classification and summary, 2nd Edition. HIH publication no. (CDC) 84–8237. Atlanta, GA: U. S. Department of Health and Human Services, Centers for Disease Control.

Burnett, S. L.; Beuchat, L. R. (2001). Human pathogens associated with raw produce and unpasteurized juices, and difficulties in decontamination. Journal of Industrial Microbiology and Biotechnology, 27, 104–110.

Caporael, L. R. (1976). Ergotism: The satan loosed in Salem? Convulsive ergotism may have been a physiological basis for the Salem witchcraft crises in 1692. Science, 192, 21–26.

Centers for Disease Control and Prevention. (2013). Surveillance for foodborne disease outbreaks – United States, 2009–2010. Morbidity and Mortality Weekly, 62, 41–47.

Cory, J. E. L.; Allen, V. M.; Hudsen, W. R.; Breslin, M. F.; Davies, R. H. (2002). Source of salmonella on broiler carcasses during transportation and processing: Modes of contamination and methods of control. Journal of Applied Microbiology, 92, 424–432.

Costerton, J. W.; Stewart, P. S.; Greenberg, E. P. (1999). Bacterial biofilms: A common cause of persistent infections. *Science*, 284, 1318–1322.

Dewaal, C. S.; Hicks, G.; Barlow, K.; Alderton, L.; Vegosen, L. (2006). Foods associated with foodborne illness outbreaks from 1990 through 2003. Food Protection Trends, 26, 466–473.

Dey, M., Mayo, J. A., Saville, D., Wolyniak, C. and Klontz, K. C. (2013). Recalls of foods due to microbiological contamination classified by the U.S. Food and Drug Administration, fiscal years 2003 through 2011. Journal of food Protection, 76, 932–938.

Endrikat, S.; Gallagher, D.; Pouillot, R.; Quesenberry, H. H.; LaBarre, D.; Schroeder, C. M.; Kause, J. (2010). A comparative assessment for *Listeria monocytogenes* in prepackaged versus retail-sliced deli meat. Journal of Food Protection, 73, 612–619.

Erbguth, F. J. (2004). Historical notes on botulism, *Clostridium botulinum*, botulinum toxin, and the idea of the therapeutic use of the toxin. Movement Disorder, 19, S2-S6.

Frank, J. F. (2001). Microbial attachment to food and food contact surfaces. Advances in Food and Nutrition Research, 43, 319–370.

Gomabas, D. E.; Chen, Y.; Clavero, R. S.; Scott, V. N. (2003). Survey of *Listeria monocytogenes* in ready-to-eat foods. Journal of Food Protection, 66, 559–569.

Hanning I. B.; Nutt, J. D.; Ricke, S. C. (2009). Salmonellosis outbreaks in the United States due to fresh produce: Sources and potential intervention measures. Foodborne Pathogens and Disease, 6, 635–648.

Hedberg, C. W.; Osterholm, M. T. (1994). Changing epidemiology of food-borne diseases: A Minnesota perspective. Clinical and Infectious Diseases, 18, 671–682.

Hedberg, C. W.; Palazzi-Churas, K. L.; Radke, V. J.; Selman, C. A.; Tauxe, R. V. (2008). The use of clinical profiles in the investigation of foodborne outbreaks in restaurants: United States, 1982–1997. Epidemiology and Infection, 136, 65–72.

Hedberg, C. W.; Smith, S. J.; Kirkland, E.; Radke, V.; Jones, T. F.; Selman, C. A. (2006). Systematic environmental evaluations to identify food safety differences between outbreak and nonoutbreak restaurants. Journal of Food Protection, 69(11), 2697–2702.

Heyndrickx, M.; Vandekerchove, D.; Herman, L.; Rollier, I.; Grijspeerdt, K.; De Zutter, L. (2002). Routes for salmonella contamination of poultry meat: Epidemiological study from hatchery to slaughterhouse. Epidemiology and Infection, 129, 253–265.

Hoffman, R. E.; Greenblatt, J.; Matyas, B. T.; Sharp, D. J.; Estebanm, E.; Hodge, K.; Liang, A. (2005). Capacity of state and territorial health agencies to prevent foodborne illness. Emerging and Infectious Disease, 11, 11–16.

Howie, J. W., (1968). Typhoid in Aberdeen, 1964. Journal of Applied Bacteriology, 31, 171–178.

Huffman, R. (2002). Current and future technologies for the decontamination of carcasses and fresh meat. *Meat Science,* 62, 285–294.

Iwamoto, M., Ayers, T., Mahon, B. E., and Swerdlow, D. L. (2010). Epidemiology of seafood-associated infections in the United States. Clinical Microbiology Reviews, 23, 399–411.

Jay, J. M.; Loessner, M. J.; Golden, D. A. (2005). Modern Food Microbiology, 7th Edition. Springer: New York.

Leistner, L. (2000). Basic aspects of food preservation by hurdle technology. International Journal of Food Microbiology, 55, 181–186.

Lin, C. M.; Takeuchi, K.; Zhang, L.; Dohm, C. B.; Meyer, J. D.; Hall, P. A.; Doyle, M. P. (2006). Cross-contamination between processing equipment and deli meats by *Listeria monocytogenes.* Journal of Food Protection, 69, 71–19.

Lynch, M.; Painter, J.; Woodruff, R.; Braden, C. (2006). Surveillance for foodborne disease outbreaks – United States, 1998–2002. Morbidity and Mortality Weekly Report, 55, 1–34.

Mead, P. S.; Griffin, P. M. (1998). *Escherichia coli* O157: H7. *The Lancet,* 352, 1207–1212.

Montville, T. J.; Matthews, K. R. (2005). Food Microbiology an Introduction. ASM Press: Washington D. C.

Neal, J. (2013). Comparative analysis of training delivery methods for new employees cleaning and sanitizing retail deli slicers: An exploratory study. Food Control, 29, 149–155.

Neal, J. A. Cabrera-Diaz, E. and Castillo, A. (2012). Attachment of *Escherichia coli* O157:H7 and *Salmonella* on spinach (*Spinacia oleracea*) using confocal microscopy. Agriculture, Food and Analytical Bacteriology, 2, 275–279

Painter, J. A.; Ayers, T.; Woodruff, R.; Blanton, E.; Perez, N.; Hoekstra, R. M.; Griffin, P. M.; Braden, C. (2009). Recipe for foodborne outbreaks: A scheme for categorizing and grouping implicated foods. Foodborne Pathogens and Disease, 6, 1259–1264.

Sapers, G. M. (2001). Efficacy of washing and sanitizing methods for disinfection of fresh fruit and vegetable products. Food Technology and Biotechnology, 39, 305–311.

Scallan, E.; Hoekstra, R.; Angulo, F.; Tauxe, R.; Widdowson, M.; Roy, S.; Jones, J.; Griffin, P. (2011). Foodborne illness acquired in the United States-Major pathogens. Emerging and Infectious Disease. 17, 7–15.

Seo, K. H.; Frank, J. F. (1999). Attachment of *Escherichia coli* O157:H7 to lettuce leaf surface and bacterial viability in response to chlorine treatment as demonstrated by using confocal scanning laser microscopy. Journal of Food Protection, 62, 3–9.

Sivapalasingam S.; Friedman, C. R.; Cohen, L.; Tauxe, R. V. (2004). Fresh produce: A growing cause of outbreaks of foodborne illness in the United States, 1973 through 1997. Journal of Food Protection, 67, 2342–2353.

Tarver, T. (2013). Modernizing food safety from farm to fork. Food Technology, 67, 25–32.

United States Department of Agriculture Food Safety and Inspection Service. (2011). Irradiation and food safety answers to frequently asked questions. Available at: http://www.fsis.usda.gov/Fact_Sheets/Irradiation_and_Food_Safety/. Accessed 7 April, 2013.

Vorst, K. L.; Todd, E. C. D.; Ryzer, E. T. (2006). Transfer of *Listeria monocytogenes* during mechanical slicing of turkey breast, bologna, and salami. Journal of Food Protection, 69, 619–626.

CHAPTER 4

WHAT HAPPENS DURING FOOD ESTABLISHMENT INSPECTIONS IN THE UNITED STATES

BARBARA ALMANZA, PhD, RD
Professor, Purdue University

CONTENTS

Food establishment or restaurant inspections are a routine part of the food safety process in the United States and date back to at least the 1930s or 1940s with the first federal food code relating to restaurants (Almanza, Ismail, and Mills, 2002; Food and Drug Administration, 2012). The Food and Drug Administration (FDA) Food Code "establishes definitions; sets standards for management and personnel, food operations, and equipment and facilities; and provides for food establishment plan review, permit issuance, inspection, employee restriction, and permit suspension" (FDA, 2009). The FDA Food Code is currently updated every four years with the latest version published in 2009 (as of May 2013). The Food Code is not federal law or regulation, but does represent the best advice for a uniform system of regulation in regards to regulations set up by state and local regulatory agencies for food establishment inspections (FDA, 2009). According to the FDA (2011), 49 out of 50 states have adopted some version of the FDA Food Code. The version of the FDA Food Code used in state regulations depends on the frequency of legislative changes within that state. More specifically, four states had adopted the 1993, 1995 or 1997 (4%) as July 2011. In addition, ten states use the 1999 Food Code (13%), 11 states use the 2001 (38%), 20-one use the 2005 (39%), and three (as well as Puerto Rico) have adopted the 2009 Food Code (2%).

This variation in food codes from one state to another creates some variations in inspections. For example, although many states use a system that monitors and reports the number of critical and noncritical violations, other states do not. The state of Mississippi uses a letter grade system. In their system, an "A" grade is received if the food establishment had no critical violations (MSDH, 2013). A grade of "B" is received if critical violations were found, but corrected during the inspection. "C" indicates that critical violations were found but some or all of them were not corrected during the inspection. It is also given if critical violations were repeated from the last inspection.

Delaware (which is one of the three states that has adopted the 2009 FDA Food Code) monitors Priority, Priority Foundation, and Core violations in addition to a category that they call COS (Corrected on Site) for violations that are corrected during the inspection (DHSS, 2012). In their code, Priority and Priority Foundation violations are defined as those practices that are more likely to increase the risk of foodborne illness, whereas

Core violations involve important retail practices such as facilities maintenance and general sanitation, but are less likely to lead to foodborne illness. Finally, New Hampshire (NHDHHS, 2012) uses a system of colors to indicate the inspection results. Green inspections signify no priority item violations or that the priority item violation(s) was immediately and permanently corrected at the time of the inspection. Yellow inspections are those where a priority item violations was identified, but not correctly immediately and/or permanently at the time of the inspection. Red inspections indicate that there is an imminent health hazard or the food establishment was operating without a license, or that the license had expired.

These and other common types of inspection systems are outlined later in this chapter. In spite of these variations, the potential "violations" (whatever term is used by that state) that are monitored by a state's inspectors are generally the same, if not identical. All states will inspect the personnel handling, food, equipment and facilities, use of poisonous or toxic materials (such as cleaning chemicals), and their compliance with other regulations in the food code (such as requirements for manager certification, posting of inspection results, etc.). The following information is based on the federal food code (FDA 2009) and describes the typical process as regulated by most states. Temporary food establishments (found at fairs, festivals, or farmers' markets) that operate for a period of no more than 14 consecutive days in conjunction with a single event are regulated, but may be subject to different procedures for obtaining a permit and the inspection process. Not-for-profit food establishments are generally not subject to the inspection requirements in most states, but may be subject to certain regulations in some states or jurisdictions. Finally, cruise ships that dock at U.S. ports are also inspected somewhat differently in that they are inspected by the U.S. Centers for Disease Control and Prevention. Inspections for cruise ships are even more comprehensive in that they also include water disinfection practices in addition to food.

4.1 HOW NEW RESTAURANTS OBTAIN A PERMIT

Regulations for operation of a food establishment are extensive and typically begin prior to the opening of the restaurant. First, the owner or "officer of legal ownership" for the proposed food establishment submits to the local (city or county) regulatory authority the plans and specifications of the proposed food establishment for review and approval. The plans and specifications should include the intended menu; volume of food to be stored, prepared, sold or served; food establishment layout, mechanical schematics, construction materials and finish schedules; equipment types, manufacturers, models, locations, dimensions, performance capacities, and installation. This submission is required for: construction of a new food establishment, conversion of an existing structure to use as a food establishment, or remodeling or changing the type of food operation (as required in the state or local government code).

When construction is close to completion and is at least 30 days prior to the planned opening, the owner then contacts the local regulatory authority to submit a written application for a permit and arrange for a preoperational inspection. The permit application includes detailed information about the operation such as: owner name and contact information, ownership type, type of food establishment (mobile, temporary, or permanent), menu, preparation and handling methods, whether consumers will be an immune compromised or highly susceptible to foodborne illness population (such as a hospital, daycare, or nursing home), supervisor/manager contact information, certification in safe food handling (if required), and a signed statement as to the accuracy of the provided information. The preoperational inspection is then conducted to verify that the food establishment is constructed and equipped according to the approved application. Because of the need to coordinate this inspection with the completion of construction, this inspection is known in advance (unlike the routine inspections). It is one of the few inspections that is, in fact, not unannounced. If the preoperational inspection is passed, the permit is then issued and the food establishment may open.

Generally, an additional inspection is conducted soon after opening to assess the operation and handling of food as this was not possible to observe prior to opening. After this inspection, the food establishment is then placed into the routine inspection schedule as enforced by the local regulatory authority.

Permits are required to be posted in the food establishment in an area where consumers may see it. Permits are not transferable to a different person or establishment. If a business is sold, the new owner will need apply for his or her own permit to operate. Similarly, if an owner closes one establishment and opens a new establishment, they will need to apply for a new permit for that location. Finally, if an owner is lucky enough to own more than one restaurant, they will need to have a permit for each of those locations. The cost of the permits is based on local or state regulations and may take into account the size of the operation and whether it is a temporary or permanent type of food establishment. Permits are required to be renewed annually.

4.2 THE INSPECTION PROCESS

Once the restaurant is in operation, they may undergo three general types of inspections or visits by health inspectors as regulated by the local or state health code regulations. These are routine inspections, follow-ups to routine inspections, and complaint inspections. Routine inspections are the regular inspections done in all food establishments once they have been permitted. Typically these are done at least once every six months, although other systems may be used in some jurisdictions to determine the frequency of inspection and are discussed in a later section of this chapter. Follow-ups to routine inspections are typically conducted when serious or critical and/or repeated violations are found to ensure that needed corrections are being done. Complaint inspections or visits may also occur and result when a consumer or other person has called the health department with a food safety concern about a food establishment.

Follow-up on complaint calls is taken very seriously by the local regulatory authority. Occasionally prank or frivolous complaints are made (by disgruntled former employees of a food establishment for example). To avoid this situation, health departments may ask for the name of the complaint caller, although the name of the caller is typically kept anonymous by the local regulatory authority so that those people with concerns feel free to share their complaint. Even further, many states have set up communication networks to be able to respond 24/7/365 to food safety emergencies through local law enforcement or other communication systems.

4.3 FREQUENCY OF INSPECTIONS

Even though the local regulatory authority may use a six-month schedule for routine inspections, the actual frequency may depend on several factors.

These may include the workload of the local regulatory authority, the inspection history of the food establishment, the size of the operation, and the type of operation. In the state of Indiana for example, there are 92 county health departments, one city health department, and the Indiana State Department of Health in Indianapolis for a total of approximately 200 inspectors. Some counties have several inspectors, while other more rural counties must share inspectors with another county. Budget cuts and expanding job responsibilities have made the job of the local health inspector more challenging. The number of food establishments for which a health inspector is responsible may therefore vary greatly by number and by geographic area. Consequently, the frequency of inspection may be impacted by these factors.

Inspection history may also make a difference. Food establishments with a history of frequent or repeated violations may find themselves inspected more frequently. Differences in inspection frequency may also occur with size of the operation with large operations being inspected more frequently. Similarly, certain types of food establishments with immune compromised or more foodborne illness susceptible populations (hospitals, daycares, schools, or nursing homes for example) may be inspected more frequently as compared to other types of operations (NHDHHS, 2013).

Some regulatory authorities have set different requirements for frequency of inspection. New York City for example, conducts an unannounced onsite inspection once a year (NYCDOHMH, 2010). Delaware inspects most establishments (as many states do) twice a year (DHSS, 2012).

By contrast, in order to make more efficient use of resources, some regulatory authorities are using a risk-based system to determine the appropriate frequency of inspection (Frash, Almanza, and Stahura, 2003). The benefit of such a system is that health departments may prioritize their resources to those food establishments posing higher levels of food safety risk. In Marion County in Indiana, for example, risk assessment is determined during an initial inspection in which a health inspector will look for evidence of high, medium, and low risk classification criteria (Frash, Almanza, and Stahura, 2003). Criteria include methods of handling potentially hazardous (also referred to as temperature controlled for safety) foods and evidence of repeat valid complaints or repeat critical violations or other sanitation problems. Areas of potentially hazardous food (PHF) handling that are observed include how foods are cooled, whether they are prepared more than 12 h in advance of serving, reheating of PHF, catering off-site, vacuum-packaging, and serving immune compromised customers on a regular basis. Establishments that are determined to be high risk are inspected on a 120-day cycle, medium risk on a 180 day cycle, and low risk on a 360 day cycle.

New Hampshire also uses risk to determine inspection frequency such that high-risk establishments are inspected once per year and low risk are inspected only once every two to three years (NDDHHS, 2012). In their system, schools are also inspected yearly. Finally, the state of Mississippi also uses a system of risk to determine frequency, but the range is from one to four times a year (MSDH, 2013). In their system a cigarette/food/news stand may be inspected only once per year whereas an establishment with a highly susceptible to foodborne illness population such as a nursing home would be inspected four times per year. In addition, a food establishment that cooks to order may be inspected twice a year, whereas a food establishment with advanced preparation techniques might be inspected three times a year.

Frequency of inspection has been found in some studies to be a factor in food safety compliance. Research conducted in three provinces in Canada (Mathias, et al., 1995) showed that the longer the time since the last inspection, the lower the inspection score. Their results also showed that if inspections were conducted at least annually, inspection scores were not significantly different from each other. On the other hand, a more recent study done in Canada (Newbold, et al., 2008) found the opposite. In their study, no difference was found in food safety compliance of high-risk operations when they were inspected more frequently.

Although states almost always conduct inspections on an unannounced basis, at least one study has suggested that there may be an advantage to occasionally conducting announced inspections (Reske, et al., 2007). Results of this Minnesota study showed that during announced inspections, the number of critical violations was reduced by one-half and the number of noncritical violations was reduced by two-thirds. Although the violation decrease would be an expected outcome from an announced inspection, the results of this study also found that subsequent unannounced inspections also showed fewer violations. More specifically, there were

30% fewer violations related to "person-in-charge demonstrates knowledge of foodborne disease prevention" and 60% fewer violations related to cross-contamination. In addition, restaurant operators expressed favorable attitudes toward the selected use of the announced restaurant inspections.

4.4 WHAT IS ASSESSED DURING A HEALTH INSPECTION

Inspections comprehensively assess food safety in the food establishment. Of course, inspections assess kitchen and "back of the house" areas such as storerooms, cleaning areas, dish rooms, coolers and freezers, food preparation, and food holding areas. In addition, inspections will also generally include the outside premises (parking lot and dumpster and/or recycling areas) and the front of the house (server stations, dining rooms, and restrooms) as these also may impact the operation of the food establishment in providing safe food for customers.

Methods for inspecting differ widely, in part because health inspector training is not standardized in the United States and is regulated by the local or state jurisdiction. In many states, health inspectors shadow more experienced health inspectors as part of their training. This creates diversity in the inspection process because of the diversity of training methods and the "trainers." In fact research studies have shown that individual scores are significantly impacted by the individual health inspector (Lee, Nelson, and Almanza, 2010) and might even be useful in determining additional training needs of health inspectors or as a way to standardize inspection scores (Lee, Nelson, and Almanza, 2012). Therefore, individual health inspectors may vary in their approach to the inspection although they all evaluate the food establishment using the same regulations.

The issue of health inspector training is being addressed with a newer inspection form (Food Establishment Inspection Report, Form 3-A) in the 2009 FDA Food Code. This form is particularly useful in training health inspectors to assess all areas covered by the Food Code without overlooking anything. In this inspection form, 54 areas of inspection are specified and the health inspector is expected to place a check mark for each area as to whether the food establishment is "IN" our "OUT" of compliance with further information detailed about the type of violation observed on additional pages of the inspection form.

Although inspections may therefore vary, generally, health inspectors start their observations as they approach the food establishment, noting the condition of the outside premises. Once they enter the establishment, they will approach the person in charge, show their credentials, explain the purpose of their visit (such as routine inspection, complaint follow-up, etc.), and request permission to conduct the inspection. This is important for the food establishment's person in charge because they need to be aware of who is in their operation at all times in order to make certain that unauthorized individuals are not allowed access to food or food products. Although the person in charge may technically refuse to allow the health inspector to conduct the inspection, this could also result in a revocation (permanent withdrawal) of their permit. Therefore, it would be extremely rare for the person in charge to refuse entry to the health inspector. Once they have received permission, the health inspector will then conduct the inspection. The person in charge may accompany the health inspector if they wish and are generally encouraged to do so by the health inspector.

There are several advantages to the person in charge accompanying the health inspector. They include the opportunity to answer questions asked by the health inspector on the spot in regards to their policies on handling food products when observations are being made and the opportunity to ask questions to help improve food handling under circumstances unique or new to their operation. For example, some operations may lack a utility room and have questions on how to safely fill mop buckets or dispose of mop water. Finally, research has also shown that health inspectors are more likely to not write down violations if the person in charge has a good attitude and is interested in food safety, corrects violations on the spot during an inspection, and accompanies the health inspector (Johnson, 2012).

Because there are so many areas to assess during an inspection, many health inspectors will inspect an establishment using a set pattern or sequence so they do not forget to comprehensively assess all areas. For example, they may first go outside to the dumpster area after which they return inside to wash their hands. This accomplishes several purposes: first their hands are cleaned and they have set an example for the establishment and second they may better assess the hand washing area for evidence of hot water, soap, and paper towels. Next, they may move to the dry storage areas and then to the dish and equipment cleaning areas. They may then move to the various food preparation areas and observe or assess employee behaviors, food temperatures, equipment, facilities, etc. Finally, they may move to food server areas and then on to the front of the house where they inspect any self-service areas, as well as table cleaning and setting, the menus (to check for the consumer advisory for raw or undercooked animal foods), and even the restrooms (both male and female).

Because of differences in the complexity of the food operations from one establishment to another, the length of time needed for an inspection varies widely. Relatively simple operations, such as fast food restaurants where food is only reheated from the frozen state, that have minimal violations may only require 20 min for an inspection. Other restaurants where food is prepared "from scratch" that have multiple violations can easily require more than two hours.

After completion of the inspection, the health inspector will finish the report on-site. After the report has been written, the health inspector will then give a copy of the report to the person in charge and the person in charge will be asked to sign the report as an acknowledgement that they have received it. Follow-up to the inspections are also planned if there are significant violations, and dates are given by which the violations are expected to be corrected. Generally this will be less than ten days for more serious violations. Some jurisdictions have additional procedures. For example, in Tippecanoe County in Indiana, food establishments may write a response to the inspection, which will be filed with the report.

Occasionally, additional procedures may be required. A suspension is a temporary withdrawal of a permit to operate. Suspensions may be done without warning, notice, or hearing. They are imposed when there is an immediate and substantial public health risk and require operations to cease immediately. Examples of situations that may require closure (if the food establishment has not already closed voluntarily) include: a back-up of sewage into the restaurant, an inability to hold foods safely (no electricity for hot holding or refrigeration), an inability to clean needed dishes such as might be caused by a lack of hot water, or a substantial insect or rodent infestation. Reinstatement of one's permit is done if violations are corrected and the health authorities determine that there is no more threat to the public health.

A revocation is a permanent withdrawal of an establishment's permit and may be imposed for serious and repeated violations or for a refusal to allow inspections. A written notice is sent to the establishment after which operations must cease. The establishment cannot reopen following a revocation. Revocations may be used as a final consequence of a formal hearing process. Formal hearings are an administrative procedure that may be used to determine whether a food establishment should continue to operate. In Tippecanoe County in Indiana, the health officer for the county presides over a judicial process. Lawyers representing the Tippecanoe County Health Department provide evidence in regards to the food safety concerns with a food establishment. Lawyers or owners of the food establishment are also allowed the opportunity to provide evidence to refute the food safety concerns. At the conclusion of the hearing, the health officer makes a decision and notifies the food establishment in writing of the decision that may include the closure (revocation) of the restaurant.

A hearing may also be requested by the food establishment if they believe that they have been unfairly inspected, denied an application for a permit, or for what they consider other adverse administration determinations. Requests for a hearing must be made by the establishment within a specified time from the administrative action (most commonly it is within ten days).

An informal hearing or meeting with the health department is also sometimes used as a preceding step to the formal hearing. Its purpose is to encourage food establishments to comply with the food code and thereby avoid the more difficult and expensive formal hearing process.

Many areas of the country publish or offer information about health inspection reports to the public. The method varies greatly. For example, in some areas newspapers publish the inspection reports on a daily, weekly, or monthly basis. Alternatively, some health departments make these available through their websites. In other jurisdictions, cities such as New York City, Los Angeles, and San Diego and at least one state (Alabama) require restaurants to post their scores in a prominent location in their establishment such as the door or front window. St. Charles County in Missouri uses an even more modern piece of technology to offer information about their restaurants' recent inspection reports. Since 2012, they have incorporated a QR code with the food service certification licenses so that customers with smart phones and a scan application can see a restaurant's recent health ratings by scanning the QR code with their smart phone. These licenses are required to be placed in highly visible areas such as front doors, drive-through windows or near checkouts (Scott, 2012).

4.5 DIFFERENT INSPECTION FORMATS

States differ in their inspection format. Some of the most common types of inspections are outlined next. They include a traditional system, a system assessing critical and noncritical violations, a system employing three categories of violations (priority, priority foundation and core violations), a risk based inspection format, and a HACCP (Hazard Analysis of Critical Control Points) inspection.

Other variations are used by individual states. New York City has a unique system (NYCDOHMH, 2012) that they use in inspecting 24,000 restaurants per year. Violations according to New York City's regulations, fall into three categories: public health hazards (such as failing to keep foods at the correct temperature) which are worth a minimum of seven points and must be corrected before the inspection ends; critical violations (such as serving salads without properly washing the ingredients first) which carry a minimum of five points, and a general violation (such as not properly sanitizing utensils used in cooking) which are scored at a minimum of two points. Each restaurant has two opportunities to earn an "A" grade in every inspection cycle. If they do not earn an "A" on the first inspection, they are scored, but not graded. An inspector will return (usually) within a month to reinspect. Grades are assigned as follows: a score between 0–13 points earns an "A" grade, 14–27 points results in a "B" grade, and those with 28 or more points earn a "C."

In the traditional system (described in earlier versions of the food code, but not used commonly today), food establishments are assessed on a possible 100-point scale. Demerits from one to five points are given for each violation found. For example, minor violations may be worth one or two points. These minor violations are generally required to be corrected as soon as possible or at least by the time of the next inspection. Major violations are worth four or five points and are generally required to be corrected as soon as possible or no later than ten days after the inspection.

A failing score was given in older versions of the FDA Food Code as less than 60 points after violations were subtracted from the 100 possible points, although some states set different levels. In the state of Indiana for example, prior to the adoption of a newer inspection system, a failing score was considered to be less than 75 points.

Disadvantages with the traditional system were that there was no penalty for repeated violations and that inspectors could not cite multiple violations of the same type with any greater weight. The inability to penalize repeat violations meant that some establishments might continue to have the same violation, including potentially serious temperature violations, and still maintain a relatively high inspection score in spite of their poor compliance in regards to that violation. The inability to cite multiple violations with a greater weight meant that establishments with: a) one PHF that was found to be in the temperature danger zone; *b*) a holding unit such as a refrigerator with all PHFs inside that unit held at the wrong temperature; and *c*) all holding units or refrigerators with all PHFs held at the wrong temperature, might all receive the same number of violations (one violation worth five points). Clearly, these disadvantages were frustrating for inspectors as the establishments in these examples represent different levels of food safety risk. More recent editions of the FDA Food Code have therefore recommended other types of inspection formats.

One of these is a system that assesses critical and noncritical violations. This system does have some analogies to the previous system in that the noncritical violations are largely what were defined as the minor violations (worth one or two points) in the traditional system and the critical violations are generally what were defined as the major violations (worth four or five points). On the other hand, the critical and noncritical violation format includes more narrative to describe each of these violations. This improves communication with the food establishment and is therefore helpful in getting violations corrected.

Repeated violations are noted for both critical and noncritical violations. In addition, multiple violations of the same type are also indicated. In contrast to the previous example given for the traditional system, food temperature violations might be assessed by temperature holding unit, so that one refrigerator that was not holding foods at the correct temperature might be given one critical violation, two refrigerators that were holding foods incorrectly would be given two critical violations, etc.

One disadvantage, at least from the perspective of the person in charge of the food establishment, is that the failing score in terms of the number of critical and noncritical violations is generally not specified in the food code, but is based on the experience and professional judgment of the health inspector. Although health inspectors are able to expertly assess the risk that violations pose, the interpretation of risk is likely to be less clear to the person in charge so that the interpretation of passing score may appear ambiguous.

The most current inspection format described in the most recent FDA Food Code (FDA, 2009) is the Criticality system. In this system there are three categories of violations that are referred to as Priority, Priority Foundation, and Core. Priority items are defined by the FDA as those actions or situations that would directly contribute to the elimination, prevention or reduction of hazards associated with foodborne illness or injury to an acceptable level and where there is no other provision that more directly controls the hazard. Priority Foundation items support, facilitate, or enable compliance with Priority Items. FDA examples of Priority Foundation Items include personnel training, infrastructure or needed equipment, Hazard Analysis of Critical Control Points plans, documentation or record keeping required of the food establishment, and required food labeling. Core items relate to basic sanitation, as defined by the FDA, and include general sanitation, operational controls, sanitation standard operating procedures (SSOPs), facilities or structures, equipment design, or general maintenance.

Although often viewed as a proactive food establishment management system to assure the safety of hazardous foods at critical control points during their receiving, storage, and handling, Hazard Analysis of Critical Control Points or HACCP is also used in some jurisdictions as an inspection tool. A HACCP inspection will differ in the way that the inspector assesses the food establishment. For example, during a HACCP inspection, the inspector will observe the flow of food in the operation. They may start in the receiving area and evaluate the temperature and appearance of the delivery truck and how food is checked as it enters the establishment. Next, they will observe how it is placed into storage, and then how it is handled during food preparation. In this way, the inspector continues so that they can evaluate fully evaluate the safety of the food as it moves through the operation from the time it is received to the time that it is served. This is central to the difference between HACCP inspections and other types of inspections that only capture a snapshot of the establishment's performance at the time of the inspector's observation. As might be expected, HACCP inspections generally require more time because of the need to observe the flow of food through the establishment rather than simply observing operations occurring at one point in time.

Risk-based inspections are another variation of the inspection process. The purpose of risk-based inspections, as defined by the FDA (2009) is to focus inspections on the control of foodborne illness risk factors and the assessment of the amount of active managerial control an operator has on foodborne illness risk factors, thereby making a great impact in a limited timeframe on reducing foodborne illness.

These have been particularly useful when many establishments must be inspected in a limited time period or with limited resources. For example, during Super Bowl XLVI held in Indianapolis in February of 2012, a staggering number of inspections were required in the very short time period prior to and during the Super Bowl. Inspections included large catered events, Indianapolis bars and restaurants, gourmet chefs cooking for big named celebrities and athletes, and a huge number of temporary establishments offering food to visitors. Foods were inspected in at least 19 venues around the city including the Indianapolis State Fairgrounds (Fan Jam), the NFL Experience in the downtown area, the Patriot and Giants team hotels, Lucas Oil Stadium, the NCAA, the Indianapolis Zoo, the Indiana State Museum, and Gleaners Food Bank Warehouse (ISDH, 2012). Even further, 28 inspections were conducted at high profile events such as the Taste of the NFL, the NFL Owners Party and the NFL Commissioners Ball with the number of people attending these events ranging from 100 to 4,000 attendees. The signature event for the NFL was the Taste of the NFL, a charity event in which 38 chefs from states with an NFL team participated (ISDH, 2012).

Because of the tremendous inspection requirements, extensive planning, coordination, training, and additional resources were required for Super Bowl XLVI. As is typical for risk-based inspections, the most critical and likely to-occur violations were prioritized for assessment so that the huge quantity of temporary and permanent food establishments that provided food venues for the Super Bowl both inside and outside Lucas Oil Stadium could be inspected in an extremely efficient process. Limited evaluations were given to violations that would not be likely to occur and/

or were of minor significance. Participating health inspectors conducting these risk-based inspections referred to their responsibilities as "extreme" inspecting (Danao, 2012).

Finally, as mentioned previously there is an inspection format (outlined in the 2005 and later FDA Food Codes) that requires health inspectors to assess all 54 areas of the food establishment for either being in-compliance or out-of-compliance. This system is particularly useful in training health inspectors because it forces them to assess all potential violation areas and develops their professional skills in comprehensive assessment.

4.6 HOW TO BE BEST PREPARED FOR AN INSPECTION AND HOW TO USE THAT INFORMATION

Food inspections are a valuable tool in assuring the safety of foods in food establishments. In the United States there is a long and dynamically evolving set of regulations pertaining to food establishments that dates back almost 100 years. Guidelines or recommendations are set forth at the federal level in the FDA Food Code. Regulations are set forth by government agencies at the state level in their state food codes and enforcement (and sometimes additional regulations) are imposed by local government agencies.

Differences in inspection formats from one area of the country to another therefore may occur. Even further, health inspectors may also differ in their approach to the food establishment inspection based on their training and education. These differences make it difficult to understand the U.S. food establishment inspection system. Chain restaurants for example, may find it hard to understand regulations when they have establishments in more than one state (Allen, 2000). This chapter seeks to explain these differences and the reasons for some of those differences. When operating a food establishment or conducting research in a particular state, it is essential to get a copy of the state or local food code to fully understand the regulations. These are generally available at no charge through state or local health department websites (or whatever government agency is responsible for food establishment inspections).

Although sometimes viewed as an adversarial relationship, research (Johnson, 2012) has shown that a professional relationship with the health inspector that is supportive of the goals of the inspection can be mutually beneficial in improving the inspection results. Accompanying the health inspector during the inspection and correcting violations (if they should occur) during the inspection has been shown to be important in achieving high inspection scores. Beyond this, it would, of course, be better to not have violations occur at all. This is likely to be achieved with a proactive managerial process that might include: manager certification in safe food handling, employee food safety training, strong manager support of food safety practices in the establishment, and regular self-assessments or audits of food safety practices. In addition, some restaurants hire food safety consultants to audit their operations or place added financial incentives to motivate their operations to maintain high food safety standards and achieve good inspection scores.

It has been said that a good inspection score is worth its weight in gold. This certainly may be true for food safety inspections. Achieving an excellent food safety inspection score means that customers are safer, the food establishments attract more customers, and health inspectors find that the inspection process can be used to bring food establishments to higher levels of food safety. Health inspections are an important tool in bringing about these important goals.

KEYWORDS

- **FDA Food Code**
- **HACCP inspection**
- **risk based inspection format**
- **violations**

REFERENCES

Allen, R.L. (2000). Making the grade: Standardized health inspection is key to food excellence. *Nation's Restaurant News, 34*, 43.

Almanza, B.A., Ismail, J., and Mills, J. (2002). The impact of publishing food inspection scores. *Journal of Environmental Health, 66*(9), 10–14.

Danao, S. (2012). Above and beyond in Super Bowl XLVI inspections. *FoodBytes, 13*(1), 3. [Indiana State Department of Health Newsletter], Retrieved from http://www.in.gov/isdh/files/Food_Bytes_Fall_Spring_2012_Final.pdf

DHSS. (2012). *Inspection services: Retail food establishments.* Dover, DE: Delaware Department of Health and Social Services. Retrieved from http://www.dhss.delaware.gov/dhss/dph/hsp/feir.html

Food and Drug Administration. (2009). *Food Code 2009.* College Park, MD: United States Department of Health and Human Services, Public Health Service, Food and Drug Administration. Retrieved from http://www.fda.gov/Food/FoodSafety/RetailFoodProtection/FoodCode/FoodCode2009/default.htm

Food and Drug Administration. (2011). *Real progress in food code adoptions.* College Park, MD: United States Food and Drug Administration. Retrieved from http://www.fda.gov/Food /GuidanceRegulation/RetailFoodProtection/FederalStateCooperativePrograms/ucm108156.htm

Food and Drug Administration. (2012). *FDA Food Code 2009: Previous editions of codes recommended by the United States Public Health Service for regulating 0perations providing food directly to the consumer.* College Park, MD: United States Food and Drug Administration. Retrieved from http://www.fda.gov/Food/FoodSafety/RetailFoodProtection/FoodCode/FoodCode2009/ucm188148.htm

Frash, Jr., R., Almanza, B., and Stahura, J. (2003). Assessment of food safety risk. *International Journal of Hospitality and Tourism Administration, 4*(4), 25–44.

ISDH. (2012). Super victory for food protection. *FoodBytes, 13*(1), 3. [Indiana State Department of Health Newsletter], Retrieved from http://www.in.gov/isdh/files/Food_Bytes_Fall_Spring_2012_Final.pdf

Johnson, A.C. (2012). *Factors that affect what is written down a retail food establishment inspection report.* (Master's thesis). Purdue University, West Lafayette, IN.

Lee, J.E., Nelson, D.C., and Almanza, B.A. (2010). The impact of individual health inspectors on the results of restaurant sanitation inspections: Empirical evidence. *Journal of Hospitality Marketing and Management, 19*, 326–339.

Lee, J.E., Nelson, D.C., and Almanza, B.A. (2012). Health inspection reports as predictors of specific training needs. *International Journal of Hospitality Management, 31*, 2, 522–528.

Mathias, R.G., Sizto, R., Hazlewood, A., and Cocksedge, W. (1995). The effects of inspection frequency and food handler education on restaurant inspection violations. *Revue Canadienne de Sante' Publique, 86*(1), 46–50.

MSDH. (2013). *Food code and food facility permits.* Jackson, MS: Mississippi State Department of Health. Retrieved from http://msdh.ms.gov/msdhsite/ static/30,5301,77,333.html

Newbold, K.B., McKeary, M., Hart, R., and Hall, R. (2008). Restaurant inspection frequency and food safety compliance. *Journal of Environmental Health, 71*(4), 56–61.

NHDHHS. (2012). *Food establishment inspection results.* Concord, NH: New Hampshire Department of Health of Human Services. Rctrieved from http://www.dhs.nh.gov/dphs/fp/inspectionresults.htm

NYCDOHMH (2012). *How we score and grade.* New York City, NY: New York City Department of Health and Mental Health. Retrieved from http://www.nyc.gov/html/doh/downloads/pdf/rii/how-we-score-grade.pdf

NYCDOHMH. (2010). *What to expect when you're inspected: A guide for food service operators.* New York City, NY: New York City Department of Health and Mental Health. Retrieved from http://www.nyc.gov/html/doh/downloads/pdf/rii/blue-book.pdf

Reske, K.A., Jenkins, T., Fernandez, C., VanAmber, D., and Hedberg, C.W. (2007). Beneficial effects of implementing an announced restaurant inspection program. *Journal of Environmental Health, 69*(9), 27–34.

Scott, J. (2012, March 26). Check out restaurants before you eat. *St. Charles Patch,* [St Charles County, Missouri]. Retrieved from http://stcharles.patch.com/articles/check-out-restaurants-before-you-eat-9366c2aa

CHAPTER 5

LEGISLATIVE PROCESS, REGULATORY TRENDS, AND ASSOCIATIONS

KEVIN R. ROBERTS

Associate Professor, Kansas State University

BRENDA H. HALBROOK

Director, Office of Food Safety, USDA Food and Nutrition Service

AND JEANNIE SNEED

Professor, Kansas State University

CONTENTS

An overview of the legislative process and history of food laws in the United States (U.S.) is important to understanding the complex structure of today's food safety systems in the U.S. This chapter will discuss the legislative process within the U.S., provide a brief history of food laws and regulations, and examine the regulatory agencies that oversee and enforce existing food laws. Finally, trade and industry associations associated with food safety research and developing and maintaining industry food safety standards are discussed.

5.1 THE LEGISLATIVE PROCESS

The U.S. government was designed by the founding fathers to have three separate, yet equal, branches of government: legislative, executive, and judicial. The legislative branch writes and enacts laws, the executive branch enforces those laws, and the judicial branch interprets and reviews laws. The design ensures that no one branch has too much power.

The legislative branch has two chambers, the House of Representatives and the Senate. All responsibilities granted to the legislative branch are laid out in Article I, Section I, of the U.S. Constitution. Each chamber has fundamentally equal power in their roles and functions. However, any legislation dealing with revenue must originate in the House of Representatives and only the Senate can conduct impeachments. The Senate also provides advice and consent for treaties and nominations made by the Executive Branch.

5.1.1 INTRODUCTION OF LEGISLATION

A bill may be introduced in either chamber but must be sponsored by a member of the chamber in which it is introduced. In both the House and the Senate, once a bill is introduced, it is sent to a committee where most work on the bill is completed. Committees are chaired by a member of the majority party and comprise majority and minority party members equal to the ratio of the full chamber (The Library of Congress, n.d.).

The number of bills each committee receives during a legislative session means the committee chair has considerable power in setting the agenda for the committee and will often decide which issues the committee will take up. Once the full committee has received a bill, committee members conduct hearings on the bill or will markup the bill (The Library of Congress, n.d.).

Hearings, while not required for a bill to advance in the legislative process, provide a setting in which external constituents can provide short remarks on the bill. Generally, these remarks are accompanied by a detailed written version of the feedback on the bill. These hearings provide committee members an opportunity to assess the proposed legislation (The Library of Congress, n.d.).

Once the committee chair chooses to bring a bill to the committee, it will be placed on the committee agenda for markup. The committee can then consider changes to proposed legislation by offering and voting on amendments and/or completely changing the text of the bill. Once the committee has completed its markup, the bill is referred to the chamber for consideration. Subcommittees are often an important part of the markup process. The role of subcommittees varies significantly from chamber to chamber and even from committee to committee within the same chamber. However, bills can only be referred to the chamber for consideration by the committee itself, never directly from a subcommittee (The Library of Congress, n.d.).

5.1.2 LEGISLATION AND THE HOUSE OF REPRESENTATIVES

If a House of Representative's committee proposes a bill, the bill goes to the floor for debate. The House considers bills using a number of different procedures; members can propose amendments during the floor debate. Most bills are considered under the "suspension of rules" procedure, which limits the total debate of the bill to 40 min without permitting any amendments to the bill. However, to pass any legislation under suspension of the rules, a two-thirds majority is needed (The Library of Congress, n.d.).

Other than suspension of rules, legislation can be considered under rules designed for that specific bill. The House Rules Committee, dominated by the majority party, governs all special rules; each rule must be discussed on the house floor and voted on. The rules for any specific bill can be as different as the bills themselves. According to the Library of Congress,

> *"Common provisions found in a special rule include selection of the text to be considered, limitations on debate, and limits on the amendments that can be offered on the floor. For instance, sometimes the committee reports a rule that places few restrictions at all on amending, which can result in dozens of amendments being offered on the floor during consideration. In other cases, the special rule will allow only specific predetermined amendments to be offered, or even preclude floor amendments all together."* (The Library of Congress, n.d., para. 2)

Once the rules are established and amendments proposed, the bill must be voted on by the entire House. Generally, a brief debate will again be allowed, and the bill will be voted on by all members in attendance (The Library of Congress, n.d.).

5.1.3 LEGISLATION AND THE SENATE

To consider a bill on the floor of the U.S. Senate, the Senate must first agree to bring forward the legislation. Senators may offer amendments to the proposed legislation only once it is brought to the floor. In many cases, the Senate will act using unanimous consent agreements, which will spell out the rules for any piece of legislation, but all senators must consent to have these rules put in effect. Outside of this, the only way debate of a bill can be limited is through the cloture rule, which requires a three-fifths vote of the Senate, often called a supermajority. Just as in the House, once the rules are established and amendments have been made, the bill must be voted on by the entire Senate (The Library of Congress, n.d.).

5.1.4 RESOLVING CONFLICTS

Before any bill can be sent to the President for approval or veto, the same bill must be approved by both the House of Representatives and the Senate. Once a bill is passed by either chamber, the official document is prepared and presented to the other chamber for approval. In many cases, the bill is simply agreed to in the other chamber and will move onto the President (The Library of Congress, n.d.).

In some cases, one chamber will amend legislation proposed by the other chamber. If this occurs, the chamber that originally passed the bill must again vote to accept the amended bill. They may also respond with other changes. For legislation to pass, one chamber must agree to the bill as proposed by the other chamber. In some cases, a Conference Committee is formed, comprising members from both the House of Representatives and the Senate. The Conference Committee negotiates a compromise between the two chambers. If a compromise is reached, a conference report is prepared and sent to both chambers for approval. The conference report must be agreed to, without changes, by both chambers for the legislation to move on to the President (The Library of Congress, n.d.).

5.1.5 LEGISLATION AND THE PRESIDENT

Once both chambers have agreed on a piece of legislation, it is sent to the President. As spelled out in Article 1, Section 7, of the U.S. Constitution, the President has ten working days, excluding Sundays, to sign the bill or veto it. If the President fails to sign or veto the bill within those ten days, or if the President signs the bill, the legislation is enacted into law.

If the President chooses to veto the bill, the bill is sent back to the chamber in which it originated. A veto can be overridden if each chamber can gather support from two-thirds of voting members. Overriding a Presidential veto is rare (The Library of Congress, n.d.).

Note that while all legislation must originate from a congressional representative, the President does have influence over legislation, even before it arrives on his desk for signature. Because the Presidential veto is difficult to override, Congress will typically consider the President's position on proposed legislation. The President can also suggest legislation but cannot be the one who brings it to the chamber (The Library of Congress, n.d.).

As you can see, writing and passing legislation is a long and arduous process, with many roadblocks in the process. Clearly, each piece of legislation must have a member of the Senate or House of Representatives who believes in it strongly and will champion the cause. Moreover, the process is often highly political. As we explore the history of food safety legislation, you will notice long gaps of time when no legislation was passed.

Often as we look at our nation's history, catastrophic events have caused a ground swell of support from the public for legislation, and that has forced Congress to act.

5.1.6 HISTORY OF FOOD LAWS

In the early days, after the U.S. declared its independence from Great Britain in 1776, laws protecting the safety of food, drugs, and other harmful substances were unheard of. By 1862, President Abraham Lincoln had the newly developed Department of Agriculture (USDA) and appointed a chief chemist to serve in the Bureau of Chemistry there. The history of legislation and regulation of food in the U.S. begins at this point.

It was not until 1880, however, that the Bureau of Chemistry, under the leadership of Chief Chemist Peter Collier, began to focus on food as well as drugs (U.S. FDA, 2009b). At that time, few government controls protected the public from unethical manufacturers. While health professionals, food packers, and agricultural businesses wanted Federal legislation, staunch opposition still existed to any laws regulating the food industry (U.S. FDA, 2009b). In 1883, Harvey Wiley, M.D., was appointed Chief Chemist in the Bureau of Chemistry. He believed that "foods rather than drugs were a greater harm to the public at the time" (U.S. FDA, 2009a). For about 25 years, more than 100 bills intended to protect the public from the food supply were introduced into Congress. All were defeated (U.S. Food and Drug Administration, 2009a; U.S. Food and Drug Administration, 2009b).

During this time, Wiley took his crusade directly to the American public, speaking at women's, civic, and business clubs all over the U.S. He enlisted magazines such as *Ladies Home Journal* and *Good Housekeeping* to print cartoons and editorials in an attempt to sway public opinion and force Congress to act on one of his pure food bills (Janssen, 1981).

In 1906, the public was convinced of the need for regulation. Support from women all over the country forced Congress to pass the Pure Food and Drug Act, which prevented the manufacture, sale, or transportation of adulterated or misbranded foods, drugs, medicines, and liquors.

Early that same year, *The Jungle*, a novel by American journalist Upton Sinclair was published. *The Jungle* described the deplorable conditions for workers in the Chicago meat-packing industry and described the meats that they processed. Sinclair wanted to highlight the plight of the workers but set off a reaction that reached the office of President Theodore Roosevelt. Sinclair is quoted as saying, “I aimed at the nation’s heart, but by accident I hit it in the stomach.”

The Jungle served as the impetus behind the Federal Meat Inspection Act, which prohibited the sale of adulterated or misbranded meat products for food and ensured that animals were slaughtered and meat processed under sanitary conditions. President Roosevelt signed the Federal Meat Inspection Act on the same day as the Pure Food and Drug Act in 1906 (Janssen, 1981). Over the next 100-plus years, these two acts would be expanded through other legislation to further protect the American public.

The Bureau of Chemistry enforced the 1906 laws until 1927, when it was reorganized and named the USDA Food, Drug, and Insecticide Administration. In 1931, it was renamed the Food and Drug Administration, still under USDA (Janssen, 1981).

Further protection was inspired when, in 1937, a pharmaceutical company released a new liquid form of sulfanilamide, known as the “elixir of sulfanilamide,” a drug used to treat streptococcal infections. This new form of the drug was responsible for the deaths of over 100 people in 15 states from Virginia to California. The elixir was found to contain diethylene glycol, a deadly poison that is often used in antifreeze today. This event, and the resulting public outcry, prompted Congress to pass the 1938 Food, Drug, and Cosmetic Act, which gave the FDA the authority to issue food safety standards. The act was signed into law by President Roosevelt on June 25, 1938 (Ballentine, 1981).

In 1940, to help separate consumer interests from producer interests, the Food and Drug Administration (FDA) was moved from the USDA to the Federal Security Agency, which would become the Department of Health, Education, and Welfare in 1953. This federal agency is now known as the Department of Health and Human Services (DHHS) (Janssen, 1981).

Over several years between 1949 and 1958, several Congressional committees generated three amendments that fundamentally changed the character of the U.S. food and drug laws: the Pesticide Amendment (1954), the Food Additives Amendment (1958), and the Color Additive

Amendment (1960). The Food Additives Amendment of 1958 established the additives that were generally recognized as safe (GRAS) and the process for scientific procedures to demonstrate that a substance was safe for its intended use. With these laws on the books, it could be said for the first time that no substance could legally be introduced into the U.S. food supply unless it had been determined safe. By requiring the manufacturers to do the research, a problem of unmanageable size was made manageable.

By the 1960s, the market had grown increasingly complex, and states were conducting their own inspections of meats and meat production facilities. Thus, in 1965, the USDA Meat Inspection Division and the Poultry Division were merged into the Agricultural Research Service. In 1967, the Federal Meat Inspection Act was amended by the Wholesome Meat Act, and 1968, the Poultry Act was amended by the Poultry Products Inspection Act, to meet new inspection challenges. By 1977, the USDA Food Safety and Quality Service was formed to manage these two acts and to grade and inspect meat and poultry. In 1981, the Food Safety and Quality Service became the Food Safety and Inspection Service of USDA.

In 1970, Congress passed the Egg Products Inspection Act (EPIA) that provides for the mandatory continuous inspection of liquid, frozen, and dried egg products to ensure that they were wholesome, otherwise not adulterated, and properly labeled and packaged to protect the health and welfare of consumers. In 1995, the USDA Food Safety and Inspection Service (FSIS) assumed responsibility for inspecting pasteurized liquid, frozen, or dried egg products. The FDA is responsible for the safety of eggs still in the shell.

One of the most significant recent regulations for improving food safety was reducing pathogens; Hazard Analysis and Critical Control Point (HACCP) Systems were issued in their current form by the FSIS on July 25, 1996. This regulation required slaughter facilities to implement a HACCP-based program to assure the safety and integrity of meat and poultry products. An April 1997 guidebook (www.haccpalliance.org/alliance/haccpmodels/guidebook.pdf) was published to support HACCP in slaughtering and processing meat.

The Food Safety Modernization Act was signed into law on January 4, 2011, representing another significant change to food laws in the U.S. The law ensures that the U.S. food supply is safe by shifting the focus to preventing food contamination rather than responding to contamination after the fact (U.S. FDA, 2013).

5.2 REGULATORY AGENCIES

5.2.1 U.S. DEPARTMENT OF AGRICULTURE, FOOD SAFETY AND INSPECTION SERVICE (FSIS)

The FSIS is the public health agency in the USDA responsible for ensuring that the commercial supply of the nation's meat (excluding game meats, such as venison), poultry, and egg products is safe, wholesome, and correctly labeled and packaged.

The FSIS enforces the Federal Meat Inspection Act (FMIA), the Poultry Products Inspection Act (PPIA), and the Egg Products Inspection Act. These laws require federal inspection and regulation of meat, poultry, and processed egg products prepared for use as human food. The FSIS also verifies and enforces industry compliance with the Humane Methods of Slaughter Act, which requires that all livestock inspected under the FMIA are humanely handled.

In FY 2010, the agency employed more than 9,800 personnel, including more than 8,000 in-plant and other front-line personnel, protecting public health in approximately 6,200 federally inspected, privately owned meat, poultry, egg product, and other slaughtering or processing plants. FSIS inspectors must be present at all times during the slaughter of food animals. Generally, an inspector will be at a processing plant once per shift during processing, although they may be present more often.

The FSIS inspects each animal carcass and sets food safety standards for products through inspection and enforcement. Only federally inspected and approved products can be transported from state to state or exported.

FSIS inspection personnel verify that an establishment maintains proper sanitation procedures, follows its HACCP plan, and complies with all FSIS regulations for slaughter and processing as well as humane handling and slaughter requirements. If an establishment violates these requirements, inspectors can take regulatory action, including suspending inspection. To verify that HACCP plans are working properly, the FSIS also takes microbiological samples that must meet standards established by the FSIS. They also test for chemical residues in meat, poultry, and egg products.

5.2.2 FOOD AND DRUG ADMINISTRATION (FDA)

The FDA is responsible for overseeing the safety and security of most of the U.S. food supply and thus protecting public health, a vital part of FDA's mission and a primary task of FDA's Center for Food Safety and Applied Nutrition (CFSAN). Under the leadership of the FDA Office of Foods, CFSAN, the Center for Veterinary Medicine, and the foods-related portion of the Office of Regulatory Affairs, the agency's unified food program protects and promotes the public health through the following:

- ensuring the safety of foods for humans, including food additives and dietary supplements, by setting science-based standards for preventing foodborne illness, and ensuring compliance with these standards;
- ensuring the safety of animal feed and the safety and effectiveness of animal drugs, including the safety of drug residues in human food derived from animals;
- protecting the food and feed supply from intentional contamination; and
- ensuring that food labels are truthful, containing reliable information consumers can use to choose healthy diets.

While some functions of the modern FDA can be traced back to the late-1830s and the Department of State, the agency is currently within the DHHS (see sidebar). In addition to protecting the food supply, the FDA assures the safety of human and veterinary drugs, biological products, medical devices, and cosmetics. Specifically, FDA oversees the safety of whole shell eggs, seafood, milk, grain products, and fruits and vegetables. The FDA assures that foods it regulates are not just safe, but wholesome, sanitary, and properly labeled. It oversees and regulates approximately 80% of the domestic food supply with the FSIS regulating the other 20%, meat from livestock, poultry, and some egg products. Some meats, like venison and other game meat, are regulated by the FDA, not the FSIS. In addition, the FDA regulates bottled water, food additives, and infant formulas, as well as livestock feeds and pet foods. The FDA, through the U.S. Public Health Service, also develops the Model Food Code.

Location of FDA and its Predecessors in Federal Government (SIDEBAR)

Year	Designation and Location	Statute
1839	Patent Office, Department of State	5 Stat. 353, 354 (1839)
1849	Chemical Laboratory of the Agricultural Division in the Patent Office, Department of the Interior	9 Stat. 395 (1849)
1862	Chemical Division, Department of Agriculture	12 Stat. 387 (1862)
1889	Chemical Division, U.S. Department of Agriculture	25 Stat. 659 (1889)
1890	Division of Chemistry, USDA	26 Stat. 282, 283
1901	Bureau of Chemistry, USDA	31 Stat. 922, 930 (1901)
1927	Food and Drug Insecticide Administration, USDA	44 Stat. 976, 1002 (1927)
1930	Food and Drug Administration, USDA	46 Stat. 392, 422 (1930)
1940	FDA, Federal Security Agency	54 Stat. 1234, 1237 (1940)
1953	FDA, Department of Health, Education, and Welfare	67 Stat. 631, 632 (1953)
1979	FDA, Department of Health and Human Services	93 Stat. 668, 695 (1979)

From "Location of FDA and its Predecessors in Federal Government," U.S. Food and Drug Administration. Available at: http://www.fda.gov/AboutFDA/WhatWeDo/History/Overviews/LocationofFDAanditsPredecessorsinFederalGovernment/default.htm

5.2.3 THE MODEL FOOD CODE

Realizing that effective disease prevention requires applying sanitation guidelines from production through consumption, in 1934, the Public Health Service developed a recommended food code for the retail food industry. Several other versions of the code have been released over the years (Table 1). Hundreds of jurisdictions and many federal agencies have adopted some version of the model food code (U.S. Public Health Service, 2009).

The model Food Code is not a federal law nor is it a regulation. It represents the agency's best advice, based on science, to ensure a food supply that is safe for human consumption. The code is updated every four years, with supplements to the existing edition coming out between editions. States and other jurisdictions can decide to implement the code in its entirety or in part (U.S. Public Health Service, 2009).

5.3 OTHER GOVERNMENTAL AGENCIES

Other government agencies that play a key role in the safety of the U.S. food supply include the Centers for Disease Control and Prevention (CDC) and the National Oceanic and Atmospheric Administration (NOAA). In addition, the National Advisory Committee on Microbiological Criteria for Food (NACMCF) provides scientific advice on food safety issues for federal food safety agencies and stakeholders.

5.3.1 CENTERS FOR DISEASE CONTROL AND PREVENTION

The CDC is a nonregulatory agency within the DHHS that heads federal efforts to gather data on foodborne illnesses, investigate foodborne illnesses and outbreaks, attribute illness to specific foods and settings, and monitor the effectiveness of prevention and control in reducing foodborne illnesses. The CDC also helps state and local health departments build their epidemiology, laboratory, and environmental capacities for surveying foodborne disease and responding to outbreaks. The CDC also captures information through FoodNet, which tracks information on diagnosed foodborne illnesses as trends over time and serves as a benchmark for reducing nationwide incidents of foodborne illness. The CDC works closely with the FSIS and the FDA in coping with foodborne illness outbreaks and assisting in tracking outbreaks to their sources.

Another important function of CDC in relation to the hospitality industry is the vessel sanitation program (VSP). This program is designed to assist the cruise ship industry in preventing and controlling the "introduction, transmission, and spread of gastrointestinal illnesses on cruise ships." To achieve this mission, CDC inspects cruise ships, monitors illnesses, investigates outbreaks, trains cruise ship employees, and provides health information to the industry and the traveling public. Inspections scores and corrective reports from each ship inspection are available on the CDC website (www.cdc.gov/nceh/vsp/). CDC also provides many resources to the industry including operations manuals and construction guidelines as well as fact sheets on a number of related topics.

TABLE 1 Overview of Food Code Editions in the U.S.

1934	*Restaurant Sanitation Regulations*
1935	*An Ordinance Regulating Food and Drink Establishments*
1938	*Ordinance and Code Regulating Eating and Drinking Establishments*
1940	*Ordinance and Code Regulating Eating and Drinking Establishments*
1943	*Ordinance and Code Regulating Eating and Drinking Establishments*
1957	*The Vending of Foods and Beverages – A Sanitation Ordinance and Code*
1962	*Food Service Sanitation Manual, including A Model Food Service Sanitation Ordinance and Code*
1965	*The Vending of Food and Beverages – A Sanitation Ordinance and Code*
1976	*Food Service Sanitation Manual Including A Model Food Service Sanitation Ordinance*
1978	*The Vending of Food and Beverages Including A Model Sanitation Ordinance*
1982	*Retail Food Store Sanitation Code*
1993	*Food Code, 1993*
1995	*Food Code, 1995*
1997	*Food Code, 1997*
1999	*Food Code, 1999*
2001	*Food Code, 2001*
2003	*Supplement to the 2001 Food Code*
2005	*Food Code, 2005*
2007	*Supplement to the 2005 Food Code*
2009	*Food Code, 2009*
2011	*Supplement to the 2009 Food Code*

Adapted from "Food Code, 2009," U.S. Food and Drug Administration. Available at: http://www.fda.gov/downloads/Food/GuidanceRegulation/UCM189448.pdf

5.3.2 NATIONAL OCEANIC AND ATMOSPHERIC ADMINISTRATION

Many state and federal agencies including the FDA and the Department of Commerce work together to ensure that the seafood we buy is safe, wholesome, and properly labeled. NOAA, located in the Department of Commerce, oversees fisheries management in the U.S. Under the 1946

Agricultural Marketing Act, the NOAA Seafood Inspection Program inspects fish, shellfish, and fishery products. The NOAA Seafood Inspection Program is often called the U.S. Department of Commerce (USDC) Seafood Inspection Program and bears the USDC mark.

The Seafood Inspection Program is a voluntary, fee-based inspection service to fishing boats, processing plants, and retailers to ensure compliance with all seafood regulations from whole fish to processed products. Inspectors verify label accuracy, including country of origin, net weight, and species identification. NOAA's seafood inspectors see about one-fifth of the seafood consumed in the U.S. every year and find some kind of fraud in at least 40% of all products submitted to them voluntarily. If a seafood business passes inspection, it is considered an "Approved Establishment" and may carry a U.S. Grade A or U.S. Department of Commerce-inspected label on its products.

5.3.3 NATIONAL ADVISORY COMMITTEE ON MICROBIOLOGICAL CRITERIA FOR FOOD

The NACMCF was established in 1988 and provides valuable scientific advice on food safety issues for federal food safety agencies and stakeholders. The Committee was established under the USDA Departmental Regulation 1043–28 by the Secretary of Agriculture after consulting with the Secretary of DHHS about the need for a joint USDA and DHHS advisory committee; the National Academy of Sciences recommended an interagency approach to microbiological criteria because a number of different federal, state, and local agencies are responsible for food safety, all of which could benefit from the scientific evidence provided by this committee. Several federal departments support NACMCF: the DHHS (FDA and CDC), the USDA (FSIS), the Department of Commerce (NOAA), and the Department of Defense (Veterinary Services Activity).

NACMCF provides scientific advice and recommendations to the Secretaries of DHHS and USDA on public health issues related to the safety and wholesomeness of the U.S. food supply. One primary concern involves developing microbiological criteria, the review and evaluation of epidemiological and risk assessment data, and developing or reviewing methods for assessing microbial hazards in foods. One notable achievement involves the standardized principles of Hazard Analysis and Critical Control Point (HACCP) systems (*see* the HACCP report at this link: http://www.fda.gov/Food/GuidanceRegulation/HACCP/ucm2006801.htm). NACMCF reports frequently serve as a foundation for regulations and programs to reduce foodborne illness and enhance public health. The following are links to three reports of particular interest for retail food safety:

- FINAL REPORT: Hot Holding Temperatures (Jan 25, 2002), http://www.fsis.usda.gov/wps/portal/frame-redirect?url=http://www.fsis.usda.gov/OPHS/NACMCF/2002/rep_hothold1.htm#.UcIa3pxdDq4
- Background Materials on Hot Holding Temperatures (PDF Only), http://www.fsis.usda.gov/wps/wcm/connect/af7fd153–6afa-46ce-9f7f-328b8e778fb6/hotholdcharge.pdf?MOD=AJPERES
- Final Recommendations on Bare Hand Contact with Ready-to-Eat Foods (Sep 24, 1999), http://www.fsis.usda.gov/wps/portal/frame-redirect?url=http://www.fsis.usda.gov/OPHS/NACMCF/1999/report_barehand.htm#.UcIbkJxdDq4

5.3.3.1 PROFESSIONAL AND TRADE ASSOCIATIONS

A number of professional and trade associations also assist in assuring the safety of food. These include the National Environmental Health Association (NEHA), International Association for Food Protection (IAFP), Institute of Food Technologists (IFT), National Restaurant Association (NRA), and the Conference for Food Protection (CFP). Each association makes a unique contribution to food safety.

5.3.3.2 NATIONAL ENVIRONMENTAL HEALTH ASSOCIATION

The NEHA (www.neha.org) is a professional society with more than 4,500 members drawn from both the public and private sectors, and includes academia, and the uniformed services. Most members are employed by state and county health departments. The mission of NEHA is "to advance the environmental health and protection professional for the purpose of providing a healthful environment for all."

NEHA publishes the *Journal of Environmental Health* ten times each year. The journal keeps members up to date on current issues and new research in environmental health. In addition, the journal provides information about products and services and employment opportunities. The journal covers a wide scope of environmental issues, including air quality, drinking water, food safety and protection, hazardous materials, occupational safety and health, terrorism and biosecurity issues, vector control, wastewater management, and water quality. All research article submissions are peer reviewed.

NEHA also provides 10 credentials to recognize professional competence in the field. Three are directly related to retail food safety: Certified in Comprehensive Food Safety (CCFS), Certified Professional-Food Safety (CP-FS), and Registered Environmental Health Specialist/Registered Sanitarian (REHS/RS). The remaining seven credentials are for areas like wastewater, hazardous substances, healthy homes, and radon.

NEHA has a strong focus on education, including many on-going training programs and a national network of trainers. Training focuses on food handling for employees and managers, HACCP at the retail level, and HACCP for processors. Many of their comprehensive training programs are available on line. These programs provide training for certificates in most of their training programs.

NEHA also has a strong public policy advocacy program to address issues that concern members. The organization has adopted several position papers that can be found on the NEHA web site.

5.3.3.3 INTERNATIONAL ASSOCIATION FOR FOOD PROTECTION

The IAFP (www.foodprotection.org) is a professional organization with more than 3,600 members drawn from educators, researchers, government officials, microbiologists, food industry executives, and quality control professionals involved with food from growing to processing and preparation. The mission of the IAFP is "to provide food safety professionals worldwide with a forum to exchange information on protecting the food supply."

IAFP publishes two journals, the *Journal of Food Protection* and *Food Protection Trends,* both of which are peer reviewed. The *Journal of Food Protection* is a monthly publication that contains scientific research and authoritative review articles on a variety of topics related to food safety and quality. Research published in this journal tends to focus on pathogenic microorganisms. *Food Protection Trends* is a bi-monthly publication with articles on applied research, applications of current technology, general interest topics related to food safety, and updates on government regulations.

IAFP provides a variety of other resources. Of particular interest to practitioners in retail food safety are the food safety and food allergy icons. They include 14 food safety icons covering basic food safety principles such as temperature control, hand washing, no bare hand contact, and cooking, as well as food allergy icons for 23 different foods. These icons can be downloaded from the IAFP website and used for signage, training materials, recipes, menus, food packaging, or other places where food safety or allergy reminders would be helpful.

IAFP provides many educational opportunities for its members, including an annual meeting, European Symposia, and a number of international meetings. In addition, the organization sponsors a number of webinars that can be viewed either live or archived on their website. Webinar topics vary and address current issues in food safety.

5.3.3.4 INSTITUTE OF FOOD TECHNOLOGISTS

The IFT (www.ift.org) is a leading organization for connecting food scientists worldwide. The mission of IFT is to "advance the science of food." The stated long-range vision is "ensure a safe and abundant food supply contributing to healthier people everywhere."

IFT publishes the *Journal of Food Science* 12 times each year, both in hard copy and on-line. This journal publishes peer-reviewed original research and critical reviews. The association also publishes the *Journal of Food Science Education*, focusing on improving food science education at all levels. This on-line, peer-reviewed journal allows for the sharing of novel teaching approaches to food science education. IFT also publishes *Comprehensive Reviews in Food Science and Food Safety*, which is a peer-reviewed online journal. Published bi-monthly, it provides in-depth reviews on specific topics. In addition, the association publishes *Food Technology* magazine, which addresses current issues of interest to the IFT membership.

IFT offers a Certified Food Scientist (CFS) credential, which recognizes the knowledge and skills of food scientists. Continuing education requirements to maintain the credential help assure that the CFS remains current. The credential recognizes that an individual meets criteria of competence and provides some assurance to employers that individuals meet IFT standards for performance.

IFT has a strong education focus. In addition to having meetings at the national, state, and local levels, they sponsor research summits to provide in-depth exchanges of information among the world's leading food scientists.

IFT also supports a public policy advocacy program for issues that affect food science. The association monitors meetings and public hearings to provide its members with opportunities to make public comments. The association provides testimony, statements, and commentary from its expert members.

5.3.3.5 NATIONAL RESTAURANT ASSOCIATION

The NRA (www.restaurant.org) represents almost 500,000 restaurant businesses through its national and state associations. The mission of NRA states, "We exist to help our members—the cornerstone of their communities—build customer loyalty, rewarding careers and financial success." To achieve this mission, NRA represents and advocates for foodservice interests with policymakers at the local, state, and national levels, provides tools and systems to improve operations, and offers networking, education, and research resources.

NRA developed one of the first food safety training courses and certification programs, ServSafe®. ServSafe® Manager Training is an accredited program with the American National Standards Institute/Conference on Food Protection. In addition, NRA offers ServSafe® Food Handler programs that meet state specific requirements.

NRA also is actively involved in public policy issues, monitoring legislation that may affect the industry. In addition, the NRA has an active lobby to influence laws and regulations. Finally, they sponsor the Restaurant PAC, a political action committee that helps elect congressional candidates who are educated in issues of interest to the restaurant industry.

5.3.3.6 CONFERENCE FOR FOOD PROTECTION (CFP)

The CFP (www.foodprotect.org) is a nonprofit organization started in 1971. The website states that the CFP "provides a representative and equitable partnership among regulators, industry, academia, professional organizations, and consumers to identify problems, formulate recommendations, and develop and implement practices that ensure food safety." The CFP meets biennially to provide opportunities for the various stakeholders to meet and discuss issues. While the CFP has no regulatory authority, it does provide a powerful voice for nongovernmental stakeholders by influencing model laws and regulations.

The CFP has three councils: Council I addresses laws and regulations; Council II addresses administration, education, and certification; and Council III addresses science and technology. Issues related to retail food, foodservice, retail food stores, or vending may be submitted to CFP, and must include a description of how the issue affects the retail food industry, the public health significance of the issue, and suggested solutions or rationale. Upon submission, the Issues Committee will review and clarify. If the issue is accepted, then it is assigned to one of the three councils. The issues submitted to each council are posted to the CFP website at least 40 days prior to the biennial meeting. During the biennial meeting, the council will deliberate about the issues submitted and a decision will be made to accept as written, accept as amended, or for no action taken with a reason. On the final day of the biennial meeting, each council chair presents recommendations to the Assembly of Delegates, who vote to accept or reject the councils' recommendations.

The CFP, and their process for introducing issues to government, has a significant effect on the FDA Food Code. At the 2012 biennial meeting, nontyphoidal Salmonella, reduced oxygen packaging, improving ground beef food safety in restaurants and food service, harmonizing time/temperature charts in Food Code with FSIS Guidance, and revision of the listeria retail guidelines were reviewed, in addition to other issues.

CFP is also involved in food safety education and certification. CFP, as an independent voluntary organization, has identified the essential components of a food protection manager certification program and established a mechanism to determine if organizations who seek to provide certificates meet these established standards. CFP works with the American National Standards Institute (ANSI), which is the accrediting organization for the CFP Standards for Accreditation of Food Protection Manager Certification Programs. Currently, four organizations have accredited ANSI-CFP Programs: 360training.com, Inc. for Learn2Serve® Food Protection Manager Certification Program; National Registry for Food Safety Professionals for the Food Protection Manager Certification Program and the International Certified Food Safety Manager program; the National Restaurant Association for ServSafe® Food Protection Manager Certification Program; and Prometric Inc. for the Food Protection Manager Certification Program.

5.4 SUMMARY

Food safety legislation and regulations have had a major impact on the safety of the U.S. food supply. Even so, food safety continues to evolve. In the United States, numerous government agencies and professional associations work to advance the science of food safety and propose policies in their efforts to maintain a safe food supply. With the Food Safety Modernization Act and the active research programs related to food safety, many changes are on the horizon. Check the government agency and professional association websites periodically to learn about the latest developments related to food safety.

Without a doubt, food safety has improved in the United States over the last 100 years. While it is difficult to predict the future, issues likely to be addressed in the short-term include changes in training and certification programs, expansion of inspections, emerging pathogens and contamination issues as well as issues related to the protection of food products from deliberate contamination. Hospitality educators and researchers can play a significant role in providing leadership to improve food safety in the hospitality industry.

KEYWORDS

- **Center for Food Safety and Applied Nutrition**
- **Food Safety and Inspection Service**
- **HACCP plan**
- **International Association for Food Protection**
- **National Environmental Health Association**

REFERENCES

Ballentine, C. (1981). *Sulfanilamide Disaster*. Retrieved June 18, 2013, from Taste of Raspberries, Taste of Death: The 1937 Elixir Sulfanilamide Incident: http://www.fda.gov/AboutFDA/WhatWeDo/History/ProductRegulation/SulfanilamideDisaster/

Janssen, W. F. (1981). *The story of the laws behind the labels*. Retrieved June 17, 2013, from Part I: The 1906 Food and Drugs Act: http://www.fda.gov/AboutFDA/WhatWeDo/History/Overviews/ucm056044.htm

Swann, J. (2005). *Apothecary's cabinet*. 100 Years of the 1906 Food and Drugs Act: The formation and early work of the Drug Laboratory, USDA Bureau of Chemistry.

The Library of Congress. (n.d.). *The Legislative Process*. Retrieved June 17, 2013, from http://beta.congress.gov/legislative-process/

U.S. Food and Drug Administration. (2009a, June 18). *Harvey W. Wiley*. Retrieved June 17, 2013, from http://www.fda.gov/AboutFDA/WhatWeDo/History/CentennialofFDA/HarveyW.Wiley/default.htm

U.S. Food and Drug Administration. (2009b, February 09). *Milestones in Food and Drug Law History*. Retrieved June 06, 2013, from http://www.fda.gov/AboutFDA/WhatWeDo/History/Milestones/ucm081229.htm

U.S. Food and Drug Administration. (2013). *About FSMA*. Retrieved June 18, 2013, from http://www.fda.gov/Food/GuidanceRegulation/FSMA/ucm247546.htm

U.S. Public Health Service. (2009). *2009 Food Code*. College Park, MD: U.S. Food and Drug Administration.

PART 2

OPERATIONAL ISSUES

CHAPTER 6

FOOD SAFETY AND THE LAW: UNDERSTANDING THE REAL-LIFE LIABILITY RISKS

DENNIS STEARNS, JD

Professor from Practice, Seattle University School of Law, of Counsel, Marler Clark, LLP, PC

CONTENTS

6.1 INTRODUCTION

For as long as there has been the exchange of food among people, whether by barter or sale, there have been concerns about the quality and safety of the food exchanged. Some of the earliest laws on record involve food, including laws that empowered supervisors to patrol markets to prohibit the sale of adulterated goods, and laws like the 1202 Assize of Bread, which required that a loaf be sold for a fair price and accurate weight, with violators subject to being "drawn upon a hurdle…through the greatest of streets, where the most people are assembled, and through streets which are most dirty, the false loaf hanging from his neck."[1] It can therefore be said that food was one of the earliest subjects of legal regulation, because it was a subject of concern for the poor and powerful alike. When it comes to adulterated food, deadly pathogens do not discriminate in terms of who is injured or killed.

Not surprisingly, the rise of professional foodservice—what has come to be known as the hospitality industry[2]–accompanied the rise of commercial industry. As has been noted by numerous historians, one of the many effects of the industrial revolution was the movement of rural workers into cities where the reliance on professional food sellers increased significantly, partly out of necessity given the long hours worked.[3] And, of course, as the reliance on professionally prepared food increased, so did reliance on

[1]Leslie Hart, *A History of Adulteration of Food Before 1906*, 7 Food Drug L.J. 5, 9 (1952). A second offense earned the baker at least an hour in the pillory. *Id.* For other interesting articles on the history of food law, see James F. Bush, *"By Hercules! The More Common the Wine, the More Wholesome!" Science and the Adulteration of Food and Other Natural Products in Ancient Rome*, 57 Food Drug L.J. 573, 576 (2002) (describing the adulteration of food in ancient Rome, for example, the substitution of inferior local products for more expensive imported ones, and the mixing of spoiled product with fresh); George M. Burditt, *The History of Food Law*, 50 Food Drug L.J. 191, 197 (1995) (noting that Cato and other prominent Romans reported or made warnings about the quality of food and wine, including the adulteration of bread with chalk); and Peter Barton Hutt and Peter Barton Hutt II, *A History of Government Regulation of Adulteration and Misbranding of Food*, 39 Food Drug L.J. 2, 27-34 (1984) (surveying the development of scientific methods to detect food adulteration, including the pioneering work of Fredrick Accum, the first scientist "to investigate food adulteration in earnest" and, who in 1820, published *A Treatise on Adulterations in Food, and Culinary Poisons* that revealed widespread adulteration of food in London).

[2]The term "hospitality industry" is broad in its meaning and a bit cumbersome to use repeatedly in the text. For that reason, this chapter will focus on the sale of prepared food at retail and most often uses the term "restaurant" to refer to these sellers. As used, however, the term can include sellers as diverse as someone with a stall at a farmers market, a hospital foodservice operation, or a hotel dining room.

[3]John Burnett, Plenty and Want, A Social History of Diet in England from 1815 to the Present Day 54 (Scolar Press: London rev. ed. 1979) (describing, among other things, a situation still familiar to many today—that is, workers having "little time or energy left for cooking").

the food being prepared safely. The same can be said with regard to the industrialization of food preparation itself. As meat and produce began to be shipped farther, and its preparation accomplished on ever-larger scales, the need (and demand) for government inspection grew. Anyone who has read Upton Sinclair's novel, *The Jungle*, likely understands how a great public disgust with unsanitary practices in the meat industry was translated into one of the first federal laws governing the manufacturer and interstate shipment of food—the Pure Food Act, which was enacted in 1906, and gave rise to a regulatory environment that exists today.[4]

Despite how far back in time issues of food and law seem to reach, to understand the subject today, history seems really to have begun in the early months of 1993—that is, when the Jack in the Box *E. coli* O157:H7 outbreak became news. Beginning in the middle of November 1992, and through to the end of February 1993, there were more than five hundred lab-confirmed *E. coli* O157:H7 infections and at least four related deaths, making it the largest reported outbreak in the history of the United States.[5] But the lab confirmation of infections did not tell the whole story; hundreds more had been made sick, the majority of them children.[6] All were infected as a result of a decision to go and eat a burger at a Jack in the Box restaurant, a burger that proved to be unsafe.

Reports of the outbreak and its victims dominated the news for well over a year, spurring changes everywhere, from how consumers viewed the safety of ground beef and fast food, to how federal government regulated the manufacture and inspection of meat products.[7] The media focus on issues of food and food safety has only increased over time, making

---

[4]The regulations are much changed, as are the enabling laws, like the Federal Meat Inspection Act, and the Food, Drugs, and Cosmetics Act. The idea of inspection, testing, and enforcement through recalls and impounding of adulterated food, still exists today. The inspectors of the early twentieth century might not understand Hazard Analysis, Critical Control Point (HACCP) principles, but they would understand the intent of such principles—the need to reassure the public that food has been prepared under sanitary conditions, making it less likely to cause illness and death.

[5]Denis Stearns, *Preempting Food Safety: An Examination of USDA Rulemaking and Its* E. coli *O157:H7 Policy in Light of* Estate of Kriefall ex rel. Kriefall v. Excel Corporation, 1 J. Food L. and Pol. 375, 390 (providing details of the outbreak, including its epidemiology and the resulting investigation).

[6]*Id.* at 390 nn.71, 73-74 and accompanying text.

[7]*Id.* at 390-97 (summarizing the changes spurred by the outbreak).See also Marion Nestle, Safe Food: Bacteria, Biotechnology, and Bio-Terrorism 73-85 (Univ. Cal. Press, 2003) (describing the outbreak and how it spurred a number of significant regulatory changes, including the requirement of "safe-handling" instruction on meat products).

every outbreak that occurs widely reported news.[8] The recent reports of horse meat being found in a wide variety of food products, [9] and the indictment of several corporate executives for selling *Salmonella*-contaminated peanuts, are but two current examples of the media's continued focus food and food safety.[10] But still the outbreaks continue, with seemingly no end in sight.[11]

Given the continuing public attention to issues of food safety, it is that much more important for responsible foodservice operators to understand the continuing legal risks faced if foodborne illness is not prevented. As will be emphasized again in the conclusion to this chapter, the best defense to a lawsuit is not facing it in the first place. Although it is not clear the extent to which the risk of liability is an incentive to improved food safety, those who ignore the risk of liability would not appear to be those most likely to pay close attention to the risk of unsafe food. For that reason, this chapter is designed as a comprehensive primer on the risk of legal liability in the context of the manufacture and sale of food. The first section will introduce the reader to some history and key legal principles, using

---

[8]During my fifteen years as a partner at Marler Clark, every time there was a foodborne illness outbreak the firm would quickly receive calls from media outlets seeking comments and interviews. Nearly from the beginning, we had a full-time person whose job it was to handle media contacts and to arrange for interviews with our attorneys and our clients.

[9]*See, for example,* Alan Cowell, *A Hint of Horse Meat Has a Nation Squirming More Than Its Neighbors*, N.Y. Times, Feb. 25, 2013 at A8 ("For weeks, the land has been seized with a spreading, Europe-wide scandal over discoveries of equine DNA in processed meals sold under household brands packaged as exclusively bovine—spaghetti Bolognese, lasagna and burgers among them.").

[10]*See, for example,* Sabrina Tavernise, *Charges Filed in Peanut Salmonella Case*, N.Y. Times, Feb. 21, 2013 at B5 (reporting that "a 76-count indictment...charged Stewart Parnell, 58, the former owner of the Peanut Corporation of America...with criminal fraud and conspiracy, for his role in what they said was a scheme to ship peanut products known to be contaminated..."). While at Marler Clark, I worked on a large number of the lawsuits arising from this particular outbreak.

[11]A study that the CDC issued in 2013 estimated that over nine million persons each year suffer a foodborne illness due to a major pathogen, a category that includes *E. coli* O157:H7. *See* John A. Painter, *et al.*, *Attribution of Foodborne Illnesses, Hospitalizations, and Death to Food Commodities by Using Outbreak Data, United States, 1998–2008*, 19 Emerging Infect. Dis. (No. 3) 407, 407 (Mar. 2013). The study also confirms how and why outbreaks are so important to understanding the extent of foodborne illness because "linking an illness to a particular food is rarely possible except during an outbreak." *Id.* The extent to which there has been progress in improving food safety is, predictably, the subject of considerable dispute, mostly because most reported outbreaks vastly underrepresent the true extent of foodborne illness, which is never linked to a particular source. *See* Denis Stearns, *On (Cr)edibility: Why Food in the United States May Never Be Safe*, Stanford L. and Pol. Rev. 245, 249 and n. 12 (2010) (explaining how the profitability of food depends in part on the ability to avoid investment in improved safety while causing significant amounts of foodborne illness that is never traced to its source).

real outbreak cases and lawsuits as examples. The second section takes a deeper look at some specific issues, again using real cases as examples. The approach here is not so much to scare as to educate. But if the real cases produce a dose of reinforcing fear, that would certainly not be a bad thing, especially if it leads to consistently safer food.

6.2 UNDERSTANDING THE RISK OF LEGAL LIABILITY: ORIGINS AND KEY PRINCIPLES

There are three basic concepts that must be understood if one is to understand how the law applies to food and food safety. First, the exchange of food for money is a "sale of goods" governed in part by principles of *contract law*. Second, the preparation of food for sale is governed by "standards of care" imposed under principles of *tort law*. Third, most food sold is a "product" governed by principles of *products liability law*, which is a kind of hybrid law containing both contract and tort principles.

In a typical law school, learning the basics of contract and tort law can take an entire year (two semesters) each, while products liability law is introduced briefly in torts class, with a subsequent semester devoted to the topic as an elective. To say that what follows is but a brief introduction to these topics would be an understatement. But, that said, by leaving out the many exceptions to every rule—there is *always* an exception it seems in the law—an understandable overview of legal liability can still be provided. Just keep in mind that the laws of each state can be remarkably different, and that a rule is only as clear-cut as the facts to which it is being applied.

6.2.1 THE CONCEPT OF DAMAGES AND LIABILITY

Just as one serious consequence of selling unsafe food is a person falling ill, another consequence is that person filing a lawsuit against the one that sold the food. The one who files the lawsuit is the *plaintiff*, and the one against whom the lawsuit is filed is the *defendant*. As is likely already understood, the reason for filing a lawsuit is to seek compensation for damages, which is why such damages are often called compensatory.

Such damages are of two kinds: economic and noneconomic.[12] Economic damages—sometimes helpfully referred to as "out-of-pocket" damages—include things like medical bills and lost wages, that is, things that would not have needed to be paid except for the fact of the foodborne illness. Non-economic damages include things like pain and suffering, emotional distress, and loss of enjoyment of life. Perhaps not surprisingly, economic damages are easier to calculate and prove than noneconomic.[13]

Although a person filing a lawsuit is seeking to recover damages, another way of putting it is that the person is seeking to hold the one that caused the injury **liable** for damages. Viewed from the perspective of the one being sued, the filed lawsuit therefore poses a **liability risk**. Therefore, just as there are food safety risks to be understood and controlled, so can the sale of unsafe food create liability risks that must be understood if they are to be controlled. But when there is no such control, a liability risk can turn into a liability reality in the form of a lawsuit. And that is when a lawyer is hired to defend a restaurant, with the first step being to analyze the likelihood that the lawsuit will result in the plaintiff's recovery of damages.[14]

To analyze the likely liability a restaurant faces, an attorney reads the *complaint*, the legal document that begins the lawsuit. The complaint sets forth the relevant facts as *allegations* and the legal basis for plaintiff's *claims* (or *causes of action*). These claims are often referred to as *theories of liability* because each one represents what is called a legal theory—that is, the basis upon which the plaintiff alleges the defendant is liable for the

[12]In addition to there being multiple exceptions for every rule, the law is notorious for calling the same thing by multiple names. For example, in the law, economic damages are also called "special" damages, while non-economic damages are called "general" damages.

[13]Because the award of non-economic damages has in the past been controversial, giving rise to phrases like "runaway jury" or "jackpot justice," some states have imposed limitations on the recovery of non-economic damages. For example, in Colorado, an award of non-economic damages cannot, under any circumstances, exceed $500,000. *See* Colo. Rev. Stat. 13-21-102.5(3).

[14]A few words on the hiring of attorneys are merited here. Most restaurants buy liability insurance to protect against the risk of a damage award. As a result, when a lawsuit is filed, the restaurant knows it will not need to pay damages itself (except for any "deductible" amount that must be paid). In addition to protecting the restaurant from paying damages, the insurance also covers the cost of hiring an attorney, although it is the insurance company that selects the attorney and controls the defense. For this reason, the insurance company can decide to settle a claim even if the restaurant thinks it did nothing wrong.

damages caused. Despite needing only one liability to recover, it is common for plaintiffs to allege (or "*plead*") more than one theory, doing so for different strategic reasons.[15] The plaintiff is not, however, entitled to recover on each liability theory.[16] And in many states, a plaintiff must elect to proceed to trial on a single theory.

Turning now to the theories that can be used in cases involving unsafe food, there are three: warranty, negligence, and strict liability, each of which will now be discussed in turn.

6.2.2 BREACH OF WARRANTY

Although there was for many decades controversy over whether warranty law was more contract or tort, there is now no disputing that it continues to exist in both places, depending on what kind of warranty we are talking about. In its simplest terms, a warranty is a promise that a thing sold possesses certain qualities or can perform in a described way. A warranty can be as simply as someone saying, "Here is your loaf of bread" as the baker hands the loaf across the counter to the customer. But in the law, nothing can stay simple long. For example, does the term "bread" mean that the loaf was made with only flour, water, yeast, and salt, without the addition of chalk or alum to bulk up its size or weight? Thus came the need for the many laws passed from the 11th century on that attempted to define what was meant by the term "bread," like the previously mentioned Assize of Bread that King John enacted in 1202.[17]

But setting the challenging definitional issues aside for the moment, the fact that a warranty is a promise is what is most key here. And where a promise has been broken it can be alleged that the seller breached a warranty to the seller, making the seller liable for the damages caused by that

[15]Some of the strategic reasons for pleading different liability theories will be discussed below when specific theories are described and explained.

[16]That would be what is called a "double recovery," which is strictly prohibited. The justification for the recovery of damages is to "make the plaintiff whole," which is to say, to compensate for what has been lost. To allow a double recovery would be to put the plaintiff in a better position than prior to injury.

[17]A job perhaps more entertaining than being a bread inspector was that of the "ale testers," who were tasked to "at all times try, taste, and assize the beer and ale to be put on sale, whether the same be wholesome for man's body…." Hart, *supra* note 1, at 10.

breach. Consequently, breach of warranty is one theory of liability upon which a plaintiffs can seek to recover damages. The challenge, though, with this liability theory is that the buyer who was injured is required to have been the recipient of the promise. Historically, this requirement was called the rule of privity, a rule that stated one could not recover for breach of warranty without having been "in privity" with the one who is alleged to have made the promise.

The famous legal case to which this rule is often ascribed is called *Winterbottom v. Wright.*[18] Decided in England in 1842, the case involved the driver of a mail coach, Mr. Winterbottom, who was maimed for life when a defective axle caused the coach to collapse. The driver sued the company that had promised to keep the coach in proper working condition, alleging in his complaint that the company had breached the promise. The court agreed that the promise had in fact been breached, and the breach had caused the driver severe injury. But the court stressed that not every wrong has a remedy, [19] with one of the judges stating "[u]nless we confine the operation of contracts such as this to the parties who entered into them, the most absurd and outrageous consequences to which I can see no limit would ensue."

The rule of privity stood as a bar to recovery for decades, frustrating plaintiffs at every turn. Interestingly, the case that finally created a lasting and influential exception to the rule involved food—canned tongue that had spoiled, and that had been sold at a restaurant.[20] The plaintiff was the restaurant, suing the manufacturer of the canned tongue on the grounds that it had caused a customer to become ill, and the restaurant to face not just a lawsuit, but also damage to its reputation and sales. The restaurant had not, however, bought the canned tongue from the Armour meat company, but from a local supplier instead. Consequently, like Mr. Winterbot-

[18]10 M. and W. 109, 152 Eng. Rep. 402 (Ex. 1842).

[19]The court used a then well-known (among lawyers) Latin phrase—*damnum absque injuria*, meaning "loss without injury." A few lines later in the opinion, one of the judges defends the court's seeming lack of sympathy as follows: "it is, no doubt, a hardship upon the plaintiff to be without a remedy, but, by that consideration we ought not to be influenced. Hard cases, it has been frequently observed, are apt to introduce bad law." Of course, the irony is that the Winterbottom case went on to be considered by many—certainly not all—to itself be "bad law."

[20]*Mazetti v. Armour and Co.*, 75 Wash. 622, 135 P. 633 (1913).

tom, the restaurant had not been the direct recipient of a promise (or "in privity"). This time, however, the lack of privity did not make the lawsuit fail. As the court explained, "Remedies of injured consumers ought not to be made to depend upon the intricacies of the law of sales. The obligation of the manufacturer should not be based alone on privity of contract. It should rest as was once said, upon the demands of social justice." And so, in this case, justice was said to prevail, doing so because the court ruled that food carried with it an implied promise, and therefore the buyer or consumer of the food could recover based on a theory of breach of *implied warranty*.

Having arisen in food cases, the theory of implied warranty continues to this day, applicable to both food and other products. This theory is usually considered part of contract law—specifically, Article 2 of the Uniform Commercial Code, which governs the sale of goods. Although the details of Article 2 are beyond the scope of this chapter, it is enough for a general understanding to know that key term is "*merchantability*."[21] Unlike an express warranty, where the parties expressly agree to terms (often written, but not necessarily), no agreement is needed for there to be an implied warranty; it is imposed as a matter of law. The idea of merchantability is that goods of a given kind must be "fit for the ordinary purposes for which such goods are used."[22] When it comes to food, it probably goes without saying that food contaminated with a pathogen is not fit for its ordinary purpose—eating. But sometimes the issue is not so clear.

For example, a restaurant defended itself against a breach of implied warranty claim by arguing that it is unreasonable for a consumer to expect raw clams to be free of bacteria like *Vibrio vulnificus* and that, in any case, the bacteria should be considered "natural," not "foreign," meaning the

---

[21]There is also an implied warranty of fitness for a particular purpose, but this warranty is usually applicable to commercial sales only, as between those in business, and not sales between a business and a consumer. An example of an implied warranty of fitness for a particular purpose is where apples are sold to a pie manufacturer, where the seller knows what the apples are meant to be used for. If the apples turn out to be unfit for the making of pies, the buyer could allege a breach of the implied warranty of fitness for a particular purpose and recover damages as a result.

[22]*See* U.C.C. Article 2, Section 314(2)(c). There are five other definitions of "merchantability," including that the goods "conform to the promises or affirmations of fact made on the container or label if any." *Id.* at Section 314(2)(f). The full text of Article 2 can be found online at the Cornell University School of Law's Legal Information Institute at http://www.law.cornell.edu/ucc/2/overview.html

claims were of merchantable quality.[23] The court rejected foreign-versus-natural test as the sole means of determining merchantability, explaining that "the problem with the foreign-natural distinction is that there is no truly logical basis for treating the two classes of injurious substances differently."[24] Summing up the superiority of what is called the consumer-expectation test, the court quoted from an Illinois Supreme Court case on the issue as follows:

In an era of consumerism, the foreign-natural standard is an anachronism. It flatly and unjustifiably protects food processors and sellers from liability even when the technology may be readily available to remove injurious natural objects from foods. The consumer expectation test, on the other hand, imposes no greater burden upon processors or sellers than to guarantee that their food products meet the standards of safety that consumers customarily and reasonably have come to expect from the food industry.[25]

Having adopted the consumer-expectations test, the court went on to conclude that it was not reasonable for the consumer to expect that a raw and unprocessed product like clams will be free of a naturally occurring bacteria, especially where the general public is aware that there are health risks inherent to consuming raw shellfish.

The court's ruling was not the end of the lawsuit, however, the court further ruled that the plaintiff should be allowed the opportunity to prove that the improper storage and handling of the raw claims, at temperatures alleged to have been above 45°F, caused bacterial growth, thus making

---

[23]*Clime v. Dewey Beach Enterprises*, 831 F. Supp. 341 (Dist. Del. 1993). In this case, the plaintiffs sued a restaurant for injured suffered as a result of eating raw clams contaminated with *Vibrio vulnificus*, a not uncommon bacteria found in clams and oysters. For most healthy individuals, an infection (if it results) will cause only mild symptoms. In this case, however, the plaintiff was an alcoholic with cirrhosis of the liver, making him susceptible to much more serious injury.

[24]*Clime*, 831 F. Supp. at 348. The court further observed that the "distinction is also somewhat artificial from a practical perspective. A small, but unforgiving, pearl from an oyster can cause as much damage as a "foreign" piece of metal when a consumer bites down on it." *Id.* (citing several cases).

[25]*Id.* at 348 (quoting *Jackson v. Nestle-Beich, Inc.*, 589 N.E.2d 547, 551 (Ill. 1992)). The quoted case involved someone injured by biting into a chocolate covered pecan and caramel candy that had a pecan shell in it, cracking the plaintiff's tooth. As stated by the court, the reasonable consumer expectation test "provides that, regardless whether a substance in a food product is natural to an ingredient thereof, liability will lie for injuries caused by the substance where the consumer of the product would not reasonably have expected to find the substance in the product." *Jackson*, 589 N.E.2d at 548 (citing as an example *Zabner v. Howard Johnson's,* 201 So. 2d 824 (Fla. App. 1967), a case involving a walnut shell in maple walnut ice cream).

the clams not merchantable for that reason. Because this warranty claim rested on facts subject to dispute (the restaurant denied improper storage), the court ruled that it was for the jury to decide the facts and reach a conclusion on the merchantability question. For similar reasons, the court stated that the plaintiff could also try to convince the jury that the alleged improper handling showed negligence on the part of restaurant. Consequently, for this plaintiff, negligence was an alternate theory of liability, and it is also our next one to discuss.

6.2.3 NEGLIGENCE

The commonplace concept of negligence is unlikely to be a puzzle to many. To be negligent is to act carelessly—that is, without using sufficient care. The legal concept of negligence is not so different at its core; however, turning the concept of negligence into a rule that can be applied in a nonarbitrary way to a diverse set of facts is complicated task. Indeed, the task is so complicated—and subject to criticism and controversy—that it is not an understatement to say that courts have struggled with devising the rules of negligence for well over a century (and further back if you count English law). But that said, the basic rules of negligence are today straightforward if applied to simple facts.

Take, for example, one of the earliest cases in the United States to establish the rules for proving a case of negligence, *Brown v. Kendall.*[26] The case involves two dogs that began fighting, prompting the owners to each begin hitting the dogs with sticks. In the process, Kendall accidentally hit Brown in the eye, seriously injuring him. The case ended up on appeal, and it is from this decision that many of the key rules still applied today can be found. These rules are as follows: (1) negligence involves a nonwillful act and, thus, differs from an allegation that one has intentionally caused harm; (2) the one suing (the plaintiff) has the burden of proving that the person sued (the defendant) was "at fault" for willfully committing an act that caused injury; and (3) fault is determined by asking if the defendant

[26]60 Mass. (6 Cush.) 292 (1850).

acted with "ordinary care and prudence" given the existing circumstances. It is this idea of fault that becomes the "reasonable person standard."[27]

As one might expect (especially in the law), the reasonable person standard is not without controversy, even though, as a rule of law, it is well established. What is maybe most controversial about this rule is the use of an objective standard, a standard that is not based on what a particular person was able to do under the circumstances, or what a particular person in fact knew at the time. Instead, the reasonable person standard is an objective, not a subjective, test that compares the defendant's conduct to an external and objective standard. This comparison is to what a reasonable person, under the same circumstances, *should have* known and *should have* done. Thus, it is no defense to a claim of negligence to answer that one was unaware of a risk if the plaintiff can show that a reasonable person would have been aware of the risk and acted to prevent the injury. And that is the question in food cases: Did the restaurant exercise *reasonable care*—that is, act as a reasonable restaurant would have acted under the same circumstances?

To understand the question of reasonable care, two related things should be kept in mind. First, it is generally assumed that a restaurant that acts with reasonable care will be a restaurant that does not cause its customers to fall ill from eating its food. Second, when a jury is called upon to apply the reasonable care standard, the conduct of the restaurant is viewed in hindsight, which is not a view that is often favorable to decisions that appeared much more reasonable at the time made. The Jack in the Box outbreak provides an excellent example of how defending against negligence claims can become a near-impossible task for a restaurant when actions are viewed after the fact.

From one perspective, the Jack in the Box case is fairly simple. Its supplier had delivered frozen ground beef patties, which were prepared

---

[27]Under modern rules, a claim of negligence is typically broken into elements: (1) duty, (2) breach, (3) injury, and (4) causation. Thus, for example, if I was to allege in a complaint that Restaurant-X was liable to me for negligence, I would allege that Restaurant-X owed me a duty to serve safe food, that it breached the duty by serving unsafe food, and that as a result of serving me unsafe food, Restaurant-X caused me injury. Because the issue of "duty" is complicated in theory, but not in practice when it comes to food safety cases, this issue will not be discussed in this chapter.

in restaurants as they had been for years. The FDA Model Food Code at the time recommended that ground beef be cooked to a minimum internal temperature of 140° Fahrenheit, a recommendation each restaurant followed, relying on electronic timers that had been programmed based on extensive cooking tests done by food scientists at the corporate office. But during the weeks in which customers were later found to have been infected, the ground beef patties used were heavily contaminated with *E. coli* O157:H7, and the cooking process had been insufficient to eliminate the deadly pathogen. Despite relying on its supplier to provide it with "safe and wholesome" ground beef, just as the USDA regulations required, Jack in the Box had received ground beef that everyone would later refer to as "tainted," "poisoned," and certainly "unsafe." Consequently, one might reasonably see the supplier of the ground beef patties as most at fault and the reason for the outbreak. But it did not turn out that way.

At the time, Jack in the Box was operating no differently than other restaurants. Everyone relied on timers to achieve a desired doneness for cooked meat products. No one, except maybe restaurant inspectors, used thermometers to double-check internal temperatures. Therefore, a defense attorney could credibly argue that Jack in the Box had acted in a way that was consistent with prevailing *industry standards*, also sometimes called "*custom*."[28] Just as a plaintiff can show that a defendant deviated from custom (or fell below industry standards) to prove a lack of reasonable care, so too can a defendant show compliance to show a jury that it acted with reasonable care. But, either way, the fact of a custom's existence, and compliance or deviation from it, is but one of many things for the jury to consider in determining if there was negligence. It must also be remembered that an industry standard, even if uniformly followed, can still be deemed to be unreasonable, and so not a defense. As the famous Judge Learned Hand wrote in a case that nearly all student study in their first year of law school, "Courts must in the end say what is required; there

[28]Generally, to be admissible, evidence of "custom" must show that a practice is well-established and broadly followed, making it likely that member of the industry should know of the custom.

are precautions so imperative that even their universal disregard will not excuse their omission."[29]

Much like custom can help define the standard of care, *applicable regulations and rules* can also define it.[30] For example, at the time of the Jack in the Box outbreak, California, Nevada, and Idaho, all required ground beef to be cooked just as the Model Food Code recommended. Therefore, no one could say that Jack in the Box had violated the law in cooking its ground beef patties. But in Washington, about a year earlier, the regulation had been changed, raised from 140° to 155° Fahrenheit.[31] Although Jack in the Box initially claimed to have received no notice of the change in the Washington law, that proved to be partly untrue. And that is how the "fairly simple" case not only became quite complicated, but the potential blameworthiness of Jack in the Box began to increase in a way that gave the plaintiffs a decided strategic advantage.

One reason that Jack in the Box could not assert in court that it was unaware of the rule change was because it had "*constructive knowledge*" of the change. Unlike "*actual knowledge*," which looks at evidence of what managers and employees in fact knew, constructive knowledge is what *should have been known* based either on available information, or what information could have been obtained as a result of a reasonable investigation. Thus, for example, if written notice of a regulatory change had been sent to a company, but never read by anyone (perhaps it was misplaced), the company is still said to have possessed constructive knowledge of the change.[32] Similarly, if a company knows that states make changes to a

---

[29]*T.J. Hooper*, 60 F.2d 737, 740 (2d Cir. 1932). In the *T.J. Hooper* case, a tugboat operator was sued for negligence as a result of it having lost two barges of coal in a storm. The boats had not been outfitted with radios and, thus, could not receive storm-warnings. The defendant argued that it had fully complied with existing custom because most tugboat operators at the time did not outfit boats with radios. Rejecting this argument, the court held that not adopting available new technology could be proof of negligence.

[30]In a minority of states, proven violation of a safety statute or regulation can itself be a theory of liability called "*negligence per se*." In states where this is a viable legal theory, the standard imposed by the statute or regulation replaces the reasonable person standard, meaning that violation of the statute or regulation is itself sufficient proof of negligence. In a majority of states, the violation of a safety statute or regulation is treated like the violation of custom—that is, as evidence of negligence, but not conclusively.

[31]As based on the 2009 FDA Model Food Code, Washington State now requires both time and temperature monitoring, allowing, for example, ground beef to be cooked to an internal temperature of 140° for 3 min, or 150° for 1 min, or to 158° and removed. WAC 246-215-03400(1)(b).

[32]A company is also said to have constructive knowledge of all that its employees know. Thus, for example, if several employees noticed that the standard setting on a grill resulted in hamburgers being undercooked, the company would be said to have knowledge of the undercooking even if the employees never reported what they knew to the corporate office.

food code, and that information is available for the asking, a company is said to possess constructive knowledge of any changes, even if no one had ever asked. In sum, a restaurant is evaluated against the standard of reasonable care based on what should have been known and done.

Most would agree that ignorance of law, willful or otherwise, should not be rewarded. Thus, Jack in the Box could ultimately not deny that it should have known the regulations for each of the states in which it operated, making it no defense to say that Washington State had done a miserable job getting the word out.[33] But none of that mattered from either a legal or public relations perspective. Instead, Jack in the Box had no choice but to defend against allegation that it had "made a conscious decision to disregard Washington law," doing so with a "total disregard for the health of its customers."[34] Whether these allegations would be deemed true was ultimately for a jury to decide; however, the eventual settlement of all pending cases made sure that risk of an unpredictable jury verdict was one risk that that never became a reality.

The reason the Jack in the Box case provides an excellent real-life example of the proof of a negligence claim is that it shows a restaurant company making decisions that were all arguably reasonable at the time made. Jack in the Box operated its restaurants in a way that was consistent with industry standards, and it followed the FDA Model Food Code in how it cooked its ground beef patties. Jack in the Box also bought ground beef patties from a USDA inspected supplier that was subject to the same safety laws as every other meat supplier in the United States. Nonetheless, there is no question that a great majority of juries, if asked to apply the "reasonable care" standard, would reach a verdict in favor of the plaintiff, especially in Washington where the cooking standard for ground beef had recently been changed. That Jack in the Box had little in the way of actual knowledge of this change would not matter at all. A jury would say that it should have known, and that Jack in the Box was one hundred percent negligent as a result.

[33]Indeed, of all of the counties in Washington, it appears that only one had even starting enforcing the new cooking-requirement. Food-service workers seeking their food-card were, at the time of the outbreak, still taking tests where 140° Fahrenheit was marked the correct answer to the question about minimum internal temperatures for ground beef.

[34]Elaine Porterfield and Adam Berliant, "Jack In The Box Ignored Food Safety Regulations, Court Documents Say," The Spokesmen Review, June 17, 1995.

But being one hundred percent negligent is not the same thing as being one hundred percent liable. And for that principle, the Jack in the Box case also provides a good example. Even though Jack in the Box faced liability under a negligence theory (among other theories, including strict liability, which will be discussed next), it was not the only defendant facing potential liability. The other defendants included the supplier of ground beef patties (The Vons Companies) and the various suppliers of raw beef that were used to make the ground beef patties. Had a case gone to trial and all the way to a verdict, the jury would have been asked to allocate fault as between the defendants, each of which would then have paid its percentage of the total verdict. Although not all states operate in this way—called "comparative fault" after the Uniform Comparative Fault Act—all states have some means by which a company held liable can either shift, or later recover, a portion of the liability to others who share responsibility.[35] Of course, being able to have another company contribute to a verdict's payment is no protection against being sued in the first place and all of the costs that come with that, including bad publicity and the resulting damage to reputation and sales.

6.2.4 *STRICT LIABILITY*

As shown, proving negligence is not too difficult with the benefit of hindsight, assuming that enough of the facts are known or can be discovered. But negligence acts are more often subtle or unseen, especially in food cases. Cross-contamination can occur even when great care is used by a majority of a restaurant's employees; an employee can be infected with a communicable disease and not show any symptoms at all; and contaminated ingredients can be delivered to a restaurant, with such contamination invisible to even the most well-trained eye. That is not to say that negligence is impossible to prove in food cases; many an outbreak report has detailed operator errors of a shocking kind. But many restaurants that

[35]The shifting or allocation of responsibility can be done in a number of ways, including ahead of time by way of agreement. Although the agreements were relatively rare at the time of the Jack in the Box outbreak, most restaurants now require suppliers to "indemnify" them for any liability resulting from the ingredients supplied, and to "hold harmless" from lawsuits, including paying for defense costs.

are identified as the source of an outbreak are well run, with well-trained employees, and admirable food safety systems in place. The question thus arises: How can a restaurant that is not negligent still be held liable for the damages caused by foodborne illness? The answer is the doctrine of strict liability—liability imposed without proof of fault.

In general terms, strict liability is not complicated, especially as applied to retail foodservice operations, like restaurants. To state a legally viable claim of strict liability, three things (or elements) must be alleged: (1) the restaurant manufactured and sold a product; (2) the product was defective when sold;[36] and (3) the defect caused a plaintiff to be injured. Put even more simply, for strict liability, there must be a product, a defect, and an injury. Now, proving all three of these things is not always easy. But in the case of an outbreak linked to a restaurant, such proof is rarely too difficult either.

6.2.5 PROVING THE MANUFACTURE AND SALE OF A PRODUCT

When most people think of manufacturing, they think of factories and things like cars, refrigerators, and lawn mowers. People are less prone to think of a restaurant as a manufacturer, but that is precisely what it is legally speaking. In many states, the issue of whether a company is a manufacturer matters, because strict liability does not apply to nonmanufacturing retailers, those who are said to "act as mere conduits in the chain of distribution."[37] But that was not always true.

As originally adopted by the American Law Institute in 1965, the Restatement of Torts (Second) §402A applied strict liability to everyone in the chain of distribution. Consequently, it did not matter if you were

---

[36]Three are three categories of defect: (1) manufacturing defects, (2) design defects, and (3) defects of warning, instruction, or marketing. With food cases, the defect is almost always in manufacturing, or that food is not reasonably safe as constructed. Although the rules is stated variously, the crux remains the same—that is, because of some kind of accident, the product did not end up as expected, or does not comply with the manufacturer's own specification. For example, a subway sandwich maker does not intend for a sandwich to contain *Listeria* in addition to cold-cuts, shredded lettuce, and other ingredients.

[37]*Almquist v. Finley School District*, 57 P.3d 1191, 1197 (Wash. Ct. App. 2002) (explaining that strict liability should apply to only those who exercise "actual control" over the product versus those who pass the product along unchanged like a distributor or grocer).

nothing more than the shipping company transporting boxes of canned peas, or the grocer who put the peas on the shelf for sale; if a can of peas ended up making someone sick, the plaintiff was free to sue the grocer, who then would need to file what is called an "indemnity action" or "contribution claim" against the other companies allegedly at fault to be reimbursed for what was paid in damages to the plaintiffs. Initially, chain-of-distribution liability seemed to be the most fair because it was thought that the plaintiff should not be shouldered with the responsibility of tracking down a manufacturer when others in the chain of distribution were in a better position to do so. It was also thought that retailers were best positioned to put pressure on upstream suppliers to make safer products, while also able to enter into contractual agreements to allocate the risks, or buy insurance as protection.[38] But as retailers more and more were subject to product liability lawsuits, pressure built to offer some protection against liability, which then prompted changes to the law.

While retailers can most everywhere be held liable for negligence, in a near-majority of states, strict liability no longer applies unless you are a manufacturer.[39] Therefore, the issue of whether one is a manufacturer can be of great legal significance in determining the risk of liability faced. In general, foodservice operations are manufacturers with regard to most products sold. Where the issue can become contentious, however, is when there is more than one potential manufacturer involved. For example, in a case involving an *E. coli* O157:H7 outbreak linked to tacos served to elementary school students, the district argued that it was not a manufacturer because another company was indisputably the manufacturer of the contaminated ground beef used to make the tacos. Both the trial court and

---

[38]These rationales for applying strict liability to retailers was first announced in the much-studied, much-debated case of *Vandermark v. Ford Motor Co.*, 391 P.2d 168 (Cal.1964). In an oft-quoted passage from the case, the court explained that, "Retailers like manufacturers are engaged in the business of distributing goods to the public. They are an integral part of the overall producing and marketing enterprise that should bear the cost of injuries resulting from defective products.").

[39]*See* J. Sinunu and A. Kott, *Protection For Retailers: Developments In Strict Product Liability*, 23 Westlaw Journal: Product Liability, Issue 5, at 3 and n. 21 (June 2011) ("Twenty-four states have since enacted some form of legislative protection from strict product liability for non-manufacturing sellers."). Most of the seller-protection statute includes exceptions, however. For example, in Washington State, a non-manufacturing seller can still be held strictly liable if "the court determines that it is highly probable that the claimant would be unable to enforce a judgment against any manufacturer." RCW 7.72.040(2)(b).

the appellate court rejected this argument, deciding that that the school district's actions fit squarely within the definition of manufacturer because it had a design for cooking the meat, its recipe, and the district's cooking process fell neatly within the dictionary definitions for produce, make, fabricate and construct.[40]

In other cases, the question of whether a restaurant is a manufacturer for a given product may not be so clear as the cooking of taco meat and its assembly into tacos. For example, a restaurant could have a self-serve salad bar where all of the items on it are purchased prewashed and precut and, thus, ready to use, with the only step taken being to open the packages and pour the contents into containers on the salad bar. In a *Salmonella* outbreak case linked to a buffet-restaurant in Georgia, a trial court decided that strict liability did not apply because the restaurant was not a manufacturer, only a seller. Such a decision is likely an anomaly, however. Still, the case shows that the issue is not always clear-cut, and some plaintiffs can be forced to prove negligence to recover damages.[41] It is for that reason, among others, that complaints always include claims for strict liability and negligence, and sometimes breach of warranty too.

6.2.6 PROVING THE PRODUCT WAS DEFECTIVE

One peculiarity of unsafe food cases is that there is rarely direct evidence of the product and its contamination. And there is a simple reason for that: food must be eaten if it is going to make someone sick. As a result, persons claiming to be injured by unsafe food must prove the fact of contamination by way of circumstantial evidence. The legal doctrine from which this means of proof derives is call *res ipsa loquitor* (race ip-suh low-qui-tore), or "the thing speaks for itself." This doctrine developed to allow plaintiffs to prove the fact of a negligent act where direct proof would otherwise be impossible, thus dooming the case. And, of course, the rule is from an old

[40]*See Almquist*, 114 Wn. App. at 405 ("to make means to bring a material thing into being by forming shaping, or altering material; to fabricate means to form into a whole by uniting parts and to construct means to form, make, or create by combining parts or elements").

[41]For more on the differences between strict liability and negligence, see the first of the three case studies below, *The Case of the Contaminated Ingredients* at __.

English case, *Byrne v. Boadle.*[42] In this case, the plaintiff was injured when a barrel of flour rolled out of a warehouse and fell on him. The trial judge had been convinced to dismiss the case for lack of evidence of any particular negligent act. On appeal, the court overturned the dismissal, observing that "There are certain cases of which it may be said *res ipsa loquitor*, and this seems one of them." The court further explained (showing eminent common sense, I might add), "A barrel could not roll out of a warehouse without some negligence, and to say that a plaintiff who is injured by it must call witness from the warehouse to prove negligence seems to me preposterous."

Like the barrel that fell on the head of Mr. Byrne, which it would not have done in the absence of some negligence, food that is contaminated would not be so unless a mistake of some kind had occurred. That the mistake occurred despite the exercise of all possible care is of no relevance at all to a finding of liability. All that must be proved is that the food was in fact unsafe. And this is where the fact of the injury comes in to be used as proof of the unsafe (or defective) condition of the product. This manner of proof is often referred to as the "*malfunction doctrine*," and it usually applies in food cases.[43] When direct evidence of a specific defect is unavailable, a plaintiff can attempt to prove a product was defective by showing (1) the injury-causing incident does not ordinarily occur in the absence of a defect, (2) any defect most likely existed when the product left the defendant's control, and (3) the injury is not likely explained by other reasonably possible causes. As will be explained in greater detail in the section that follows, only the likelihood of other possible causes is commonly subject to dispute when a plaintiff has a confirmed case of foodborne illness that investigators have linked to an outbreak. On the other hand, nonoutbreak cases are notoriously difficult to prove based on the use of circumstantial evidence alone. Therefore, even with the mal-

[42] 2 H. and C. 722, 159 Eng. Rep. 299 (Exch. 1863).

[43] Although direct evidence of food contamination is not common, it is not unheard of either. For example, in cases involving a sushi restaurant in Arkansas, more than one Mexican food restaurant, and a national fast-casual restaurant (like an Applebee's), I have had clients who have taken home leftovers that ended up testing positive for a pathogen. Similarly, outbreak investigations sometimes find ingredients in restaurants that test positive, allowing a link to the persons who got sick eating at the restaurant.

function doctrine as a tool, its successful use is mostly restricted to persons injured as part of an outbreak.

6.2.7 PROVING CAUSATION-IN-FACT[44]

Whether asserting strict liability or negligence, a plaintiff must prove that the injury complained of would not have occurred "but for" eating the unsafe food. The "but for" causation analysis is often more confusing in theory than in practice, making an example from a challenging real-life case helpful here. An eight-year-old girl and her cousin of the same age share a hamburger, each eating half. Both fall ill within two to three days. The cousin has a confirmed *E. coli* O157:H7 infection that, fortunately, is not severe, and she recovers quickly. The eight-year-old girl suffers symptoms like her cousin, going on to develop hemolytic uremic syndrome (HUS), a deadly complication almost always associated with *E. coli* O157:H7 infections in children. Although both of the children consumed other food items known to be at risk for contamination with *E. coli* O157:H7, the only food that the two children had in common was the hamburger. Therefore, based on these facts a convincing argument could be (and was) made that, *but for* consuming the hamburger, and *but for* its consumption, neither of the children would have been infected with *E. coli* O157:H7 and injured.

A plaintiff need not prove causation beyond a reasonable doubt—the standard applicable to criminal trials. Neither must a plaintiff proof causation on a statistically significant basis—that is, to a 95 percent confidence level. Instead, causation only need be proven on a more probable than not basis, which is sometimes also referred to as the "preponderance of the evidence" or "balance of the evidence" standard. Put simply, this standard requires any amount more likely than not. Thought of as a scale, even the slightest of tips in favor of something being the cause is enough to meet the burden of proof under this standard. It is only when some other cause

[44]Proving cause "in fact" (or factual causation) is not the full extent of the causation question in the law. There are also the concepts of "proximate cause" of which "legal causation" is a part. These concepts are complex and policy-driven at times, making them ill-suited for summary treatment. Consequently, this section is focusing on causation-in fact, which is where a food case would usually fail.

is more probable, or if one or more causes are equally probable, that the standard is not met.[45]

Proving that another cause of illness is equally probable is a standard defense to a circumstantial case. Such a defense can often put a restaurant in a difficult position, because it is usually a supplier that is a candidate for finger pointing (or, *allocation of fault*). For example, in an outbreak of hepatitis A infections linked to a Chi Chi's restaurant in Pennsylvania, investigators were able to identify contaminated green onions as the cause of infection. Over 500 of the restaurant's customers were infected and three died.[46] There was, as a result, no real question that the restaurant was liable under a strict liability theory for having caused the infections through its use of contaminated green onions. The trace-back proved to be less decisive in revealing other potentially responsible companies. Sysco delivered the green onions to the restaurant, but it had no involvement beyond transportation.[47] There was also a fruit and produce wholesaler with which Chi Chi's had a contract to buy the green onions. There were multiple suppliers of green onions to the wholesaler, and three growers in Mexico identified as potential points of origin. But the trace-back evidence never got beyond the wholesaler because each of the suppliers were of equal probability for having provided the contaminated green onions that made it to the restaurant. And there were no records kept by Sysco or the restaurant that could have helped identify the other potentially responsible parties. Consequently, the injured customers focused on recovering damages from the restaurant, which fortunately had sufficient insurance, while Chi Chi's sued Sysco and wholesaler only, alleging breach of warranty.

---

[45]When there are multiple causes that have contributed to causing the same injury, some courts use what is called the "substantial factor" test, which asks whether one or more defendants "contributed in a material way" to causing the injury. The jury is then asked to determine the extent of each defendant's contribution and the verdict is allocated according to percentages. In practice, the "but for" causation test is not really any different, because it has never been the rule that there can only be one "but for" cause.

[46]CDC, Hepatitis A Outbreak Associated with Green Onions at a Restaurant --- Monaca, Pennsylvania, 2003, 52 MMWR Weekly, Vol. 47, 1155-57, November 28, 2003. Chi Chi's ended up filing for bankruptcy and never resumed business again as an ongoing restaurant operation.

[47]Which is not to say that Sysco escaped liability because Chi Chi's filed a breach of contract claim against it seeking to be indemnified (paid back) for having supplied it with green onions not fit for use in the manufacture of food. Much litigation ensued as between the insurance companies for Sysco and the other suppliers in the chain of distribution for the green onions, with millions at stake.

6.2.8 DEFENSES TO LIABILITY

When it comes to food cases, the best defense is to sell only safe food. To say that may seem silly, but it is said based on over 20 years of experience in litigating food cases. For the first 15 years that the Marler Clark law firm operated, it represented thousands of victims of foodborne illness and filed hundreds of lawsuits. During that time, only three cases went to trial, and only one all the way to a jury verdict.[48] That means nearly all of the other cases settled. That is not to say that everyone who contacted the firm became a client—most did not. And the main reason that someone did not become a client was because there was a lack of evidence of the cause of the illness. Even someone with a confirmed *Salmonella* infection will not necessarily be attributable to a given outbreak or exposure at a given restaurant (based on food history). Consequently, for those food cases that are successfully defended, the most common reason for the success is the ability to disprove causation.[49]

Disproving causation—or, more accurately, convincing the jury that the plaintiff did not succeed in proving causation by a preponderance of the evidence—is a defense that works equally well against a strict liability and a negligence claim. There are more available defenses against a negligence claim, like showing that the restaurant acted in a reasonable manner, consistent with the applicable standard of care. But such a defense is of no real help if the plaintiff also has asserted a strict liability claim, which does not require proof of fault. Therefore, it bears emphasizing again that strict liability in every respect lives up to its description as "strict," making

---

[48]The case that went all the way to verdict, and then on appeal, involved the lawsuits arising from the *E. coli* O157:H7 outbreak at the elementary school in eastern Washington. After a nearly month-long trial on the issue of liability (the trial on damages was to follow), the jury was asked to allocate fault as between the school district and the supplier of the contaminated ground beef. The jury allocated 100% of the fault to the school district for its improper cooking.

[49]Although covering the timespan prior to when Marler Clark became the first law firm to focus exclusively on food cases, a study published by the USDA Economic Research Service provided support for the assertion here that most food cases that go to trial fail on causation grounds. *See* Jean C. Buzby, *et al.*, Product Liability and Microbial Foodborne Illness, Agricultural Economic Report No. (AER-799), at 18 April 2001 ("Despite their greater reliance on medical experts, most consumer plaintiffs failed to convince juries that defendant firms were legally responsible for causing their illness.").

defenses against negligence mostly irrelevant, except in perhaps keeping the award of damages down.[50]

6.3 FURTHER UNDERSTANDING THE RISK OF LEGAL LIABILITY: MORE REAL-LIFE CASE STUDIES

The law is not only used to impose liability after the fact of an injury; the law can also be used to shift the responsibility for that risk to the one who, at least theoretically, is in the best position to control the risk. Of course, for that shift to certainly occur, the risk needs to have been identified ahead of time, which is what did not happen in the following three case studies, leading to some tragic results.

6.3.1 STRICT LIABILITY VERSUS NEGLIGENCE: THE CASE OF THE CONTAMINATED INGREDIENTS

One of the more common liability scenarios is where a restaurant receives an ingredient that is contaminated or otherwise unsafe. When such an ingredient is added in unchanged, with no cooking or other intervention, there are few grounds for alleging negligence. For example, imagine a restaurant like Jimmy Johns that uses alfalfa sprouts to make subway sandwiches. If those sprouts are contaminated with *Salmonella*, there is little chance for the restaurant to prevent susceptible customers from falling ill. Would it therefore be a complete defense to liability to point out that (1) the restaurant did not do anything to the sprouts except put them on a sandwich, and (2) there was no way to know that the sprouts were contaminated, given that *Salmonella* cannot be detected by sight or smell. The short answer is: No, the restaurant could still be liable.

First, the question of negligence is not so cut-and-dried to allow a restaurant to escape liability because it neither caused, nor actually knew of,

[50]In evaluating a case for settlement purposes, the relative culpability or wrongfulness of a given restaurant's acts are definitely taken into account because, in general, juries tend to award more when it appears that a restaurant is more blameworthy in causing an illness. On the other hand, when it can be shown that a restaurant had a well-run operation with only a minor misstep attributable, for example, to an employee error, a jury is much less likely to award more than what seems minimally fair. Having had to explain this to clients many times, I usually refer to this jury-effect as the "good citizen discount."

the contamination. With something like sprouts, the risk of contamination has been known for some time. For example, a study published in 2001 noted, "at least 12 reported sprout-related disease outbreaks involving a total of more than 1,500 cases have been reported since 1995."[51] And that was just the beginning. Outbreaks linked to contaminated sprouts happened with such regularity that it no longer took a creative attorney to argue that a restaurant using sprouts was negligent just in choosing to do so. In other words, certain foods can acquire enough of an outbreak record to be deemed inherently risky.[52] For example, in the lawsuits arising from a 1996 unpasteurized apple juice outbreak, a coffee-shop retailer was added as a defendant on the grounds that it had sold the juice, pouring it into a cup before serving it. In arguing that it should be dismissed from the lawsuit, the retailer argued that it had not made the juice, did not know that it was contaminated, and could not as a matter of law be held negligent. Opposing the dismissal, the plaintiffs argued (in a brief that I wrote) that there had been numerous outbreaks linked to unpasteurized apple juice, making its sale potentially negligent. The court agreed.[53]

In addition to using an ingredient that is known to be risky (or should be known to be), the second way negligence can arise from the unknowing use of contaminated ingredients is through the careless selection of a supplier. Just like certain ingredients can have track records that should put a user on notice of the risk being taken by the ingredient's use, suppliers have track records too—especially if the industry involved is subject to inspection. I can barely count the times over the years where in reviewing

[51]Mary E. Proctor, *et al.*, *Multistate Outbreak of Salmonella Serovar Muenchen Infections Associated with Alfalfa Sprouts Grown from Seeds Pretreated with Calcium Hypochlorite,* 39 Journal of Clinical Microbiology, Vol. 10, 3461, 3461 (Oct. 2001). These outbreak figures are from the United States only and, thus, omit one of the largest outbreaks of all time—also linked to sprouts. This historic outbreak involved *E. coli* O157:H7, and occurred in Japan in 1996, sickening 9,441 people, mostly school children

[52]In February 2012, Jimmy John's permanently stopped selling sprouts on its sandwiches, this after the fifth sprouts-related outbreak of illnesses linked to sandwiches sold at the restaurant chain. Eight months later, Kroger, the nation's largest grocery store chain, also announced that it would stop selling sprouts. *See* Elizabeth Weise, *Kroger stores stop selling sprouts as too dangerous*, USA Today, Oct. 20, 2012.

[53]It should be emphasized that such a ruling is not a finding that the retailer was negligent; it was a legal conclusion that the plaintiffs had alleged sufficient facts to have earned the opportunity to prove negligence at trial. In this particular case, a settlement was eventually reached with the maker of the juice.

the inspection records of a given manufacturing plant I said to myself: "Who would buy from these people?" For example, in an outbreak linked to a supplier of peanuts and peanut products, one food company had inspected the supplier's plant and its testing records and then decided not to purchase from the supplier. Another equally large food company relied upon a private third party inspector that the supplier hired itself to prepare a report. After passing this inspection, the other food company began buying peanuts to use in its own products. A multistate *Salmonella* outbreak linked to peanuts was announced by the CDC, leading to the supplier's bankruptcy filing and, ultimately, criminal charges. Yet, despite the clear culpability of the peanut supplier, the company that had decided buy peanuts on the basis of a third party inspection was still subject to lawsuits and significant defense costs and liability risks, based on both negligence and strict liability theories.

6.3.2 UNREASONABLE RISKS AND DUTIES OF CARE: THE CASE OF THE DEADLY SALAD BAR

One case that arguably merits an entire chapter of its own is the 2000 Sizzler Steakhouse *E. coli* O157:H7 outbreak that occurred in Milwaukee, Wisconsin. Over 100 people were sickened as a result of eating contaminated ready-to-eat items from a salad bar, and a three-year-old girl, Brianna Kriefall died as a result of her *E. coli* infection.[54] The litigation went on for over a decade and involved multiple appeals, one that made its way all the way to the United States Supreme Court.[55] But what makes this case such an instructive example is how, with the benefit of hindsight, the

[54]*See Kriefall v. Sizzler USA Franchise Inc.*, 816 N.W.2d 853 (Wis. 2012). The Kriefall case settled for $10.5 million dollars.

[55]The issue that made its way to the U.S. Supreme Court, which ultimately decided against taking the case to review, was whether tri-tips contaminated with *E. coli* O157:H7 could be deemed to be unsafe and defective under state law if the USDA regulations against this pathogen were specific to ground beef only. The meat company, Excel, convinced the trial court that federal law preempted (trumped) state law on this issue and, thus, the trial court dismissed all strict liability claims against Excel. The Wisconsin Court of Appeals reversed the trial court's decision, reinstating the lawsuits against Excel. It was this decision that was appealed to the Wisconsin Supreme Court and then the U.S. Supreme Court. For more on this issue and these cases, *see* Denis Stearns, Preempting Food Safety: An Examination of USDA Rulemaking and its *E. coli* O157:H7 Policy in Light of Estate of Kriefall ex rel. Kriefall v. Excel Corporation, 1 J. Food Law and Pol. 375 (Fall 2005).

outbreak seemed so easy to prevent—if only the risks had been identified and controlled.

The two restaurants implicated in the outbreak were both franchises, long in business, and without apparent problem. As is the case with most franchise operations, the franchisor, Sizzler USA, issued operation policies, some mandatory and others not. One policy that was not mandatory was the purchase of precut, prepackaged steaks. In 1993, Sizzler restaurants in Washington and Oregon were identified as the source of an *E. coli* O157:H7 outbreak. A subsequent investigation found that knives used for cutting steaks had also been used to cut fruit and produce for the salad bar. In response to this finding, Sizzler USA switched all of its corporate restaurants to the use of precut steaks as a means of avoiding cross-contamination. Franchise restaurants did not need to make the switch, though. Then, seven years later, can you guess the cause of the outbreak at the two Milwaukee area Sizzler restaurants?

The Milwaukee restaurants had continued to purchase whole tri-tips, and to cut and tenderize the steaks in-house. Doing so was more profitable, in part because it was then possible to use the leftover trimmings to make chili. Precut steaks also cost more, although it is not clear this would be true if labor costs are also considered. By allowing its franchise restaurants to continue cutting steaks in-house, now a proven risk of cross-contamination, Sizzler USA exposed itself to a risk of liability, a risk that became reality once plaintiffs asserted negligence claims against it for the failure to use reasonable care in supervising the operations of its franchise restaurants. Of course, the two Milwaukee restaurants (both with the same owners) faced both strict liability and negligence claims. And for selling the contaminated tri-tips that were the source of *E. coli* O157:H7, Excel faced strict liability and negligence claims too. To say there was no shortage of finger pointing as between the defendants would be an immense understatement.

6.3.3 NEGLIGENCE AND PUNITIVE DAMAGES: THE CASE OF THE BROKEN HOT WATER HEATER

Our final case study is as surprising as it is instructive, showing how financial incentives and bad decision making can combine to cause a lot of avoidable sickness. This outbreak cases involved a large, national restaurant chain and its restaurant in the Midwest.[56] One of the employees had gone on vacation and somehow been infected with *Salmonella*, an infection that caused no symptoms. Asymptomatic infections are not very common, but they are not very rare either. One of the rationales for hand hygiene as an important control measure is the assumption that people can be infectious without even knowing. And that was certainly true in the case of this *Salmonella* outbreak. Not only was the employee infected, but she managed to infect several co-workers and one of the managers. All of this was risky enough, of course. But then the hot water heater broke, meaning no hot water with which to wash hands.

Faced with the decision to replace the hot water heater, the manager appears to have waited until the end of the month when, one could speculate, monthly sales totals would be wrapped up. As with many restaurants that are part of a national chain, the compensation of the managers can include bonuses paid for meeting certain sales goals. Similarly, equipment maintenance and replacement costs are allocated to the restaurant, meaning that money saved in this area adds to restaurant's profitability. Consequently, deferred maintenance can add to the bottom-line. Unfortunately, deferred maintenance can also add to the food safety risks that a restaurant faces. In this example restaurant, the dish machine apparently did not sanitize properly because the chlorine sprayer was not working correctly. This combined with a lack of hot water for at least one day, and no water at all during the time when the hot water heater was replaced while customers were still being served, created an arguably perfect environment for infected employees to spread *Salmonella* to food, plates, glasses, and eating utensils.

That a plaintiff would have a relatively easy time proving negligence in this case probably goes without saying. The more interesting—and telling—question though is whether a plaintiff could prove a claim for *punitive damages* against the restaurant. In proving punitive damages, a plaintiff must not only show negligence; the plaintiff must show what is often

[56]Although this case study is primarily based on one particular outbreak, I am not identifying the restaurant because clients that I represented at the time signed confidentiality agreements. All of the facts upon which this case study is based can be found in news articles and in the health department reports.

referred to as "malice" or "wanton and willful" conduct—that is, conduct deserving of punishment by way of an award of punitive damages. (Such damages also are called "*exemplary damages*" because they are awarded to make an example out of the defendant to deter similar conduct in the future.) To be deserving of punishment, the conduct must usually be shown to constitute a conscious disregard of a known safety risk. Here, should have known will not suffice. A plaintiff cannot ask a jury to award punitive damages unless it can be shown that a risk was known, and a conscious decision was made to take the risk anyway, putting others predictably in danger of harm. The case of the broken hot water heater is an example of such a case.

6.4 CONCLUSION

With the help of its defense attorneys (of which I was one of the leads), Jack in the Box was able to settle and dismiss all of the lawsuits filed against it. There were two keys to the success of the settlement strategy. First, the company did not contest that it would be held liable under the law of Washington and other states in which there were claims. But even being willing to admit liability does not make litigation go instantly away, especially when there are several hundred persons with claims. And, for that reason, Jack in the Box still faced over three years of intensive and intrusive litigation. I spent weeks going through every one of its files (all mostly paper back then) to respond to discovery requests—formal written requests for answers to questions and for copies of relevant documents. There were dozens of depositions too where employees were put under oath and their testimony recorded by a court reporter. When a company is sued, its insurer will pay for the defense attorneys, and for the settlement or verdict. But the time—not to mention stress—dealing with ongoing litigation, is a cost that is arguably more harmful than the payment of dollars and cents. For that reason, what I have told food companies for years is this: the best defense is to not make someone sick. Period.

KEYWORDS

- ***actual knowledge***
- ***constructive knowledge***
- ***contract law***
- ***defendant***
- ***implied warranty***
- ***liability law***
- ***merchantability***
- ***plaintiff***
- ***tort law***

CHAPTER 7

HANDWASHING

JIM MANN
Executive Director Handwashing For Life® Institute

CONTENTS

7.1 A PROCESS OFTEN OUT OF CONTROL

Handwashing, the removal of transient skin flora by cleansing or disinfection, is an agreed priority in the food industry yet compliance drives a different reality. Science has founded many new technologies but has done little to change behaviors at either the management or line worker levels. It is often neither a priority concern in the boardroom nor along food's circuitous route from farm to fork.

Handwashing is so basic to one's health yet this almost intuitively simple practice is compounded by the many variables affecting cross-contamination and the cleanliness goal. The risks arise from the natural power of pathogens and lack of process control.

Research has been largely limited to the physical sciences, led predominately by desires to make the Model Food Code more effective and science based. Risk-based research has been hampered by the lack of metrics and data to directly connect key behaviors to risk. Total solutions may well require culture change driven by behavioral science.

Contaminated hands, bare or gloved, are the major carriers of pathogens resulting in foodborne outbreaks. The food safety community estimates that 65% of foodborne illness is hand-triggered, preventable by a traditional handwash, a hand-sanitization or the donning of clean gloves (Consensus reached with 31 responses at 2013 meeting of Southwest Environmental Health Association in Laughlin NV).

The CDC estimates that 3,000 people die each year in the USA from pathogens transmitted through food prepared or served away from home. This translates to a likely majority of these deaths being attributed to hand-triggered cross-contamination. In healthcare the CDC annual death toll estimates soar to a total of 99,000 hospital/healthcare acquired infections (HAI). Healthcare infection preventionists commonly ascribe 85% of these HAI deaths to inadequate hand hygiene.

7.2 UNDERSTANDING THE RISKS

Food processing uses machines and automation to minimize hand contact, giving them better process control and lower risk of hand-triggered contamination. It is obvious that an unwashed hand can cause more damage in food processing because of batch size and the challenges of effective recalls.

Breakdowns in the hand hygiene food safety process occur regularly in restaurants, delis, convenience stores and institutional foodservice. Hands are a critical component of service for all these locations. And the process is often ill defined, poorly understood and rarely measured.

An outbreak is often the first indication that there was a system failure. It is a dangerous, costly and potentially business-ending way to learn of the likely preventable error. An operation's ability to track compliance standards creates a predictive window to catch near misses and improve the process before an outbreak is endured.

Daily restaurant routines rarely end with an outbreak. Shift upon shift, year after year an outbreak-free performance reinforces current standards of behavior regarding hand cleanliness. Poor behaviors can unintentionally be locked in as the corporate standard and ignored as operational efficiencies are continually improved.

Bacterium and virus, the two primary hand-transferred pathogen groups, differ widely in infectious dose and little is written about variables in their adhesion, release and transfer. Chemical compositions and pressure variants come into play. Skin texture, moisture levels and fingernail lengths vary widely but are rarely considered by a food worker when choosing when and how to wash.

Norovirus is far and away the leading cause of foodborne illness and is second only to the common cold in person-to-person transferred infections. Its virulence is impressive with an infectious dose of as little as 10 organisms compared to salmonella with a level greater than 100,000.

Dr. Christine Moe, a world authority on norovirus with Emory University, has demonstrated that human norovirus can live on inanimate surfaces for weeks and that commercial hand sanitizers differ greatly in their ability to kill the human norovirus (Liu et al., 2013). Previous studies had been limited to calici virus surrogates that are more easily rendered inactive. Dr. Moe's studies also show that some commercial brands of alcohol hand sanitizers are almost ineffective while others are highly effective. The differences were not attributable to differences in alcohol level.

The risk assessment process often includes a close look at the legal implications as a missed handwash could damage brand values or even terminate the business. Strict Liability is a critical legal principle that if you cause harm you are liable. It is liability without proof of fault for an injury caused by a product that is both defective and not reasonably safe. In establishing strict liability, the injured person need only prove:

1. The product was unsafe (and therefore defective), and
2. The defect caused the injury.

The focus is on the product, not the conduct of the manufacturer/food-service operator. It does not matter if every possible precaution was taken. If the product was defective, and caused an injury, the provider of the product is liable. Genetic fingerprinting of organisms makes positive identification of the source of an outbreak possible. Once the source is identified, the operator is liable.

7.3 STANDARDS FOR GOOD HANDWASHING BEHAVIORS

Operators do have standards for handwashing and high-touch surfaces. They are most commonly the default standard resulting from current practices. If the practice is poor the standard is poor. If that standard is defined as low and then not raised, business ethics are called into question.

Standards are antecedents for rewarding good hand hygiene behaviors. Without them, efficiency goals dominate and hinder handwashing enhancement programs. Ideally, a supervisor's leadership in motivating and delivering good compliance rates opens the door to salary increases and promotions.

Assessing operational handwashing is hampered from the start by the lack of meaningful metrics and even definitions. What is a clean hand? In section 1.202, the Model Food Code defines a *HAND WASHING SINK* but offers no such definition of a clean hand. In later chapters elements of handwashing processes are outlined but do not include specifications for the outcomes. Considering the range and complexity in defining these standards The Model Food Code is wise to remain less prescriptive and leave these multifaceted decisions to the operator, as it is he that is first in line regarding accountability.

Pathogen reduction is the most accepted measure of handwashing effectiveness. A 2-log reduction, 99%, is considered a basic reference point for soap and running water wash for 15 sec. This reflects the thinking summarized in the Annex of the Model Food Code but is not a codified standard.

Just as there is no one soil there is no one handwash. Effectiveness can be affected by soap choice, water temperature and hardness, friction variables, use of a nailbrush, skin condition, by length of scrub time and drying method. In cases where water is not available, effective handwashing can be achieved without water (Edmunds et al., 2010).

Careless consideration of these differences adds to the risk.

The Centers for Disease Control and Prevention (CDC) conducted an observational study using The Model Food Code as its benchmark for when to wash. They concluded 8.6 Handwashes/Employee Hour (HW/EH) were required to be code-compliant, a total of 69 Handwashes per 8 h shift. This level of handwashing is in conflict with customer service and is seen as too much to ask of employees. The regulatory community tends to agree.

The FDA's Model Food Code does not attempt to differentiate situations requiring a handwash. It simply states when to wash. Operators interpret their operations and set out risk-based standards for their training programs.

A clean, ready-to-serve hand is not codified numerically. The metric is at best a measure of activities, often limited to observation as the means to audit compliance. Intuitively, many operators, especially those with limited raw food handling, establish a minimum frequency of one handwash per employee hour, based on the likely hand contact with a raw food, a ready-to-eat (RTE) food or a contaminated surface. Time is used as a backup control factor. The one handwash per hour is an approximate conversion of the job description and specific accountabilities.

Operators with more complex menus and perhaps concerns over their own span of control or a lower tolerance for risk opt for a two to four handwashes per employee hour rate. Whatever the standard, measurement is critical when looking to motivate behavior change.

Frequently touched surfaces are often infrequently cleaned and should be considered in establishing handwashing protocols. These surfaces are the pathogen's refueling stations.

Every surface touch in a professional kitchen, restroom and service area is different and carries an unknown risk factor for contaminating the hands, bare or gloved. These hands in turn can contaminate the food or another person directly. Ranking these touches, with the HACCP, Hazard Analysis Critical Control Point, principles in mind, helps workers understand those touches that must trigger a handwash and a specific hand hygiene regimen.

Each surface has its own reason for being clean. But some surfaces are clearly more risky in contaminating hands and entering the food flow. These surfaces deserve identification, standards of cleanliness and closer monitoring than most.

Surface cleanliness in food processing establishments has an operational definition. Cleaning procedures and frequencies are established to consistently yield safe level standards. These surfaces are monitored and included in food safety audits.

The need to monitor food-processing surfaces more closely relates to batch size and the scale of a potential outbreak. Foodservice is a smaller scale food processing operation. The operational needs are very different as hands and small batches are the order of the day rather than machines controlling long production runs. Adding to the issue for foodservice are language skills, cultural differences and the fact that the public are invited into the service space, shared with the workforce. Guests are a common source of pathogens, particularly for norovirus as regularly witnessed aboard cruise ships.

Operators are no longer satisfied with the standard of "clean to sight and touch" and are commonly using ATP (adenosine tri-phosphate), to both assign metrics to the acceptable cleanliness levels and to train the cleaning staff by demonstrating and documenting what clean is. The immediate feedback of an ATP luminometer is a powerful, language-free, graphic training tool.

7.4 FACILITY FAULTS DISCOURAGE COMPLIANCE

Facility faults are common when assessing the breakdown in a hand hygiene process. Inconvenience and equipment failure are key factors that discourage good handwashing.

Handsink location is often specified without concern or sometimes even knowledge of the menu. There is an underlying current that says the workers are not going to wash anyway so why compromise the value of this prime real estate with anything less than a productive asset such as an additional beverage machine or specialty oven.

Preparing for the Plan Review and the certification to open a restaurant, kitchen designers work with a minimum number of handsinks. There are some jurisdictions that specify distances like 15 or 20 feet apart. Very creative layouts result, often just trying to minimize the number of handsinks. In reality, staff working in time-short intensive situations requiring frequent task changes are best served with a handsink within 2–5 steps, certainly not one that requires a jaunt of 20 feet.

Handsinks are also often blocked by prep tables as Operations cranks up efficiency factors or introduces new menu items. Rarely used handsinks commonly attract the staffs storage of purses or implements which otherwise require more steps.

Handwashing is easily trivialized by those involved in component specification, resulting in unreliable faucets, soap and paper towel dispensers. Decisions are too often based on price, ignoring the positive impact and time saving of quality splash-free handsinks and hygienic no-touch soap and paper towel dispensers.

Touch-free, lever-free paper towel dispensers are preferred and readily available. Air dryers lack the friction factor and their inherently slow drying makes them a poor choice for hurried food workers. The friction added by using a paper towel, particularly with a textured product, is a significant part of the hand cleaning process, offering a further 1-log pathogen reduction. Using air dryers of any type can leave a high level of suspended contaminants in place. The shorter the scrub step, the more important it is to use disposable paper towels. This is especially true in restrooms where fast handwashes are the norm and residues are naturally nasty.

The restroom too must be examined critically in setting conditions for success, as a good handwash here is the best opportunity to break the norovirus chain of cross-contamination.

7.5 ROUNDS OF CONTINUAL TRAINING

Food workers are trained and retrained. Staff turnover and language barriers are routinely cited but the lack of standards and process control squander even more of scarce training budgets. The lesser talked about two, lack of standards and leadership, are even bigger diluents of training effectiveness.

The WHY of handwashing is rarely covered effectively. Coupled with the lack of standards, any short-term gains in handwashing behaviors dissipate in days. The biggest waste of training budgets is a thorough training followed by the manager entering the kitchen without washing his or her hands.

7.6 MOVING FROM THE UNMEASURED TO HACCP HANDWASHING

Observational monitoring of handwashing has little effect on compliance because of inconsistencies and time pressures facing the managers. The Hawthorne effect also competes with a sustainable solution. Observation is expensive and inconsistent.

This no-standards environment, together with passive players and no metrics to report, sets the height of the hurdle when deciding to lower the risk of contamination via enhanced hand and high-touch surface cleanliness. The solution is in process control, starting with a risk assessment of current menus, facilities and span of management control, all in the context of the customer profile. What percentage of our customer base are children, seniors and other at-risk populations.

HACCP (Hazard Analysis Critical Control Points) provides a helpful framework to manage the risks of poor hand and high-touch surface cleanliness. For Foodservice, HACCP is the process to identify and prioritize risks and establish an integrated system of interventions. Active Managerial Control (AMC) is the implementation and supervision of food safety practices to control risk factors by the person-in-charge (PIC).

HACCP has taken the teachings of Dr. Edward Deming and effectively applied them to food safety. Dr. Deming's theories first led Japan and then the world in improving automobile quality at all price points. HACCP in the food processing industry, where the machine is king, has been an unqualified success in raising world food quality.

Applying HACCP to handwashing in foodservice, where entry-level labor is king, requires further understanding of the worker, his/her motivation and a written, reproducible process to follow, 24/7. The seven steps of HACCP include:

1. Analyze hazards.
2. Identify critical control points.
3. Establish preventive measures with critical limits for each control point.
4. Establish procedures to monitor the critical control points.
5. Establish corrective actions to be taken when monitoring shows that a critical limit has not been met.
6. Establish procedures to verify that the system is working properly.
7. Establish effective record keeping to document the system

AMC provides the framework for the USA's 30,000+ health department inspectors to assess and advise regarding local handwashing protocols, including the behavioral realities of entry-level labor. Inspection based on Active Managerial Control versus *Facility Inspections*, transforms the relationship between the operator and the regulator into one of a partnership with the common goal of public health.

Integrating handwashing into a controllable process requires standards for handwashing itself and the high-touch surface incubators where norovirus can await transfer and bacteria can grow, increasing its likelihood to reach dangerous levels. Protocols are set based on available physical and life sciences with an all-important consideration of the behavioral side of success.

Culture change is generally not a quick fix, but in many cases it is the only route to convert time-hardened handwashing behaviors into a sustainable solution. Years of accepting low compliance define today's reality; years of costly and tiresome training produce waves of temporary peaks in performance that quickly revert back to the undeclared standard.

7.7 THE HAND AND HIGH-TOUCH SURFACE RISK AUDIT

Risk-based solutions will inherently be imperfect but can be a significant improvement. It is important that those charged with risk reduction verbalize their goal as clearly as possible. Commitment by all is a key antecedent of success as approximate risks are connected with current behaviors. This audit lights the path to convert risk-based expectations into monitored, documented and rewarded behaviors.

Following the cleaning step for hands or surfaces with a kill step of sanitizing strongly reduces cross-contamination potential. Adding more time, temperature, chemical or mechanical action to the procedure can easily multiply effectiveness of single-intervention protocols. This is clearly seen in handwashing when a basic 15-second soap-water wash protocol is followed by an application of alcohol hand sanitizer. A 2-log decrease from the handwash can commonly jump to a 4-log pathogen reduction. If at-risk populations are being served, a combination of multiple interventions is recommended.

7.8 IMPLEMENTING STANDARDS

Today's best standards for hand hygiene and high-touch surface cleanliness have been developed by leading operators who are not deterred by the lack of science and nationally agreed standards. They press on and are increasingly successful in implementing risk-approximated solutions.

The language of risk when it comes to hand and surface cleanliness is a hurdle for both operators and regulators. Risk qualifiers like high and low do little without objective assessment criteria. Moreover, complexities in risk measurement often diminish or block progress.

Operators are encouraged to first set a standard on the quality of the wash prior to installing equipment to document frequency. When it's time to multiply handwash behaviors the outcomes will be at a cleanliness level consistent with the operations risk assessment. This first metric helps establish handwashing as a mission-critical event rather than these trivializing options: "Wash when convenient" and "Wash fast."

7.9 CREATING THE CONDITIONS FOR SUCCESS

Conditions for handwashing success start with the optimum kitchen layout that supports the food safety priority. This task is simplified by experienced consultants who understand how detrimental a foodborne illness can be to the operator, not only in terms of short-term losses but also the cost to uphold the public's trust and confidence in food preparation and processing techniques. Managing the risks of the unwashed hand or the unchanged glove are inherently designed into the operational process and implemented under strict management control.

An effective kitchen design is one that meets the owner's standards of food quality and efficiency. It is created with a process and the means to control the process, including good handwashing and high-touch surface cleanliness. It is a kitchen without handwashing hurdles, supporting the accountabilities of the Person-In-Charge (*PIC*) for each and every shift, year in and year out.

Handsink location is critical. Deep draw stainless steel sinks minimize splashing and avoid the standing water that encourages bacterial growth. Touch-free electronic faucets save time and water while encouraging compliance. Paired with reliable no-touch soap, hand sanitizer and paper towel dispensers, this touch-free integrated system has the ergonomics to encourage frequent use. Its ease of use creates a rhythmic reminder of the trained process.

Soap selection is important too. Its specification for effective cleaning must include easy rinsing to avoid skin damage. An unfriendly soap is a major discouragement for frequent washing. At just three handwashes per hour, staff members are often aiming at about 24 handwashes per shift.

Choice of hand sanitizer formulations must be considered relative to pathogen targets. Not all performance differences are reflected in the label claims making choices more challenging. For example, Emory University demonstrated that human norovirus was significantly harder to kill than its calici virus surrogate. Then they found one formulation that was clearly superior but due to label registration requirements, and laboratory limitations in working with norovirus, no label claims were made.

Hand drying completes the cleaning process when using a clean disposable paper towel. This friction-aided action can account for an additional 1-log reduction. When food workers make compromises in the prescribed protocols, particularly in wash time, the cleaning afforded by the drying step becomes more critical. Over-cycled paper that crumbles when wet is another deterrent to frequent handwashing.

7.10 MOTIVATION LEADS TRAINING

Changing behaviors hardened by years of frequent practice takes training and sustained repetition. The new behavior's early trials start the process of hard wiring but there are many examples where a three to six month period is needed for the new to replace the old and become the new default option.

Visual cues of success help establish the new handwashing behavior. Germ simulated tracing lotions can be applied and washed off using the desired protocol. Areas poorly washed become a roadmap lighted by a simple UV lamp. This helps understand why it takes 15 sec or more to eliminate all traces. This self-learning factor personalizes the message and helps establish the new behaviors in less time. Language-free, highly graphic video is another effective training medium for multicultural staffs.

Team learning is also important in that it engages all present and helps maintain team pride in their professionalism. Learning together helps establish and sustain the new practices with positive peer pressure.

7.11 MEASURED, MANAGED AND MOTIVATED PROCESS CONTROL

Monitoring is first and foremost a report card for Operations, measuring their degree of leadership and process control. Where is the reward for exceeding expectations? Where is the consequence for poor compliance? Without a score, the handwashing priority slips behind a long list of efficiency priorities, each with well-established metrics.

If handwashing is to ever be a sustainable priority for Operations it must be fully represented as a criterion for management progression. Handwashing scores become a basis for promotional consideration. Once this policy is clearly established, training of the staff gets less frustrating and much more productive.

Standards are first needed to assure food quality and the health of the customers. Committing to this purpose of helping others raises the handwashing priority, creating an internal compass to follow in both the quality of the wash as well as the frequency. Supervisory time is significantly reduced once the staff starts washing for an agreed cause.

Staff training is already a part of every foodservice operation at the hourly worker level. Informal observation by supervisory staff is the standard and in general has proven to be inadequate.

New monitoring technologies warrant a reassessment of handwashing and process control options. Accurate data is now available for those that believe better handwashing can significantly lower their risk of an outbreak. Poor hand hygiene remains the number one contributing factor identified in the study of outbreaks.

Observation is the overall answer but technologies are now showing up to help. Some use PIN numbers to record who is washing and who is not. Others use video, infrared and radio frequency identification (RFID). These all have their place depending on the corporate culture and budgets. All generally expose that handwashing rates are much lower than expected, often less than half of the desired level.

The first decision in choosing a handwash compliance tracking technology is the choice of team vs. individual monitoring. Both have advantages. Experience suggests this selection is highly aligned with differences in corporate cultures.

Some chains have managed this culture change from one of *production first* to *customer safety first* at a unit level, usually capitalizing on the very special leadership characteristics of an individual manager. Chain restaurants can leverage the learning of such a unit and develop a repeatable process model for their network.

This prototype facility can provide data to better value the risk of current corporate hand hygiene behaviors and determine the risk-based costs for a sustainable solution. Can it be done without raising prices? Continued nurturing can hothouse the initiative for months or years while resources are defined, acquired and aligned for rollout.

Thermal detection technology offers simplicity in implementation and maintenance. It is tailored to team measurement of time-stamped handwash frequency and provides an indicator of wash quality by recording time at the handsink – all without requiring personal badges. Manual entry of a personal code provides another simple option without the need for badges.

RFID, Radio Frequency Identification, requires a badge and is well suited for the cultures wanting to individually reward good washing behaviors. This technology can be married with video detectors and water meters to give real-time feedback on the compliance of both wash times and frequency.

7.13 HANDWASHING'S BOTTOM LINE

Handwashing is a foundation factor in defining the quality of a food product offered to the public. Breakdowns in the handwashing process endanger this intrinsic quality standard that builds consumer trust and brand value.

Changing hand hygiene behaviors is a reset of priorities where customer safety rises to the top of the list of restaurant services. Safety sets the intuitive compass for action and becomes the default setting as each employee adjusts to the day's menu and special situations such as higher than expected customer levels or the need to cover for an absent teammate.

New monitoring technologies add process sustainability and objectivity. They are motivating operators to take a closer look at managing the risks related to their poor hand hygiene and ineffective cleaning of high-touch surfaces. Metrics improve messaging and manager accountability. They provide a path to motivate sustainable behavior change. Operations can lead this track of continuous improvement from under-washing to under control.

KEYWORDS

- **customer safety first**
- **employee hour**
- **HACCP**
- **hand washing sink**
- **person-in-charge**
- **ready-to-eat**

SOURCES CONSULTED

Boone, A. A. and C. P. Gerba. (2007). Significance of fomites in the spread of respiratory disease and enteric viral disease. *Appl. Environ. Microbiol., 73*: 1687–1696.

Chen, Y., Jackson, Kristin M, Chea, Fabiola P., & Schaffner, Donald W. (2001). Quantification and Variability Analysis of Bacterial *Cross-Contamination* Rates in Common Food Service Tasks. *Journal of Food Protection, 64*(1): 70–80. Food Risk Analysis Initiative, Rutgers, the State University of New Jersey, New Brunswick, New Jersey 08901–8520, USA

Clayton, D. A.; Griffith, C. J. (April 2008). Efficacy of an extended theory of planned behavior model for predicting caterers' hand hygience practices. *International Journal of Environmental Health Research, 18*(2): 83–98.

Edmonds, Sarah L.,[1] Mann, James,[2] McCormack, Robert R.,[3] Macinga, David R.,[4] Fricker, Christopher M.,[4] Arbogast, James W.,[4] Dolan, Michael J.[4] (December 2010). SaniTwice: A Novel Approach to Hand Hygiene for Reducing Bacterial Contamination on Hands when Soap and Water are Unavailable. *Journal of Food Protection,* 73(12): 2296–2300.

Green, Laura R., Radke, Vincent, Mason, Ryan, Bushnell, Lisa,. Reimann, David W., Mack, James C., Motsinger, Michelle D., Strigger, Tammi, Selman, Carol A. (2007). Factors Related to Food Worker Hand Hygiene Practices. *Journal of Food Protection, 70*(3): 661–666.

Green, Laura R.,[1*] Selman, Carol A.,[2] Radke, Vincent,[2] Ripley, Danny[3] Mack, James C.[4] Reimann, David W.,[5] Stigger, Tammi,[6] Motsinger, Michelle,[7] & Bushnell Lisa[8]. (2006). Food Worker Hand Washing Practices: An Observation Study. *Journal of Food Protection, 69*(10): 2417–2423.

Judah, G., Donachie, P., Cobb, E., Schmidt, W., Holland, M., Curtis, V. (2010 Mar). Dirty hands: bacteria of fecal origin on commuters' hands. *Epidemiol Infect., 138*(3): 409–14.

Julian, T.R., Leckie, J.O., Boehm, A.B. (2010 Dec). Virus transfer between finger pads and fomites. *J Appl Microbiol., 109*(6): 1868–74.

Liu P., Hsiao, H. Macinga, D., Arbogast, J., Snyder, M., Moe, C.L. (2012). *Comparative Efficacy of Alcohol-based Hand Sanitizers and Antibacterial Foam Handwash against Noroviruses using The Fingerpad Method.* Emory University, Atlanta, GA, GOJO Industries, Akron, OH. http://www.foodprotect.org/issues/packets/2012packet/attachments/III 025 c.pdf

Montville, R., Chen Y., Schaffner, D. W. (2002). Risk Assessment of hand washing efficacy using literature and experimental data. *International Journal of Food Microbiology, 73*(2):305–313.

Yepiz-Gomez, M. S., Bright, K. R., & Gerba, C. P. (2006). Identity and numbers of bacteria present on tabletops and in dishcloths used to wipe down tabletops in public restaurants and bars. *Food Protection Trends, 26*: 786–792.

CHAPTER 8

TRAINING AND CERTIFICATION: IN RESEARCHING THE HAZARD IN HAZARDOUS FOODS

ROBIN B. DIPIETRO, PhD

Associate Professor, University of South Carolina

CONTENTS

8.1 INTRODUCTION

The importance of food safety in the foodservice industry cannot be overstated. The role of the foodservice industry is to provide wholesome, safe food for the public to eat while away from home. Consumers determine where they will eat based on a variety of factors such as price, location, and quality of food, but it seems clear that they should be able to assume that the safety of the food that they purchase will not be compromised by the people that prepare it. If people had to worry about every restaurant and foodservice operation regarding the safety of the food, many people would not eat out. Unfortunately, despite our need for safe food when we eat out, it has been determined that a majority of the reported cases of foodborne illness could be traced back to public eating establishments (Cotterchio, et al., 1998; Jones, et al., 2004).

The Centers for Disease Control and Prevention have stated in a recent report that most foodborne illness can be prevented. Recent data (CDC, 2013) has shown that each year one in six Americans (or more than 48 million people) gets sick, 128,000 are hospitalized, and more than 3,000 die from foodborne illness. Despite the fact that these numbers seem very high, they are actually significantly lower than the number of people impacted by foodborne illness over a decade ago (CDC, 2013). In recent years the ability of medical experts has gotten better with respect to diagnosing foodborne illnesses, there have also been changes that have caused the number of incidents to decrease. Some of the things that have changed with regards to foodborne illness and proper food safety are an increased awareness of customers to the dangers of foodborne illnesses, an increased number of measures for preventing foodborne illness, more government regulations and inspections, as well as improved industry standards that are forcing foodservice operations to do more complete training on foodborne illness and food safety. All of these changes are helping to lower the number of foodborne illnesses stemming from the foodservice industry, but more could be done in this respect.

There are statistics that show that many foodborne illnesses are caused by eating out in a foodservice operation- the numbers may scare people! Foodborne illness prevention and control is a public health priority that has been cited in causing millions of people a year to become sick. It has been determined that most of the foodborne illness outbreaks have been caused by food mishandling, often due to such issues as inappropriate preparation or handling of food by food handlers – including both managers and employees (Medeiros, et al., 2011).

The true numbers of foodborne illnesses that occur each year are hard to determine as not everyone reports these illnesses as accurately as they could. Also, doctors tend to underreport these illnesses as they may not recognize them as foodborne illness due to the fact that they may be caused from unknown emerging pathogens or bacteria that have not been determined to be an issue at the time of the illness. Another problem with knowing the true impact and scope of foodborne illness is that many individuals do not seek treatment for gastroenteritis or stomach problems. Many people have the wrong impression that when they feel sick it may be just a "bug" or the flu, when in reality, it could be foodborne illness. If more people reported their illnesses, a more realistic number could be determined that would be more representative of the actual impact of foodborne illness. The following sections will discuss the topic of food safety – what it is and why it is important, especially in the context of the foodservice industry.

8.2 WHAT IS FOOD SAFETY?

Food safety is the scientific process of ensuring that your food is safe for consumption and is kept safe throughout the process of receiving, storing, preparation, cooking, and serving it in ways that prevent foodborne illness. It is one of the most basic and important components of running a restaurant. Restaurants should not take the perspective of trying to do the minimum possible to ensure that their food is safe; they should be trying everything within their power to ensure that food is safe throughout the processes used in the restaurant.

Pathogens, or disease causing bacteria, are the primary cause of making food unsafe and causing foodborne illness. There are a variety of ways that food can become contaminated with an unsafe amount of bacteria, but improper cooking and holding and personal hygiene of the employees that are working with the food are two of the most common ways that food is contaminated. Bacteria are becoming more difficult to detect and more difficult to eliminate or reduce to safe levels. There have also been problems with the detection of new pathogens that are causing different types of foodborne illness and more severe symptoms related to those foodborne illnesses. Recent years have seen a decrease in the overall number of foodborne illness outbreaks in some developed countries, but this problem of new and stronger strains of bacteria will continue to require a global focus to keep food safe and therefore protect the public.

As the government still tries to find more ways to keep the food supply that we eat safe, they update regulations on a regular basis. There are online government resources that allow restaurant managers and owners to update themselves on the food safety policies and regulations. Online access to food safety materials and reports helps professionals stay current with information that they need. Some of the websites that can be used to determine food safety regulations and updated information on the sanitation and safety of our foods are: www.fda.gov (Food and Drug Administration) and www.usda.gov (United States Department of Agriculture), and www.cdc.gov (Centers for Disease Control and Prevention). These websites provide a plethora of information on food safety, foodborne illnesses, food safety threats, and food safety and sanitation training best practices.

There are five major factors that cause foodborne illness that impact the food safety of restaurants, and not necessarily in order of importance, they are: contaminated equipment; obtaining food from unsafe sources; improper holding times and temperature control; inadequate cooking of foods; and poor personal hygiene (FDA, 2013). All five risk factors can be reduced through the use of proper food safety training and certification of restaurant employees and managers (Murphy, et al., 2010). These factors are very much in the control of foodservice operations and restaurants; they just need to have some focus put on them. The following section highlights why food safety is so important in the foodservice industry.

8.3 WHY IS FOOD SAFETY IMPORTANT?

In addition to there being many factors that cause food safety to be compromised in a foodservice setting, there are also many groups of people that are very vulnerable to getting foodborne illness. The groups of people that are vulnerable to contracting foodborne illness also frequent the restaurant industry. The groups of people that are most at-risk when it comes to contracting a foodborne illness are: elderly, pregnant women, immune-compromised individuals, and children (especially those under five years old). These groups of people tend to have a weakened immune system that can be compromised much quicker and more severely than the general population at large. Many of the bacteria that would be ingested by the average adult person could cause a longer and more severe illness in one of the more vulnerable populations mentioned previously.

In addition to trying to protect the vulnerable populations related to foodborne illness, there are other reasons why maintaining a safe food environment is important to restaurants. There has been increased customer knowledge regarding foodborne illness and food safety. Over the past decade the government through the U.S. Department of Agriculture has mandated that meat and poultry manufacturers put food safety labels on their packaging that explain to customers how to keep the product safe by holding it properly and cooking it to the correct temperatures. These labels have increased in their details and have become a guideline for consumers.

The news reports and media outlets that discuss foodborne illness outbreaks have informed people in more detail how and where and why people are becoming ill. These news reports are immediate and discussed at length and help the public know about any situation that they need to be aware of. This gives consumers knowledge about foodborne illness, but may give them just enough information to be dangerous – yet not all of the information that they need.

During recent years, there has been an increase in the number of measures available for preventing foodborne illness. This increased information has become part of the common place labels that have been put on meats, poultry, seafood, and egg products when you purchase them at the grocery store. Having information put on the label of a grocery store item tells you what temperature to store the item, for example, "store refrigerated between 34–41°Fahrenheit." It also tells you how to cook the product, for example, "cook chicken thoroughly to 165°Fahrenheit." These simple pieces of information can help the public to better use a product and to be safe about the preparation of the item. With the increased knowledge that the public has about foodborne illness and the causes of these diseases through the increased visibility in the news media and press, it is important to arm people with procedures to help them keep food preparation in their home safe.

With the increased knowledge of the public about food safety and the increased press coverage about foodborne illness outbreaks, there has been an initiative by the foodservice industry to continue to improve their education efforts. There is more knowledge that is passed on to restaurants and other foodservice operations through the government agencies in charge of food safety across the globe. The increased information has helped restaurants do a better job of providing a safe environment. With new knowledge though comes a need for increased training of the employees and managers in order to ensure that the knowledge translates to increased performance in foodservice operations.

Governmental regulations and inspections to prevent foodborne illness have helped to identify risks in production facilities and to ensure compliance with food safety and sanitation procedures. In the United States, the U.S. Department of Agriculture (USDA) inspects meat, poultry, and dairy facilities in order to ensure sanitation. They also inspect shipments of food products that come in from other countries and are imported into the United States. The state and local regulatory agencies inspect foodservice operations on a regular basis in order to ensure safety and sanitation of those operations as well. Through the federal, state and local regulatory governmental agencies, there are guidelines put out regarding food safety and there are checks and balances that ensure that operators are performing at their best to protect the public.

The primary way that the local and state regulatory agencies try to help restaurants ensure safe environments for the public is through an inspection system of some form. There are typically agencies at all levels of the government that perform inspections during all stages of the food production process. For example, the USDA performs inspections at the beef manufacturing plants as the products are produced, and then the local government inspects the restaurants that sell the steaks and hamburgers once they are processed in a safe and sanitary environment. The federal, state and local levels of government have been created in order to review operations for sanitation and safety in order to help to eliminate food safety risks, but also to help with determining where problems occur if and when they do.

Overall an inspection produces a written record of what inspectors find in a foodservice operation. These are shared with the operation and help is given to try to get the operation into compliance with the regulations of the location. These inspections can also be used in order to determine where there should be a focus on training in the operation. There are some concerns that have been noted in the research about inspections and their reliability. Since inspections are done by a variety of people and agencies, there is the chance for inconsistency. Moreover, people can see things differently and inspections are done on a random basis; as a result there are potential problems with the way things are viewed and documented on the inspection report. Further, inspections are just a snapshot of the operation as a whole since they only occur on an every 6–12 month basis depending on the location and requirements of that state.

One last factor that shows the importance of food safety in foodservice operations is that the foodservice industry standards for training related to food safety have improved over the years. More restaurant operations are taking the initiative to help protect themselves from lawsuits and from major problems occurring by having training requirements of their own that exceed any requirements of the government. By implementing mandatory company food safety and sanitation training programs, many organizations are ensuring that they provide a positive, sanitary workplace. They train their employees and managers in order to pre-empt any government intervention into their business and to prevent foodborne illnesses or disease from occurring and being linked to their businesses.

8.4 FOODSERVICE INDUSTRY

The foodservice industry has always had a responsibility to protect the public and provide safe food, but in 1993, food safety came to the forefront of the public eye as the Jack in the Box *E. coli* outbreak became one of the most well known cases of foodborne illness in the world. This incident was caused by undercooking ground beef and unfortunately Jack in the Box had procedures in place that could have prevented the outbreak, but they were not following the proper cooking temperatures required by the health department because they believed that the beef was too tough and overdone at that temperature. This case caused more than 700 people to become ill and four children died as a result of this preventable outbreak (Golan, et al., 2004). This incident also cost more than $160 million for Jack in the Box and shows that one incident can cause devastating emotional and economical results to a foodservice operation.

The majority of foodborne illness and contamination is caused by poor handling of food in a restaurant or public foodservice operation which typically can include cross-contamination of food, inadequate cooking or holding of food, and poor personal hygiene by the food handler (Cruickshank, 1990). This is a big problem in the restaurant business where there are many people that work in the industry and turnover in jobs is high. In many segments of the restaurant industry, hourly employee turnover can be close to 100%, which means that the entire staff of a restaurant can turn over annually. This can cause a large number of different people to come in and out of a restaurant kitchen, sometimes with food safety knowledge, but oftentimes without that knowledge.

The reason that there are more foodborne illness outbreaks or food safety issues at restaurants than when preparing food at home is that there are increased risk factors that occur in a restaurant. There is high turnover and therefore employees may not be thoroughly trained, there are sometimes poor sanitary practices in storage facilities in restaurants, the environment is less easily controlled as there are more people within the operation, and bacteria is much more easily transmitted and less controlled in the restaurant environment versus the home kitchen environment (Fielding, Aquirre and Palaiologos, 2001).

There are two types of restaurants: chains and independents. Chain restaurants have multiple locations of the same type of restaurant and are owned by either the same person or company, or that are owned by multiple franchisees with consistent policies and procedures. Independent restaurants are small restaurant operations comprised of one or more restaurants that are very unique in their look, feel, and menu type. They are small businesses that are owned by one or more people, but operate on a much smaller scale and with less structure and formality than a chain restaurant. The overall restaurant industry is comprised of approximately 50% chain restaurants and 50% independent restaurants.

The type of restaurant can often help to determine the amount of resources that are available for food safety training and the type of standards that are set up for the food safety and sanitation of the restaurant. Chain restaurants tend to have more formal structure in their policies regarding food safety training and also tend to have more resources to put toward ensuring that their employees and managers are trained properly in food safety related issues. Independent restaurants often try to do the food safety training as well, but don't have as many formal policies and procedures as the chain restaurants may have. There also tends to be more consequences for a chain restaurant if they do not comply with the food safety regulations of their brand as they may have financial sanctions put on them by the corporate headquarters or they may lose the brand itself if the regulations are not followed (Murphy et al., 2010).

8.5 WHAT IS HACCP?

HACCP is the acronym that stands for the Hazard Analysis Critical Control Point and is the generally accepted process for developing a system to look at the flow of food in a foodservice organization. It is a system of tools that are designed to help identify possible causes of risks to the food safety flow in an operation, and a way to identify procedures to eliminate or minimize the risk to food safety. Many health departments (both state and local) require that foodservice operations have a HACCP plan in place in order for them to continue to operate in the public domain. Many health departments will work with foodservice operations and help them to develop a HACCP plan.

HACCP is a risk-based approach to food safety that is internationally recognized as a consistent, standard way to ensure that risks in the flow of food are minimized. HACCP is comprised of seven principles that help an organization to minimize the risks inherent in foodservice operations. Food manufacturers have been more advanced in their use of the HACCP system to prevent foodborne illness through the production of their products, but restaurants are now starting to adopt more of the principles as they try to protect themselves and the public.

The seven principles of HACCP are: perform a hazard analysis; identify critical control points; establish critical limits for those critical control points; create monitoring procedures; take corrective action; perform record keeping; and create and maintain verification procedures. Each of these steps in the HACCP process is important in ensuring that an operation is doing everything that it can to identify and minimize or eliminate hazards to food safety in their operation. In the first step, performing a hazard analysis means that the manager of the operation should look at the items that are on the menu or in the operation to determine where there are potential food safety hazards. In identifying the hazards through the menu analysis, the restaurant can determine which food items need to be assessed through the process to ensure that they are received, stored, and prepared correctly to minimize the potential problems.

The second principle is to identify critical control points in the food process. This is where points or steps in a procedure can be taken to prevent or eliminate a hazard or to reduce that hazard to an acceptable level of safety. For example checking the temperature of a cooked hamburger patty is a critical control point in the process of cooking the meat patty.

The next principle in the HACCP process is to establish critical limits for those critical control points. That is where you create a minimum or maximum value to which a hazard must be controlled. These critical limits are usually a temperature or a time that a product must meet a required standard for. An example of this would be to cook a hamburger patty to a minimum internal temperature of 155°Fahrenheit.

After that important step is done, the foodservice operation must create monitoring procedures to determine how the critical limits will be monitored. For example, if the operation is going to track a temperature on a product, it is important to know how often the temperature of the product must be taken and the correct procedures for how it is supposed to be taken.

Even if the above principles are followed correctly, there are still opportunities for mistakes or problems to be caught in the process. This is when the taking corrective action step can be done. This step ensures that the operation has a procedure in place for what to do when critical limits are not met or when there is a potential food safety hazard found in an operation. For example, if the cooking temperature requirements are not met for a hamburger patty (155° F) this procedure explains what to do in order to resolve the problem-discard the patty, reheat it to 165°, or continue cooking until it meets the critical limit. It is a way to help an operation prevent bacterial growth and/or foodborne illness.

The next step in the process is to have a recordkeeping procedure in place. This step ensures that the operation is keeping accurate records and documents. These records need to be kept in the operation for a certain amount of time (depending on your location). The records need to have accurate representation of what you are monitoring, recorded limits, and any deviations and/or actions that you have taken to ensure that the correct limit is ultimately achieved.

The last step is to create and maintain verification procedures. This is when the operation sets up a way that they can track the HACCP steps and ensure that they are working to keep food safe and minimize the risk to food in the operation. This would include specific guidelines for how to monitor the process itself, self-inspections, and sampling of products to ensure that they are safe. This should be monitored by internal employees and managers during routine inspections.

Training in the principles of HACCP is critical in ensuring that food-service operations are doing their best to minimize the risks involved in the operation. Industry awareness of these principles can help to ensure safety in the production of foods and also will help regulatory agencies understand what foodservice operations are doing to be proactive in this area. In order to achieve the goals of having a knowledgeable workforce, organizations have to train their employees and managers on the principles of HACCP and food safety regulations as a whole. By integrating the HACCP principles into the workplace, ultimately the organization is better, the food quality and safety and sanitation of the operation will be better, and the consumers will be better protected.

Studies that have been done to look at training programs that assess the effectiveness of the HACCP programs have been consistent on at least one thing—training is critical to help managers and employees assess hazards and to help control food safety issues in general. Having a HACCP based system in place and implementing training programs that follow up on these principles has been essential in reducing the number of widespread foodborne illness outbreaks in the meat and poultry industry (McCabe-Sellers and Beattie, 2004).

More consumers, employees and employers are showing increased interest in learning about food safety (McCabe-Sellers and Beattie, 2004). Employees and managers are interested in learning how they can provide a safer environment for their customers and ultimately for themselves. The following sections describe the types of training and certification that can be done for food safety issues and why this is so critical for foodservice operations to implement.

8.6 TRAINING IN FOOD SAFETY

Training is imparting knowledge or skill to someone through a variety of means that could include lecture, demonstration, or discussion. Training is a planned and organized process used to give knowledge or skills to someone through a learning experience. This information is given to someone in order to help them change or alter a behavior. In order to properly use training, the trainer must determine the objective of the training program. They need to conclude what specific skills or knowledge needs to be addressed and also they need to determine the type of training method to be used in order to meet that objective. In order to determine if training was done effectively, an assessment method should be chosen that will allow the trainer to determine if the training met the overall objective of the training program. Assessment of the training helps to give feedback as to whether the training was effective or not.

Training in food safety has been found to be one way to increase knowledge of foodservice employees and managers in sanitary food practices. Training in food safety also reduces the risks that are inherent in any foodservice operation. Because of the high turnover that tends to occur in any restaurant or foodservice operation, it is essential that training is done on a regular basis to update and improve employee and manager skills in food safety and prevention of foodborne illness. This training can help to teach new knowledge and to remind employees and managers of proper procedures or to reinforce things that they had learned before but had forgotten. Training in food safety can be done with hourly employees of the foodservice operation and/or with the management of the operation. Both types of training provide information that will help the foodservice operation protect the public and keep the products that people eat safe for consumption. The goal of doing training on food safety would be to increase the knowledge on the topic of food safety and prevention of foodborne illness and also to decrease the number of foodborne illnesses that occur in the foodservice industry.

Despite the seemingly obvious benefit of doing food safety training with both employees and managers, there is not a consistent standard for food safety training across the world. In a survey of 1,650 foodservice operations across the UK, it was determined that although there was a consistent positive feeling about food safety training in these operations, there was inconsistent application of the use of the training. In a study by Mortlock, Peters, and Griffith (2000), it was found that more than 90% of businesses had done some kind of training with foodservice employees about food safety, but only 20% of businesses had mandated manager level or supervisory level training on food safety. This is a concern in that a lack of training at the managerial level may translate into the lack of enforcement of food safety standards and a lack of follow up by the managers to regulate the food safety procedures that are supposed to be enforced.

In a review of several large chain restaurants as well as independent restaurants, the type and amount of training on food safety varies. There is no consistent standard across the foodservice industry about what type of training is required to be done in order for a restaurant to operate. Some of the restaurants required all of their managers to be certified in food safety and sanitation, and they required their employees to take a shorter food safety and sanitation course. Other restaurants do not require anything beyond reading a short policy on hand washing and sanitation in the restaurants. Some of the variation between the regulations that restaurants follow is caused by the state and local regulations that mandate a specific type and amount of training be administered by the restaurants to their employees, while other variation is due to the company mandate for food safety training.

There are a variety of ways that food safety training can be delivered and the method chosen for the training will often depend on the organization, the level of employee that is being trained, and the method determined to be most effective based on the training needs and learning style of the specific individuals. The primary ways that food safety training is done is through interactive media, audio visual materials, videos, lectures, classroom training, and physical demonstrations and performance. In order to assess the training and determine if it was effective can be done through questionnaires, monitoring of actual performance or behavior, a check list, and a Likert type scale. Studies have shown that the most commonly used and accepted training method by employees was interactive media and hands on training activities. These activities both help improve knowledge and education about a topic, and also help to change behavior and performance about a topic related to food safety.

The interactive media training is beneficial as it can be done in multiple settings and the employee can do the training at home, at the workplace, or any other convenient location for them. The initial cost of developing the training materials using interactive media or using computer simulations, etc. can be expensive, but in the long run, it is more cost effective as it can be used multiple times without the cost of a trainer actually performing the training. Also, once the program is developed and the initial costs are done, the revisions or updates can be done more effectively and inexpensively.

Hands-on training is very effective when the goals of the training are to demonstrate an action and then follow up by viewing the employee's ability to perform a task. Hands-on training is good for showing the employee the proper way to perform a task and to allow the employee to practice the performance and ask questions of the trainer. In training that had a hands-on component, there tends to be greater employee satisfaction. Doing hands on training in the actual workplace is also helpful to encourage employees to implement changes in the workplace as they can see how these changes would impact them and how they could integrate the changes into the real environment that they are working within.

The main categories of food safety training that are done in foodservice operations are: hand washing and personal hygiene, proper cooking of foods, proper storage of foods, cleaning and sanitizing of the workplace and equipment, and pest control. It is important when selecting the training method to use, to also take the time to review the content of the training. In reviewing the list of categories of training topics, it is clear to see that these topics are suited for hands on training and interactive media training as they are all good for visual training methods.

In all types of training, it was found that shorter and more concise training was most effective. In order to ensure that people are picking up the information in a consistent and meaningful way, the training needs to be in short information blocks. Bite sized training is a term that is often used to describe doing training in small sections or chunks in order to help people learn larger tasks or skills. Using the concept of bite-sized training can help employees learn smaller components of larger tasks and feel comfortable with the material before moving onto another component of training.

The use of employee training is an important way that an organization is portrayed to the public, their brand image. It helps an organization maintain internal and external competitiveness. Another reason that training is important in relation to food safety is that according to the World Health Organization, it is a way to help ensure that workers don't contaminate food. Proper food safety training also helps to eliminate or reduce food contaminants and prevents microorganism growth.

It is important in training employees and managers in foodservice operations to focus on not only the proper information to share in a training session, but also to realize that there needs to be more time spent in focusing on how that training should be delivered. Increasing knowledge is important, but ensuring that the behaviors and practices that follow in the restaurant are as important if not more important.

Proper training, assessment of the training to ensure comprehension of the material, and follow up on training to determine if actual practices have changed can help restaurants to prevent foodborne illness. Creating more effective food safety training can create stronger brand awareness by the public to specific restaurant brands that are positive examples for other foodservice organizations. The areas of most concern related to food safety training are: holding and cooking foods properly; practicing proper personal hygiene; preventing cross-contamination; procuring food from a reliable source and receiving those foods at a proper temperature. There is no one consistent way that an organization should perform food safety training, and there is no one proper certification in food safety that should be used for every individual or organization.

8.7 CERTIFICATION IN FOOD SAFETY

There are a variety of certifications related to food safety. Certification programs are more formal training programs that are regulated typically by a for-profit organization specifically designed to ensure consistent administration of the training and assessment materials. One of the most popular food safety certification programs used in the United States and in several other countries is the Serv-Safe® program that was developed by the National Restaurant Association. This training certification program is overseen by the National Restaurant Association. The Serv-Safe® program is a detailed 16 h food safety training program that includes national certification if the person passes a stringent exam with a 75% score or higher. There are many states and local governments that require one person in the foodservice operation to have this certification in order to ensure that there is someone with detailed food safety knowledge on staff at all times.

Another food safety certification program is the ANSI-CFP accreditation program. ANSI-CFP stands for American National Standards Institute Conference for Food Protection certification. This accredited Food Safety Protection Manager Training is yet another form of certification that a manager can obtain that is accepted in all states in the United States that have mandatory requirements for food manager professionals. The certification program is similar to the Serv Safe certification as it is an exam-based program that contains a wealth of information about food safety and sanitation in the workplace.

To provide an example of a more specific, locally based food safety certification program, the San Francisco Department of Public Health Food Safety Certification is another type of food safety certification that is used specifically in the state of San Francisco. It is broad enough to contain the information and guidelines for food safety information at the national level, but also includes specific regulations for the city of San Francisco in the prevention of foodborne illness. There are some other areas of the U.S. as well as other parts of the world that use specific food safety certification programs that are detailed for their locations.

The Food Handler online program is a certification program that organizations can use to help train their hourly employees. This online program is much less detailed than the 16 h Serv Safe, but gives food handlers the key prevention practices that they need to know to help provide safe food to customers. The areas covered in the certification program include personal hygiene, proper cleaning and sanitation, and proper cooking and holding of food. The online certification is currently accepted in a few states across the U.S. and helps food handlers gain information that will help them to be better employees and to better serve their customers.

Many organizations provide a training program for all of the hourly employees to at least give them the basics in food safety and sanitation. This information is beneficial for restaurants in order to help eliminate some of the liability if they do have a foodborne illness problem. Showing that employees and managers are trained and certified in food safety and sanitation will help them demonstrate the proactive nature of the organization and can help to eliminate some of the liability for problems in the restaurant. It is also just the right thing to do in being a socially responsible business.

The bottom line is that there are many companies that provide food safety training for both employees and managers. Restaurants must look at the state and local requirements for their location in order to ensure that they are following the necessary requirements. Also, many chain or multiunit restaurant companies have their own requirements for food safety training and certification that must be adhered to.

There have been mixed results related to the impact of certification of managers on the actual inspection scores of the restaurants that they operate. Wright and Feun (1986) found that the impact of managers having food safety and sanitation certification did not have a significantly positive impact on the sanitation inspection scores.

Around the world, there is not a consistent type of food safety certification program. Each country may have their own variation on food safety and the requirements that they have. The use of the HACCP program is very widespread, but also the Food Hygiene Code is a popular type of food safety certification that is used in such places as Holland and Hong Kong. The following section describes the overall impact of food safety training on restaurants.

8.8 IMPACT OF TRAINING ON RESTAURANTS

Overall, research has shown that there is a positive correlation between food safety training and certifications and increased health inspection scores (Almanza and Nesmith, 2004). Training in restaurants related to food safety has been shown to help increase sanitation inspection scores, improve the quality of food related to microbiological concerns, and has been shown to improve self-reported food safety behaviors. Unfortunately, there have been inconsistent reports on whether having the appropriate food safety knowledge increases all food safety behaviors.

Training alone in food safety and sanitation is not sufficient to stop the possibility of foodborne illness and contamination in restaurants, but training needs to be accompanied by follow up and enforcement of standards. By following up and inspecting procedures and knowledge on a regular basis, the organization can ensure that policies and procedures are implemented correctly and that their risk of having an unsafe situation is minimized.

Studies have been done in which it has been determined that in restaurants that have mandatory food safety training and certification for managers, the restaurants have had significant improvements in their health and sanitation inspection scores (Cotterchio, et al., 1998). The previously mentioned study grouped restaurants into three categories based on whether managers were given food safety training or not: mandatory training in which one manager from each location received training; voluntary training in which one manager from each location could voluntarily participate in training; and a control group where no food safety training was done. The results showed that the restaurants with mandatory training done at them had more positive food safety inspections and with longer-term results, i.e., the training helped maintain a safer environment. There has been a link between positive health inspection scores and lower cases of foodborne illness, so this result shows a propensity for there to be less foodborne illness at the restaurants that receives mandatory training in food safety and sanitation.

Training managers is critical, but also giving some food safety training to employees is important. Managers are not always the people that make the food and maintain the cleanliness and sanitation standards of the restaurant, as their hourly employees are the ones that maintain the operation of the restaurant. In a study done in the Midwest U.S., it was found that food safety training for employees helped to increase the employees' knowledge of safe food practices, and in the case of hand washing, it even helped to increase the performance of the behavior. Unfortunately, they found that knowledge alone was not enough for changing all behaviors related to food safety issues (Roberts, et al., 2008).

It is important to not only educate employees and managers on the how to of food safety, but also to help them understand the why of food safety. If people know why certain practices are done, they may have a better sense of why they should adopt a specific practice. For example, if the proper method of hand washing is shown to employees, and it is followed by a discussion about why it is important to follow those procedures, it will be more likely to be implemented by the employee. Just showing them how to wash their hands may not have as large an impact.

Organizations can assess the effectiveness of training through a variety of methods. The levels of training evaluation are: reaction, learning, behavior, results, and return on investment (Kirkpatrick, 1967; Phillips, 1997). Reaction is simply asking the people in the training session how they liked the training and the trainer. This may not be enough to determine how much they learned through the training itself. Learning is assessing the trainee about what was learned in the program and can be evaluated by testing them via a pre and post-test or by having them demonstrate a skill if necessary. Behavior evaluation is looking at what behavior is changed as a result of the training program- this can be done on the job and it actually assesses how well they are adopting the practices that were discussed in the training program. Results evaluation is looking at the business results and how that changes as a result of the training. In food safety, it would be to analyze if the conditions in the restaurant are safer or if there have been any foodborne illnesses reported after the training is done. The final assessment or evaluation method is to look at return on investment or how much has been gained versus the cost of the training program.

There is not consistency in the evaluation of training programs; most training programs are not evaluated for their effectiveness at all – so it is difficult to know if they are helping to increase knowledge of food safety issues or not. It is critical to evaluate training programs in order to know if they are being effectively used or not and to determine if actual behavior on the job is changing as a result of the training. The key to success in implementing a training program in food safety is to make sure that performance changes and customers become safer as a result of the training program. Most training programs rely on conveying information with the hope that this information will cause a change in the way that people behave as a result of the training.

One important area that can help to ensure that the training is done effectively and that behaviors are changed is to ensure that training is done on a regular basis. When food safety guidelines are changed or when new employees start, it is important to do refresher courses for food safety issues. This will help to ensure that employees are up to date with information and that there is a consistent reminder of the regulations that can help to keep food safe in a restaurant environment. Since most of the foodborne illnesses are caused by improper handling of food, including through poor personal hygiene, it is important that foodservice operations do retraining or refresher courses on a regular basis.

Measuring the impact of training over time is also important as a key to successful training program. This impact is critical as there is usually a tendency for people to perform better immediately after training. In looking at the evaluation of a training program and the impact that it makes over time, an organization is ensuring that there is not an immediate reaction to training and then it is "business as usual" in the organization over time.

Some of the benefits of proper training and follow up in the workplace are: there is an ability of managers to influence hygiene in the operation; improved satisfaction and therefore less turnover of managers (and therefore they can influence how employees are trained); it is more cost effective to train managers in depth and then have them train the employees; managers can do a better job of self-inspection and training and follow up on employees. Doing just hourly employee training has less of a long-term and broad reaching impact, but it can immediately impact the risk factors for food safety and sanitation.

8.9 SUMMARY

It can be seen through the information presented in the chapter that food safety is a critical issue that must be addressed by foodservice operations. It also is clear that training in food safety is one of the most powerful ways to improve long-term performance or behavior of employees and managers that can then prevent foodborne illness and improve customer perception and satisfaction. In research studies, it has been shown that having the presence of a food safety manager in the operation can help to improve the inspection scores of the operation. Trained managers can help the operation by noticing food safety and HACCP risks in the operation and being able to help employees recognize these issues and correct them.

In training employees to recognize the importance of food safety issues through the use of training and development, managers and organizations can ensure that they are minimizing the risk to food. Employees are the key link between the food production and the customer.

Mandatory training is helpful in improving inspection scores. In requiring food safety training, managers are more apt to take the subject seriously and put more effort into the implementation of policies and procedures that improve the overall safety of a foodservice operation.

Overall, if proper training is done and proper follow up and checkpoints exist, the food supply and foodservice operations can be safe and maintained. If critical behaviors that were discussed above are given proper focus in training programs and the assessment and follow up with these behaviors are followed up on, then the public can rest assured. It is a critical image builder to be a proactive company and provide the best training in the safest environment possible.

KEYWORDS

- **customer perception and satisfaction**
- **food safety**
- **foodborne illnesses**
- **HACCP**

REFERENCES

Almanza, B.A., and Nesmith, M.S. (2004). Food safety certification regulations in the United States. *Journal of Environmental Health*, 66(9), 10–14.

Centers for Disease Control and Prevention (2013). CDC estimates of foodborne illness in the United States. Retrieved from http://www.cdc.gov/foodborneburden/2011-foodborne-estimates.html on July 1st, 2013.

Cotterchio, M., Gunn, J., Coffill, T., Tormey, P., and Barry, M.A. (1998). Effect of a manager training program on sanitary conditions in restaurants. *Public Health Report*, July August, 113(4), 353–358.

Cruickshank, J.G. (1990). Food handlers and food poisoning: Training programs are best. *British Medical Journal*, 300 (6719), 207–208.

Egan, M.B., Raats, M.M., Grubb, S.M., Eves, A., Lumbers, M.L., Dean, M.S., and Adams, M.R. (2007). A review of food safety and food hygiene training studies in the commercial sector. *Food Control*, 18(10), 1180–1190.

Fielding, J.E., Aguirre, A., Palaiologos, E. (2001). Effectiveness of altered incentives in a food safety inspection program. *Preventive Medicine*, 32(3), 239–244.

Finch, C., and Daniel, E. (2005). Food safety knowledge and behavior of emergency food relief organization workers: Effects of food safety training intervention. *Journal of Environmental Health*, 67(9), 30–34.

Food and Drug Administration (2013). FDA Food Code 2009. Retrieved from http://www.fda.gov/Food/GuidanceRegulation/RetailFoodProtection/FoodCode/ucm188363.htm on July 1s t, 2013.

Golan, E., Roberts, T., Salay, E., Caswell, J., Ollinger, M., Moore, D. (2004). Food safety innovation in the United States: Evidence from the meat industry. US Department of Agriculture, Economic Research Service.

Hedberg, C.W., Smith, S.J., Kirkland, E., Radke, V., Jones, T.F., and Selman, C.A. (2006). Systematic environmental evaluations to identify food safety differences between outbreak and nonoutbreak restaurants. *Journal of Food Protection*, 69(11), 2697–2702.

Jones, T.F., Pavlin, B.I., LaFleur, B.J., Ingram, L.A., and Schaffner, W. (2004). Restaurant inspection scores and foodborne disease. *Emerging Infectious Diseases*, 10(4), 688–692.

Kirkpatrick, D.L. (1967). *Training and development handbook: A guide to human resource development*. New York: McGraw-Hill.

Lynch, R.A., Elledge, B.L., Griffith, C.C. and Boatright, D.T. (2003). A comparison of food safety knowledge among restaurant managers, by source of training and experience in Oklahoma County, Oklahoma. *Journal of Environmental Health*, 66(2), 9–14.

Murphy, K.S., DiPietro, R.B., Kock, G., and Lee, J. (2011). Does mandatory food safety training and certification for restaurant employees improve inspection outcomes? *International Journal of Hospitality Management*, 30, 150–156.

Phillips, J.J. (1997). *Return on investment in training and performance improvement programs*. Houston: Gulf Publishing Company.

Roberts, K.R., Barrett, B.B., Howells, A.D., Shanklin, C.W., Pilling, V.K., and Brannon, L.A. (2008). Food safety training and foodservice employees' knowledge and behavior. *Food Protection Trends*, 28(4), 252–260.

Training and Certification (2013). Food Safety State Requirements. Retrieved at http://trainandcert.com/food-safety/food-safety-state-requirements/ on July 10th, 2013.

Wright, J., and Feun, L. (1986). Food service manager certification: An evaluation of its impact. *Journal of Environmental Health*, 49(1), 12–15.

CHAPTER 9

DIFFERENCES IN THE FOOD SAFETY PERCEPTIONS OF CONSUMERS, EMPLOYEES, AND REGULATORY OFFICIALS

AMEET TYREWALA, PhD
Professor, Algonquin College

CONTENTS

9.1 INTRODUCTION

Food is an integral part of our lives; it is required by human beings for their basic survival and existence. Hence it is critical that food safety guidelines and practices be followed and observed at all times. The World Health Organization (WHO, 1988) defines food safety as "steps and measures that are taken during various stages of food preparation, like its distribution, storage, handling, and preparation to ensure it is safe to be consumed by humans." But are these steps and measure being followed? The Centers for Disease Control and Prevention (CDC, 2011) estimates the number of foodborne illnesses, hospitalizations and deaths in the USA each year. According to recent estimates 48 million people fall sick due to foodborne illness every year, 128,000 people are hospitalized, and 3,000 people die due to foodborne illnesses. These are only reported numbers, and do not include cases that go unreported where people opt for self-medication and do not visit a doctor or hospital. The U.S. Food and Drug Administration (FDA, 2009) has identified the five most important risk factors impacting food safety. They are: (1) poor personal hygiene, (2) inadequate food cooking, (3) improper food holding temperatures, (4) unsafe food sources and (5) contaminated equipment. The Centers for Disease Control and Prevention (CDC, 2011) has also identified these same five factors as majorly impacting food safety.

There are three groups that are primarily related to food safety in some way or another: Consumers, Employees and Regulatory Officials. Consumers are the ones who pay for the food when they go out to eat in restaurants, hence it is essential to know and understand their perceptions and what they expect when it comes to food safety. Employees working in the kitchen are the ones who actually receive, store, prepare and serve the food; since they are the ones who cook the food served to customers, understanding their perception towards food safety is critical. And finally health inspectors represent regulatory bodies; they know the law, have been trained to follow it and abide by it, and have been given the power and authority to implement and enforce it. This chapter will identify and present the differences in food safety perceptions of these three groups particularly with regards to hand washing in a food service setting.

9.2 IMPORTANCE OF HAND WASHING

According to the FDA poor personal hygiene is one of the top five factors impacting food safety. Hand washing is a major aspect of this. But how can a food handlers hands be a threat to food safety? Our skin is made of up two layers, the epidermis and the dermis. These layers protect the tissue beneath the skin. The epidermis is the outermost layer of the skin; although it may appear smooth it actually has a lot of cracks and crevices. These cracks and crevices in the epidermis provide a safe haven and ideal conditions for bacteria and other disease causing microorganisms to grow and multiply (Nobel, 1980; Nobel, 1992). Bacteria on the skin can be classified into two major categories: resident and transient bacteria. Resident bacteria or microorganisms are usually present in the cracks and crevices of the epidermis and the dermis. These bacteria can never be completely removed from hands. Even after washing hands resident bacteria will be present on hands. Not all resident bacteria cause disease and infections. Most transient bacteria, however, are disease-causing bacteria; they are found in cracks, crevices, webs of fingers, on fingertips and underneath fingernails. Hence it is of utmost importance that hands be washed as frequently as possible in health care settings as well as in food service settings to prevent the spread of disease-causing, transient bacteria and microorganisms like *Salmonella* spp., *Shigella* spp., *Escherichia coli* and Hepatitis A virus (Nobel, 1980; Nobel, 1992).

The FDA, 2010 has defined hand washing as "the act of cleansing hands by applying soap and water, rubbing them together vigorously, rinsing them with clean water, and thoroughly drying them. This process gets rid of dirt and germs. Every hand washing stage is important and effectively contributes to soil removal and reduction of microorganisms that can cause illness." Different organizations have different hand washing specifications and guidelines, so in order to ensure all food handlers wash hands in a uniform way and to ensure food safety, the FDA food code recommends a step-by-step procedure for washing hands. The total time recommended for washing hands is 20 sec of which 15 sec should be used for lathering and the remaining 5 sec to rinse and dry hands. Hands can be dried using a single use paper towel or a hand-drying machine.

In the 2009, Food Code the FDA has also recommended that food handlers wash their hands after the following activities:

1. After using the restroom (toilet).
2. After touching any body parts like hair, face, ears, mouth etc. other than his/her own hands and forearms.
3. After sneezing, coughing, eating or drinking, chewing tobacco or using paper towels to wipe or blow ones nose.
4. After completing a task and before moving on to the next task
5. Before wearing gloves and after removing gloves.
6. After handling dirty dishes.
7. After handling service animals or aquatic animals
8. After performing any other activity that may contaminate hands

9.3 CONSUMER PERCEPTIONS OF FOOD SAFETY

Consumers are the ones who pay to eat at restaurants and food establishments, hence it is very important to know and understand their perceptions of food safety. The restaurant industry employs around 13.1 million workers with 980,000 locations and $660.5 billion in sales (NRA, 2013). Research has shown that of every food dollar spent, 49 cents is spent at restaurants and other food service establishments. Consumers usually assume that food in restaurants is safe to eat, but what if their belief is proved otherwise? Consumers may stop eating at restaurants or reduce eating outside the home thus resulting in job and revenue losses not only for the industry but for the national economy. Hence it is critical to understand consumer perceptions of food safety.

Studies have shown that consumer confidence about food safety of a food service establishment often relies on certain factors like the taste and presentation of the meal. If the food served to consumers does not look appetizing or has a foul odor or does not taste like it normally should, consumers will not consume the food and will send it back to the kitchen. If they feel the food they were served or are being served is not safe and hygienic to eat they will never return to that restaurant again. Consumers rely heavily on visual cues to determine if the food is safe. Visual cues like clean plates and glasses, hygiene of restrooms and cleanliness of tabletop surfaces or tablecloths are some of the ways in which consumers gauge the food safety of a restaurant (Leach et al., 2001). Several studies have indicated there is a correlation between the cleanliness of restrooms and the cleanliness of kitchens. Research has shown that consumers associate the cleanliness of restrooms (toilets) with the cleanliness of kitchens, the implication being if a restaurant can't keep its toilets clean – which is a much smaller and insignificant part of the overall operation – how would they ensure food safety is being practiced at all times in the kitchen, thus raising concerns about the quality of food being served in the restaurant (Barber and Scarcelli, 2009). Hygiene of food workers in display kitchens is another key factor, which impacts consumer perceptions of food safety in restaurants. The cleanliness of employee uniforms, wearing of protective gear while preparing and serving food, the cleanliness of display kitchens, and the sanitation and cleanliness of kitchen and service equipment are among the factors, which consumers associate with food safety while in a restaurant (Henson, et al., 2006). Other factors like companions or colleagues falling sick after eating food at a particular restaurant and how crowded or busy a restaurant is, also can impact consumer perceptions about food safety. When consumers see a busy restaurant with a lot of consumers waiting to get in, the perception is that the people consuming the food like the food; this also suggests that consumers are not getting sick and that the restaurant is practicing food safety. Hence consumers will feel more inclined to eat at such restaurants due to the perception of food quality being good and food safety practices being practiced. Reviews from family and friends, and stories about their experience also impact consumer perceptions, which in turn may result in their visiting or not visiting a particular restaurant.

A basic expectation and assumption of consumers is that employees working in food service establishments will practice necessary hygiene at all times. But consumers also believe that having a regulatory or enforcing authority in place helps enforce food safety guidelines and practices. The enforcement authority in the case of food service establishments is the Health Inspector who ensures that proper food safety procedures and hygiene guidelines are being followed at all times. Consumers perceive that having food safety systems in place will ensure food safety (Wansink, 2004). Consumers like to know and want to know about restaurant inspection reports and inspection scores. The knowledge of these scores and reports helps them decide where they want to eat.

Transparency is what consumers prefer in terms of the systems, steps and actions the government is taking to enforce and ensure food safety. Some states have letter grade inspection scores, which need to be posted at the entrance of the restaurant so that consumers can see them. This gives consumers the opportunity to see how the restaurant is performing in terms of food safety and sanitation. The consumer is well informed about what restaurant he is going into, leaving no ambiguity in their mind about the food safety of the restaurant. On the flip side restaurants ensure they follow food safety guidelines in order to have a good score, which in turn attracts more customers. It's a win-win situation for both sides. When consumers were asked how long they think employees should wash hands in the kitchen, the answer given by consumers averaged to 29 sec. While the FDA recommended standard is 20 sec. If consumers had it their way they would make employees working in restaurants wash hands for 29 sec every time they needed to wash their hands (Tyrewala, et al., 2012).

Knowledge and trust are two key components, which have a significant impact on the food safety perception of consumers. There is a direct correlation between the amount of knowledge and the perception of food safety: the greater the amount of knowledge a consumer has about food safety the lesser the risk perception. When consumers were asked about their food safety perceptions regarding food manufacturers, processors, farmers markets, super markets and restaurants, they gave restaurants the lowest score indicating they still don't completely trust restaurants with regards to practicing food safety (Knight, Worosz, Todd, 2007). Having adequate knowledge about food safety does not necessarily mean consumers also practice appropriate food safety. Research has shown that when consumers were asked if they would practice food safety at their own home while cooking, most of them said they do. But when observed it was seen that consumers did not wash their hands at all the times when they should have in order to prevent cross contamination (Williamson, Gravani and Lawless, 1992). So in other words food safety knowledge of consumers does not necessarily transfer to actual food safety practices at home. Another reason for this discrepancy is that consumers perceive when they are themselves cooking food they will not get sick even if they do not practice proper hand washing practices, while research has shown that if they fall ill by eating food outside at a restaurant they perceive that the restaurant did not use proper food safety practices and hence they would not revisit that restaurant.

In some instances consumers do not have any knowledge about food practices like hand washing, nor have they ever seen or stepped into a restaurant kitchen. For consumers it is an ideal world where food service employees are supposed to practice food safety and wash their hands all the time. In a recent study consumers were asked what they thought were potential barriers for employees to wash they hands. In response consumers said lack of training, lack of resources like water, soap, paper towels and the inconvenient location of sinks were definitely not barriers preventing employees from washing their hands – although many of these were identified as barriers by employees (Tyrewala, et al., 2012). While considering how busy restaurants get, consumers perceived lack of time to be a potential barrier preventing employees from washing their hands. When asked what were the reasons employees washed their hands, consumers said employees washed hands because it was required by the law (health inspectors) to do so, because it was a management policy, because they didn't want to get sick themselves as well as they didn't want the customers to fall sick. Another important reason why employees washed hands was that they didn't want the organization to get sued by a customer who could fall sick by eating food at the restaurant due to food handlers not practicing proper hand washing and food safety. When asked if employees would wash their hands all the time when no one is watching them consumers disagreed, meaning they felt food service employees would not wash their hands when no supervisor or manager was watching them. But would definitely wash hands when they perceived they were being watched by customers, health inspectors, supervisors, managers and their colleagues (Tyrewala, et al., 2012).

Consumers perceived that employees should be able to wash hands at all times stated by the FDA food code and according to them it is very easy and practical for employees to wash hands at all the times mentioned in the food code. This perception stems from the idea that they are paying for the food being prepared, which in turn pays the employees salary, hence food handlers should practice food safety and hand washing at all times.

This is also due to lack of knowledge of the activities, work pressure, time lines and the load food service employees face in the kitchen on a daily basis. When asked about specifics like what they feel should be done when an employee cuts his/her finger while chopping in the kitchen, or washing hands before wearing or after removing gloves, or if hands should be washed between switching tasks, it was seen that consumers were neutral in their responses indicating that they did not have any knowledge about the specific food safety guidelines, raising questions about the food safety practices consumers would use while cooking at home (Tyrewala, et al., 2012). Factors like age, gender and socio economic status were found to have an impact on consumer perceptions of food safety. For example, older people do not perceive food safety to be very important and do not practice a lot of food safety at home (Altekruse, et al., 1995). On the other hand women were seen to be more conscious about food safety than men. It was also seen that people having higher socio economic status were more concerned and knowledgeable about food safety than those having a lower socio economic status. Whatever the case, understanding consumer perceptions of food safety is very important as the very survival of the restaurant industry depends on consumers.

A recent study looked at the food safety practices of consumers in a Canadian community (Nesbitt, et al., 2009). It was seen that although consumers knew washing raw fruits and vegetables was important, almost a fourth of them did not do it all them time, the perception being that not washing raw fruits and vegetables would not affect their health in any way. The elderly population was seen consuming undercooked eggs, either because the group didn't know what temperature or weren't concerned about the risk associated with the consumption of raw eggs. A group of consumers, especially those who lived on farms consumed unpasteurized milk, although it is not legally allowed to sell unpasteurized milk in Canada. The facts that it is readily available, cheap and convenient are among the reasons why some of those living in the rural community consume unpasteurized milk. This was startling considering they knew that they could fall sick by doing this; the reason they still consumed unpasteurized milk was that they perceived that they had developed the immunity to consume unpasteurized milk. Another key finding was that some consumers still preferred to thaw meat (chicken and ground beef) by leaving it out to thaw at room temperature. Consumers also perceived their judgment of touch and sight was good enough to validate the temperature of meats and to know if the meat was cooked properly or not. Very few consumers used a thermometer. Unlike what they would expect when they go to eat at a restaurant where they would expect employees and chefs to use a thermometer at all times to ensure food safety. Another interesting finding of this study was that there were more risky behaviors practiced by consumers of higher house hold incomes as compared to consumers of lower house hold incomes. This finding is actually contrary to the findings of other studies, but the reason given for this was that people with higher house hold incomes would not cook as often at home and would mostly eat out, hence they would not have the practice or the habit to practice food safety behaviors at all times. This study shed light on some important perceptions on consumers' food safety practices (Nesbitt, et al., 2009).

Studies have shown that almost 50% of consumers use the same chopping board for cutting meats and vegetables, and as they don't perceive it to be a problem with regards to cross contamination and food safety. It was seen that when consumers felt food was unfit for their own consumption they would give it to someone else for consumption with the perception that nothing would happen to the other person, this was found true about consumer perceptions at homes in Trinidad, West Indies (Boodhu, Badrie and Sookdhan, 2007). When asked how they detected if food was spoilt, almost 60% said they would taste it – under the perception that they would not fall sick even if the food were spoilt. This brings us to the concept of "optimistic bias" and " illusion of control" introduced by Redmond and Christopher in 2004. According to this study it was found that consumers had this bias that they would not fall sick when they cooked for themselves or for others even if they didn't practice all necessary food safety practices and guidelines. Also, consumers perceive a low risk of food borne poisoning occurring at home with home cooked meals especially if they prepare the meal. So even if they do fall sick by self prepared home cooked meals they usually go unreported. But if the same happens by consuming food at a restaurant then consumers have a completely different take on food safety and sanitation – which they feel needs to be practiced at all times.

9.4 EMPLOYEE PERCEPTIONS OF FOOD SAFETY

Employees working in restaurants receive, store, cook and serve food to consumers. Hence it is very important to know and understand their perceptions towards food safety. Employees working in food service establishments are by law required to be Servsafe® certified or have some kind of food safety certification, but this doesn't apply to all employees. As long as there is one manager who is Servsafe® certified in the building it is assumed that the restaurant and all its employees will follow and adhere to food safety guidelines. Research has shown that there is a strong association between the knowledge of correct hand washing practices by the person in charge and the food service workers being able to demonstrate hand washing as specified by the food code. But in some cases employees do not have the adequate knowledge about hand washing or how long one should wash their hands in spite of the manager being trained and having knowledge about hand washing practices. An observational study looked at how many times employees adequately washed their hands using soap and warm water and if they did it for the recommended 20 sec. The results were startling. A very small percentage of employees adequately washed their hands with soap and water while a larger number of employees attempted to wash their hands, but did not use soap or did not wash for their hands for 20 sec or just washed hands with cold water (Allwood, et al., 2004). This clearly indicates the need for managers to train employees and to teach them the importance of hand washing and how proper hand washing is done. This also shows a need to address the perception of employees who feel that merely rinsing hands with warm water, no using soap, washing hands with cold water and in some cases not washing hands at all will not have any impact of the safety of the food they are preparing to be served to the customer.

Today, hand sanitizers are everywhere, outside restrooms, near elevators, in public areas. The idea behind this is to promote hand hygiene and prevent disease-causing microorganisms from spreading. The advertising of hand sanitizers as the solution for keeping hands free from germs and bacteria is not completely true. Research has shown that a large number of employees perceive hand sanitizers to be a substitute for hand washing (Pragle, et al., 2007). Hence they do not wash hands but instead use hand sanitizers before switching tasks or before wearing gloves or after removing gloves and sometimes even after using the restroom. According to the FDA using hand sanitizers instead of hand washing is completely incorrect. Hand sanitizers can be used after an employee has washed his/her hands but hand sanitizers are not substitutes for hand washing. The perception that hand sanitizers can kill microorganisms on the hands, as effectively as hand washing does is what makes most employees do this. From a food safety stand point this could be very dangerous. Imagine an employees cutting raw chicken and then using a hand sanitizer to clean his hands before moving on to the next task. Training and giving employees adequate knowledge about effective hand hygiene can change this perception about hand sanitizers being used as substitutes for hand washing.

Bare hand contact is a term known to both food service employees as well as health inspectors. It basically means employees use their bare hands to handle cooked or raw food, for example while plating, cutting and deboning chicken, mixing and portioning large bowls of salad, cutting fruits and so on. Some states in the U.S. prohibit the use of bare hands by employees to handle food and insist on employees wearing gloves. Some states, however, allow bare hand contact if there are procedures in place to ensure employees wash hands properly, if they are trained and if there are adequate facilities like soap, water and paper towels to ensure proper hand washing can occur at all times. Even then research has shown bare hand contact is a major reason for the cause of foodborne illness outbreaks (Barry et al., 2004).

Employees not having the time to wash hands between tasks, cross contaminating food due to lack of knowledge and training and unavailability of resources like soap, warm water, paper towels were some of the reason sighted for employees not being able to wash hands. There have been instances when an employee has visited the restroom (toilet) but found the sink had no water or there was no soap in the dispenser or no paper towels to wipe wet hands. Subsequently if employees have bare hand contact with food there is a high probability that fecal matter could be transferred from the hands of the food worker to the food being prepared which could result in a full blown foodborne illness outbreak. In this case even if the employees want to wash their hands they could not do so due to lack of resources.

Thus many employees perceive lack of resources to be a potential barrier in performing hand washing.

Employees work in kitchens all day long. What consumers may not realize is that they are under constant pressure to complete tasks in a timely manner, ensure consistency of flavor and quality of food, and at the same time practice food safety and wash hands as recommended by the law. The law recommends times when hands need to be washed in the kitchen. When asked what potential barriers there were to prevent employees from washing their hands, an over whelming majority said lack of time is a major barrier, and that in spite of the fact that they would like to wash their hands, work demands prevent them from washing them. Another barrier mentioned by employees was the inconvenient location of hand washing sinks. Almost all employees said that sinks were located so far away and in such remote locations that even if they wanted to wash their hands as prescribed by the law, the remote or inconvenient location of the sink prevented them from doing so (Tyrewala, et al., 2012). From a consumer perspective this makes no sense, as they feel employees working in the kitchen have plenty of time. Further, walking to a sink with an inconvenient location is not a big deal according to consumers. But employees working in the kitchen know what they have to encounter everyday; hence knowing employee perception is very important.

It is important to understand the reasons why employees wash hands while working in the kitchen. A study conducted by Tyrewala, et al. (2012) found that one of the reasons employees washed their hands was that it was required by health inspectors, which in turn meant that it was required by the law. Employees also said they were more likely to wash hands when watched by a health inspector or when a health inspector was in the restaurant. Employees also said they washed hands because they didn't want consumers to fall sick and in the bargain have the restaurant get sued by the consumer for the same. This suggests that employees understand the importance of hand washing and the consequences it can have not only on them but also for the organization.

Any management policies, which need to be implemented in an organization, need to be followed not only by the employees but also by management. One of the main reasons employees said they washed their hands was because it is a management policy, they also said that they performed adequate hand washing in the presence of supervisors and managers and also looked towards managers and supervisors to be role models in performing proper hand washing and food safety behaviors as part of the management policy. Employee perception towards management policies is strongly impacted by the behavior of their managers and supervisors. If the managers and supervisors do not follow proper hand washing procedures and food safety practices as per management guidelines then employees follow suit and the managers cannot tell employees to do otherwise as they have set a poor example for them. Employees no matter how busy they are, not matter how much work pressure they have do not want customers to fall sick because of lack of adequate hand washing on their part. They understand how important consumers are for the restaurant to stay in business and for them to keep their jobs. Employees working in display kitchens practice an even higher level of hand hygiene than if they were back of the house working in a kitchen where customers didn't see them. Employees when being observed or watched by consumers tend to wash hands more frequently. The psychology behind this is that consumers watching them washing hands after each task will feel confident, safe and secure about the food they are consuming.

From a consumers perspective there is no excuse for employees not to wash their hands while handling food. So why do employees not wash their hands as often as the law recommends? What is the employee perspective? Employees have a lot of work and a very demanding schedule when they work in the kitchen. From an employees perspective it is not possible for them to finish their tasks on time as well as wash hands at all times recommended by the law.

According to them it is just not practical. If they had to wash hands according to the times recommended by law they would spend more time near the sink and less time actually doing their work. Employees feel the number of times hands need to be washed is impractical and that people making the law should actually spend time working in the kitchen to see how difficult it is to follow the recommended hand washing guidelines (Tyrewala, et al., 2012).

Most employees when asked basic questions about hand washing – like how long hand washing needs to be done or should hands be washed after handling raw protein – gave the correct answer as these are the basics and they see their colleagues around them washing hands after such tasks all the time. But there is ambiguity regarding some other situations. Although the law says that hands need to be washed before wearing gloves and after removing gloves, employees don't think it is that important to do so. The simple rationale and perception is that since they are wearing gloves they will not contaminate the food while handling it. And since they were wearing gloves while cooking, their hands didn't come in contact with food and hence they don't need to wash and clean them. But scientifically speaking this rationale falls flat; when gloves are worn they provide perfect conditions for bacteria growth like moisture, warm temperature and food. So when gloves are removed employee hands are full of bacteria and hence need to be washed. Also the FDA (2009) recommends that gloves be changed after every four hours of continuous use. Another scenario, which occurs commonly in the kitchen, is that each time an employee blows his/her nose, hands should be washed; by FDA guidelines this applies every time. But according to a recent study employees do not think it is necessary to wash hands every time. In such instances employees are either ignorant or are too time pressed.

How many of us feel it is important to wash hands after handling currency? During a study when consumers were asked this they said it was not necessary to wash hands after handling currency. When employees working in food establishments were asked the same question they gave the same answer. Although the food code doesn't specify washing hands after handling currency, many states require employees to wash hands after handling currency and before handling food. Currency is exchanged all the time and passes through many hands. Since we don't know if the hands were clean or not, it is required by employees working in food service establishments to wash their hands (Tyrewala, et al., 2012).

A number of the perceptions employees have about when to wash hands and when not to wash hands stems from the fact that they work under pressure and have time lines to meet. The number of times they need to wash hands if they follow the FDA guideline makes it difficult for them to work all together. Even if they do want to wash hands, barriers like lack of resources, lack of time and inconvenient location of sinks deter them from doing so. And finally and most importantly lack of knowledge and hands on training about hand washing guidelines results in improper hand washing and employees not washing hands at times when they should to prevent spread of food borne illnesses (Tyrewala, et al., 2012).

9.5 HEALTH INSPECTOR PERCEPTIONS OF FOOD SAFETY

Although restaurant employees know what the law is with regards to food safety they do not necessarily practice it all the time. Managers and supervisors can only exercise so much control over employee food safety practices and they don't represent the governmental bodies. Thus in the case of restaurants and food service establishments it is up to the health inspectors to enforce the law. The health inspector represents the regulatory agencies and ensures that guidelines with regards to food safety and sanitation are implemented by food service employees. Health officials inspect restaurants to check if they are following food safety guidelines recommended by the FDA. These inspections can be once a year, one in six months and sometimes-even once a month depending of the type of food service establishment and the number and type of violations the restaurant has had previously. Health inspectors usually look for critical and noncritical violations while inspecting restaurants. Critical violations are those that can directly impact people sitting and eating food in the restaurant. If there are too many critical violations there is a high probability that the health inspector might ask the restaurant to correct them immediately, or the health inspector can even shut down the restaurant. On the other hand noncritical violations are those, which would not impact consumers directly and immediately, but if not corrected in due time could pose significant health risks to consumers. Lack of refrigeration, improper hand washing facilities and blatant cross contamination are examples of critical violations while food not being labeled, and no hot water in hand washing sinks would be examples of noncritical violations.

Since health inspectors have the authority by law to shut a restaurant down and give restaurants low inspection scores if food safety guidelines are not being met, usually restaurant operators and employees fear health inspectors and try and ensure they wash hands and follow food safety guidelines as recommended by the law. Some states require inspection scores to be displayed near the entrance of a restaurant and if a restaurant receives a bad inspection score or is shut down by the health inspector it has a significant negative impact on the clientele and consumer perception about food safety of that particular restaurant. As a result consumers may not patronize that restaurant anymore, and the restaurant may go out of business. From the restaurant's perspective inspections are critical and very important. When asked consumers indicated that health inspection scores helped them make informed decisions about where to eat; further they would like to have more and easy access to health inspection scores of restaurants (Henson et al., 2006). Hence a number of health departments now have health inspection scores of restaurants uploaded on their websites for ease of access to consumers.

Consumers are very knowledgeable and informed these days, they are aware that the government has systems in place to ensure food safety. They trust that these systems work and produce the desired outcome of ensuring food safety. Health inspectors are a part of this system; because of this it is of utmost importance that there is consistency in terms of how health inspectors conduct inspections. All health inspectors are required to undergo mandatory certification and training; they also need to attend training sessions as laws are updated to be current with the FDA guidelines. Although one would expect health inspectors all across the country to be consistent due to the training they receive, studies have shown discrepancies. It was seen that some health inspectors cited a particular action of an employees as being a violation several times while some inspectors would not cite the same action as a violation every time they saw it (Lee, Almanza and Nelson, 2011). This clearly is a problem, which needs to be addressed by having health inspectors receive frequent and updated information and training to ensure violations do not go unmarked. If one of the violations is a critical violation it could lead to lead a full-blown food borne illness outbreak. Restaurants having outlets all across America have voiced the need for consistency in inspections. Due to the lack of consistency restaurant companies find it difficult to set standards as a particular employee food safety behavior may be acceptable in one state but the same action is marked as a violation in another state.

Over the course of a year health inspectors inspect several hundred restaurants – and many of them have been doing so for many years. The experience they obtain during this time is invaluable and immensely useful in that they know exactly what to look for. But this experience can also be a hindrance in ensuring consistency in inspections all over the country. When asked what influenced them and their decision on whether to mark a particular behavior as a food safety violation, 94% of health inspectors said food safety regulations and the training they received impacted their decisions, while 93% said their own personal experience, which they have had in, the past impacted their decision making. Also, 87% indicated previous inspection scores of restaurants influenced their inspection scores. Basing inspection scores on personal experiences has drawbacks to the extent that there are inconsistencies (Isaacs et al., 1999). Some health inspectors might consider an action to be a critical violation while others may write it off as a noncritical violation.

Another factor impacting the food safety perception of health inspectors was the previous score of the restaurant. Usually when restaurants receive a negative inspection score they make changes, and correct violations and mistakes with the notion that they will do better next time. But it is not correct if health inspectors use previous scores as an indicator of a restaurants food safety and let that impact the future health inspection scores (Isaacs et al., 1999). On the other hand if the same premise is applied to a restaurant getting very good inspection scores the perception of health inspectors would be that the restaurant always does a good job and hence would give them a good score. But what if an employee of that restaurant performed a behavior, which was a critical violation but was not marked for it – and many people fell sick.

The size and type of restaurant was another factor that could impact restaurant inspection scores. A full service restaurant obviously would have many more employees and perhaps more complex processes. Due to the sheer number of employees and the tasks performed, there is potential for a greater number of violations. Quick service restaurants, on the other hand, use basic cooking techniques and oftentimes have a menu that is less demanding when compared to those found in full service restaurants. Hospital and school cafeterias and kitchens also usually get higher scores as the assumption is that all food safety practices and procedures are being followed by employees working in these kitchens since they are preparing food for high risk populations like children, elderly, and the infirm that (Seiver and Hatfield, 2000). The time when the restaurant is inspected also makes a huge difference, and has a major impact on the score. A restaurant inspected during a busy lunch shift will have far more violations than if the restaurant was to be inspected in the morning when it is still quiet and there are no customers. Also the day when the restaurant is inspected matters. Full service restaurants are usually very busy over the weekend, especially during dinner service versus being inspected on a Monday evening when most restaurants are quiet. From a restaurant managers perspective it would be very inconvenient for them if inspections were conducted on a Friday evening when it is busy. From a health inspectors perspective food safety needs to be practiced and adhered to at all times – no matter the day or the time. Both these groups have a valid point to make (Seiver and Hatfield, 2000). The employee perspective is that when it gets busy it is difficult for them to wash hands at all the times recommended. Surprisingly even health inspectors believe that one of the major barriers to employees washing their hands is lack of time. Research also shows that according to health inspectors the lack of training, lack of resources (hot water, soap and paper towels) and inconveniently located sinks were major reasons why employees could not wash their hands even if they wanted to.

When asked why they think employees washed their hands, most health inspectors said employees wash their hands because they require it. The other reason was that employees washed their hands to prevent themselves from falling sick. According to health inspectors, employees do not care whether customers become ill or the organization gets sued as reasons for washing their hands correctly and at all times as recommended by the law. They also said that employees in restaurants tend to wash their hands more frequently and according to the recommended guidelines when observed by health inspectors, managers, supervisors, fellow employees and customers than when they were alone and not being observed. When asked if they think it is not practical to wash hands at time recommended by the law, health inspectors disagreed with this statement. According to them it is practical for employees to wash their hands while working in the kitchen at all times recommended by law. Health inspectors did agree that employees have many tasks in the kitchen, which they need to complete in time, but they still felt washing hands would not be a problem (Tyrewala, et al., 2012). In one study health inspectors were asked specific knowledge based questions about scenarios when they feel hand washing is necessary and when it is not. Questions were based on the times when hand washing is recommended by the FDA, for example after blowing ones nose, handling trash, handling currency, switching between tasks, handling raw and cooked foods and after cutting ones finger. Health inspectors gave correct answers to most of the scenarios but there were some scenarios where the answers given by health inspectors were not in accordance to the law (Tyrewala, et al., 2012). Clearly, differences occur in interpretation of the regulations.

9.6 CONCLUSIONS

In this chapter we have examined the viewpoints of three groups with regards to food safety in restaurants – especially hand washing: Consumers, Employees and Health Inspectors. It is important to understand the perspective of each group; it is also important for these groups to know and understand each other's perspective with regards to food safety and hand washing in food service. Consumers pay for the food they consume at restaurants; they expect the best, not only in terms of taste and presentation, but as regards the safety and wholesomeness of the food. For consumers it does not matter how much work an employee has in the kitchen or how short on time he/she is. In every scenario they expect employees to wash their hands. Surprisingly when it comes to practicing proper food handling practices at home, the standards are lower and food safety and hand washing are not as important. Moreover, most home cooks tend to think they will not make themselves sick. But consumers trust the law and expect health inspectors to toe the line and ensure food safety practices and hand washing are practiced in restaurants at all times. But studies have shown health inspection inconsistencies. The personal experience of health inspectors was seen to be a "barrier" impacting the health inspection scores of restaurants. Health inspectors' need for continuous training and updating of information was seen as a major requirement moving forward. From a health inspector perspective although they understood lack of time, lack of resources, lack of training and inconvenient location of sources to be major barriers for employees washing hands in restaurants, they still strongly felt that it is practical for employees to wash hands at times recommended by law. Although the majority of health inspectors knew the recommended times hands need to be washed according to the

FDA food code there were few who did not know this, thus pointing out a lack of continuous training and education.

Employees who are the ones actually cooking food on the line for consumers day in and day out felt very strongly that washing hands at all times as recommended by law was not practical and was almost impossible to do during peak hours when the restaurant was busy. Employees identified lack of training, resources, location of sinks and time as major barriers. Employees do understand that hand washing is important and do care for customers eating in restaurants, but these findings and perceptions of employees about the difficulty they have in hand washing at all times as recommended by law is a serious concern. Health inspectors should work in the kitchen on a busy shift and experience it themselves about how practical or impractical it is for employees to wash hands at times recommended by the law. There is no doubt that the three groups understand the importance of hand washing in restaurants, but understanding each other's perspectives gives each group valuable insight about the perspective and problems – which can be used in a constructive way to strengthen hand washing and food safety practices in restaurants.

KEYWORDS

- **dermis**
- **epidermis**
- ***Salmonella* spp.**
- ***Shigella* spp.**

REFERENCES

Allwood, P., T. Jenkins, C. Paulus, L. Johnson, and C. Hedberg. (2004) Hand washing compliance among retail food establishment workers in Minnesota. J. Food Prot. 67:2825–2828.

Altekruse, S., D. Street, S. Fein, and A. Levy. (1995) Consumer knowledge of foodborne microbial hazards and food-handling practices. J. Food Prot. 59:287–294.

Andrew J. Knight, Michelle R. Worosz, E.C.D. Todd, (2007) "Serving food safety: consumer perceptions of food safety at restaurants", International Journal of Contemporary Hospitality Management, 19(6), pp. 476–484.

Barry Michaels1, Cheryll Keller, Matthew Blevins, Greg Paoli, Todd Ruthman, Ewen Todd, Christopher J. Griffith, 2004. Prevention of food worker transmission of foodborne pathogens: risk assessment and evaluation of effective hygiene intervention strategies.

Boodhu, A., Badrie, N., and Sookdhan, N. 2007 Consumers' perceptions and awareness of safe food preparation practices at homes in Trinidad, West Indies. International Journal of Consumer Studies ISSN 1470–6423.

Centers for Disease Control and Prevention (2011). 2011 Estimates of Food borne illnesses in the United States: Findings Retrieved on February 9, 2012 from http://www.cdc.gov/features/dsfoodborneestimates/

DOI: 10.1111/j.1471–5740.2004.00088.*x*

Henson, S., Majowicz, S. Masakure, O., Sockett, P., Jones, A., Hart, R., Carr, D., Knowles, L., 2006. Consumer assessment of the safety of restaurants: the role of inspection notices and other information cues. Journal of Food Safety 26(4), 275–301.

Isaacs, Sandy; Abernathy, Tom; Hart, Bob; Wilson, Jeff, 1999. Public health inspectors in restaurants: What they do and why. Canadian Journal of Public Health; Sep/Oct 1999; 90, 5; ProQuest Research Library, pp. 348.

knowledge with home food-preparation practices. Food Technology, May, 94–100.

Leach, J., Mercer, H., Stew, G., Denver, S., 2001. Improving food hygiene standard—a customer focused approach. British Food Journal 103(4), 238–252.

Lee, J., Nelson, D., and Almanza, B. (2011) Health Inspection reports as predictors of specific training needs. International Journal of Hospitality Management 31 (2012), 522–528.

National Restaurant Association (2013). Facts at a Glance Retrieved January 9, 2013 from http://www.restaurant.org/News-Research/Research/Facts-at-a-Glance

Nelson Barber and Joseph M. Scarcelli (2009) Clean restrooms: how important are they to restaurant consumers? Journal of Food Service, DOI: 10.1111/j.1748–0159.2009.00155.x.

Nesbitt, A., Majowicz, S., Finley, R., Marshall, B., Pollari F., Sargeant, J., Ribble, C., Wilson, J., Sittler, N. (2009) High-Risk Food Consumption and Food Safety Practices in a Canadian Community, Journal of Food Protection, Vol. 72, No. 12, 2009, pp. 2575–2586

Nobel, W. C. 1980. Carriage of microorganisms on skin. In Problems in the Control of Hospital Infection. S. W. B. Newsom and A. D. S. Caldwell, eds. Royal Society of Medicine. International Congress and Symposium. Series No. 23. Academic Press. London, UK.

Nobel, W. C. 1992. The Skin Micro flora and Microbial Skin Disease. Cambridge University Press. Cambridge, U.K.

Pragle AS, Harding AK, Mack JC. Food workers' perspectives on handwashing behaviors and barriers in the restaurant environment. J Environ Health. 2007;69:27–32.

Redmond, E.C, and Griffith, C.J. 2004. Consumer perceptions of food safety risk, control and responsibility. Food Research and Consultancy Unit, University of Wales Institute, Western Avenue, Cardiff, CF5–2BY, South Wales, UK.

Seiver, O.H., Hatfield, T.H., 2000. Grading systems for retail food facilities: a risk- based analysis. Journal of Environmental Health 63(3), 22–27.

U.S. Food and Drug Administration. (2009). FDA report on the occurrence of food borne illness risk factors in selected institutional foodservice, restaurant, and retail food store facility types (2004). Retrieved February 9, 2012 from http://www.fda.gov/downloads/Food/GuidanceRegulation/RetailFoodProtection/FoodborneIllnessRiskFactorReduction/UCM224682.pdf

U.S. Food and Drug Administration. (2009). Food Code 2009 retrieved February 12, 2012 from http://www.fda.gov/Food/GuidanceRegulation/RetailFoodProtection/FoodCode/ucm181242.htm

U.S. Food and Drug Administration. (2010). Employee Health and Personal Hygiene Handbook – Personal Hygiene, retrieved February 12, 2012 from http://www.fda.gov/Food/GuidanceRegulation/RetailFoodProtection/IndustryandRegulatoryAssistanceandTrainingResources/ucm184207.htm

Wansink, B., 2004. Consumer reactions to food safety crises. Advances in Food and Nutrition Research, 48, 103–150.

Williamson, D. M., Gravani, R. B., and Lawless, H. (1992). Correlating food safety.

World Health Organization. (1988) Health Education In Food Safety. WHO/EHE/FOS/88.7.

CHAPTER 10

WHAT RESTAURANTS SHOULD DO DURING POWER OR WATER EMERGENCIES

AMEET TYREWALA, PhD
Professor, Algonquin College, Canada

CONTENTS

10.1 INTRODUCTION

During the Stone Age humans killed and ate animals for their survival. As times progressed they realized that the meat would last longer if it were stored at a cooler temperature. Hence they started storing meat in underground caves or under snow, or used ice from frozen rivers to keep it fresh longer. In the Medieval times, man started using chemicals like sodium nitrate and potassium nitrate. These chemicals were added to water, which provided a cool environment to chill containers with liquids in them. It was not until the 19th century that the concept of refrigeration came into existence and was used to keep food cold. Today restaurants all over the world use refrigerators, coolers and freezers to keep food that they prepare and serve safe and free from harmful bacteria. According to the Centers for Disease Control and Prevention (CDC) improper storage of food, unhygienic practices, cross-contamination, low quality raw materials and improper handling are the major causes for food borne illnesses (Centers for Disease Control and Prevention, Diagnosis and Management of Foodborne Illnesses, January 2001). Thus it is imperative that foods in both the raw and cooked form are stored at the correct temperature to prevent foodborne illness.

Today we live in a new and ever-changing world where the numbers of natural disasters like floods, hurricanes, snow storms, tsunamis, tornadoes and earthquakes have increased drastically over the last few decades (Telford and Cosgrave, 2006). These natural disasters are now accompanied by an ever increasing number of man-made disasters such as terrorist attacks, wars, global warming, nuclear threats, and riots, to name a few. During most of these natural and man-made disasters one of the first things that may happen is a power outage followed by a water emergency. In times of such crises, a safe food supply is critical to survival, particularly for those injured or with health risks. Restaurants and other foodservices serve millions of people around the world every day and may play an essential role in providing food during crises. This chapter will address what restaurants can do and should do when faced with a power outage or a water emergency to help ensure the safety of their food.

10.2 MOVING INTO THE 21ST CENTURY

As the world moves towards the 21st century there have been improvements in energy technology which include greener fuels, solar energy, wind energy, and so on. But while these improvements have been taking place, the world population has also been growing at an alarming rate. The current world population is 7.1 billion and growing every second. Estimates suggest that it will reach 9.6 billion by the year 2050 (United Nations Department of Economic and Social Affairs, 2013) This increase in population has resulted in an increase in the needs and demands of people, thus putting a significant strain on already depleting natural resources, utilities and other public services. This strain is particularly challenging in developing countries. China and India are ranked as the two most populated countries in the world today (United States Census Bureau, 2013) Both of these countries are poised to be global super powers in the coming decades, but both these countries also face the most acute power shortages. A recent example occurred in 2012 when a major power grid failure in north India left almost 600 million people without power for two days. In fact, rolling blackouts have become a common occurrence in most major Indian cities. A rolling blackout, also called load shedding, results when utility companies schedule systematic power cuts when the demand for power exceeds the supply. In India for example, a load shedding blackout can range from two to four hours in most months and can go as long as eight to ten hours in peak summer months. With such long blackouts the food kept in the refrigerator may become unsafe to eat. Obviously, power outages affect commercial businesses such as restaurants as well as residential areas. Restaurants need to be prepared to handle these power outages in areas where they routinely occur.

Other disasters can also dramatically affect the power supply. One of the most horrific examples occurred in 2004 when an earthquake near the coast of North Sumatra resulted in the Asian Tsunami. Waves as high as 30 meters killed 225,000 people in 14 different countries and caused tremendous loss of property and infrastructure. The first utility service to go out was the power supply when power lines were knocked down by the waves.

In addition, man-made disasters can also cause disruptions in utility services. Examples include the major disruptions caused by the terrorist attacks of 9/11 in New York and Washington D.C. which killed 3,000 people and injured thousands in 2001, and the 26/11 attacks on hotels and restaurants in Mumbai in 2008 which resulted in the deaths of 183 people. During all these natural and man-made disasters power was lost. It is still essential, however, to provide safe food during such crises to the victims of these disasters in spite of the difficulties associated with a loss of power.

10.3 THE IMPORTANCE OF REFRIGERATION AND ITS IMPACT ON FOOD SAFETY

When the power goes out, all electrical appliances and equipment including the refrigerators, coolers, walk-ins and the freezer stop working. But why is refrigeration so important to a restaurant owner? Let us understand what refrigeration is and why is it so important. The process of refrigeration takes place when food is placed inside a refrigerator. A refrigerator is a box, cabinet or room that has the capacity to remove heat by mechanical or electrical processes. Refrigeration is the process of removal of heat from a food product, body or material. When a food product is placed in a refrigerator the heat from the product is transferred to the air till the temperature of the food and the air reaches equilibrium. The cold temperature in refrigerators also slows down biological processes of the food and retards the deterioration of flavor, color and texture in the food. As a result, refrigerated fruits and vegetables maintain their freshness longer. Most importantly however, the cold temperature in refrigeration slows down and prevents the growth of pathogens, which cause foodborne illness. In addition, food stored in refrigerators should be covered. This helps to minimize possible pathogenic contamination of the food during storage. Covering the food also prevents it from losing moisture and becoming dry. This is especially true for foods stored in freezers.

Precisely following refrigeration temperature guidelines is important in maintaining product temperatures. This is important because air temperature in the refrigerator has been observed to be lower than the temperature of the actual food product (Gilbert, et al., 2007). At times, refrigerators may also have warm spots (near the door) if they are opened and closed frequently. Consideration of the refrigerator (air) temperature and the refrigerator's proper maintenance and use is therefore important in keeping foods at safe temperatures.

As a result, hotels, restaurants and other food service establishments store meats, poultry, seafood and fish, fruits, vegetables, eggs, and dairy products in the refrigerator. A good refrigerator circulates the air uniformly inside the refrigerator maintaining temperatures evenly throughout the refrigerator. But when the power goes out, there is no circulation of air, and as the air inside the refrigerator gets warmer, the temperature of the food products placed inside the refrigerator also start to increase in temperature. If this goes on for a prolonged period of time the food will become spoiled, pathogenic bacteria may grow, and food becomes unsafe to eat.

The U.S. Food and Drug Administration (FDA) has temperature guidelines for safe storage of different types of foods. According to the FDA, hot foods that are temperature controlled for safety (also called potentially hazardous foods) must be stored above 135°F (57°C). Cold foods that are temperature controlled for safety must be stored below 41°F (6°C). This is because the "temperature danger zone" is between 41°F (5°C) and 135°F (57°C). The temperature danger zone is the temperature range in which bacterial growth is at its peak. Food kept within the temperature zone for more than four hours needs to be discarded and cannot be used according to FDA guidelines. It is also recommended that when the power goes out, any food in the refrigerator should be stored for a period of only four hours after which it needs to be discarded. This principle applies only if the refrigerator door is kept closed and not opened during the four hour time period.

Bacteria need certain conditions to grow. These include a food, acidity, time, temperature, oxygen and moisture (the acronym "FATTOM" has been suggested to summarize these conditions). Temperature is one of the most critical of these conditions and is important because different microorganisms grow at different temperatures. These include: thermophiles which grow rapidly at high temperatures ranging from 113°F (45°C) to 194°F (90°C), mesophiles who thrive at medium temperatures ranging from 77°F (25°C) to 113°F (45°C) and psychrophiles which grow at lower temperatures ranging from 32°F (0°C) to 59°F (15°C). Many of the pathogenic bacteria are mesophiles, but *Listeria monocytogenese* is a microorganism that grows even at refrigerated temperatures although its growth is slower at 40°F (4.4°C) (U.S. FDA, 2013). Clearly, if a refrigerator's power goes out, the growth rate of bacteria may increase as the warming temperatures provide even more favorable conditions for their growth. Studies have shown that food stored at 10°C (50°F) will spoil four times faster than food stored at 5°C (41°F) which will spoil two times faster than food which is stored at 0°C (32°F) (Banwart, 1981).

This all goes to show how important refrigeration is in storing food and maintaining food safety. All types of refrigeration are important in temperature maintenance, but care of food in refrigerators may be prioritized in times of a crisis. Generally, most restaurants have large walk-ins or refrigerators while the size of their freezers is relatively small. This is because there is generally a much faster use or turnover of refrigerated food. Temperatures of the refrigerators are therefore very important in a crisis because of the larger quantity of food typically stored there, as well as the fact that their temperature is more likely to enter the temperature danger zone sooner than freezers. So what can restaurants do to ensure the safety of food, which is stored under refrigeration, particularly in the refrigerator? In addition, what should restaurants do for all other power needs?

10.4 RESTAURANTS IN A POWER OUTAGE

10.4.1 CLEANING AND COOKING

Restaurants usually store substantial amounts of food in their coolers and freezers to accommodate possible increases in product demand, which may occur at any point in time. Power is needed to do this, but power is also required for several other purposes, which are important in being able to provide safc food. Power is needed to heat water because hot water combined with cleaning agents helps kill germs more effectively when cleaning the restaurant or washing dishes. Power is also needed for cooking food and maintaining it at hot temperatures prior to service. Finally, power is used to keep the refrigerators and freezers running and cold, thus ensuring restaurants and food service establishments can store their foods safely for an extended period of time.

When there is a power outage, careful consideration should be given to optimize the use of scarce resources and modify operations to maintain safe food handling. For example, if hot water is not available, foodservice operations may choose to delay some tasks that may be less essential (although still important) in regards to food safety such as cleaning of floors.

In addition, they may use sanitizers that are less temperature sensitive. The CDC has guidelines for using sanitizers and disinfectants. One of the least temperature sensitive is chlorine bleach (which must be at least 55–115°F or 13–46°C) (CDC, 2011).

If hot water is not available, another practice might be to use disposable plates, glasses and cutlery which then eliminates the problem of washing dishes altogether. Hence restaurants, which face regular and frequent power cuts, should consider stocking disposable cooking pans and utensils as well as service ware, in addition to their standard plates, glasses, and cutlery.

Another suggestion in a power outage is to use cooking equipment, which only requires an alternative fuel source such as gas. Choice of gas cooking equipment should be carefully considered however. Some gas equipment still requires electricity for electronic ignitions, digital displays, thermostats, or other components. In addition, gas equipment requires adequate ventilation (hoods or ventilation systems) that may be powered by electricity. Sufficient ventilation is important to ensure employee safety and prevent build-up of carbon monoxide or other gas combustion by-products. In some cases, restaurants have used portable gas grills outdoors where there is sufficient ventilation. Alternatively, some restaurants keep a supply of canned foods, which do not require cooking in their preparation and may be used as substitutes on the menu. Finally, if the power is expected to be out for more than a few hours, some restaurants cook the highly perishable food items they have stored in the refrigerator and serve them immediately before they have a chance to spoil.

In some countries, power shortages are so common they may be even be scheduled in advance. The utility company makes an announcement in the newspapers of a scheduled power outage. Details like the dates and days when the power outage will occur, along with the listing of the times when different areas in the city will experience a power outage are mentioned in the newspaper to ensure the general public is aware of these outages and can plan accordingly. This information is even more useful to businesses and industries so that they can plan their production schedules and working hours accordingly. Foodservice establishments find this information particularly useful because they can't simply close the restaurant just because there is a six-hour power outage. A proactive restaurateur needs to think and plan ahead in case of a scheduled power outage.

If power outages are expected, some foodservice establishments keep the refrigeration units or other pieces of equipment functioning with portable generators because this ensures money is not lost due to food spoilage, in addition to its benefit in keeping food safe. Portable generators may be available to rent or buy, however advance planning is usually necessary because there may be limited availability. Generators usually provide a certain preset wattage of power. Smaller portable generators may only be able to provide wattage to power up a few small appliances or lighting for key areas for a few hours, or perhaps a small refrigerator. In order to better assist with the power outage, restaurateurs generally require generators, which are bigger and have a larger wattage output. Most generators will need gasoline. Others may use diesel fuel, propane, or natural gas. A supply of this also needs to be maintained to ensure there is enough fuel to keep the generator running for the length of time that it is needed. Restaurants located in countries or areas where frequent power outages occur may wish to buy their own generator, preferably a larger generator with a higher wattage output that the restaurant needs to handle both current and future needs for power during outages.

10.4.2 FREEZERS AND REFRIGERATORS

When the power goes out, the food in the refrigerator starts losing temperature and gaining heat. The most precise way to calculate the amount of temperature lost and heat gained is to calculate the heat transfer coefficient (U). U can be calculated experimentally by tracking the rate of heat loss through the cabinet for a given inside and outside temperature. Once U is known then the rate of heat transfer can be estimated based on the temperature of the room and the temperature of the food, using the following equation:

$$q = UA\Delta T$$

where: q is the rate of heat flow; U is the overall heat transfer coefficient; A is the surface area of the cabinet; and ΔT is the temperature difference between the temperature inside and outside the refrigerator.

Although this equation may offer a precise answer to how long food remains safe in a refrigerator or freezer without power, most restaurants do not have the necessary data to calculate it. The following discussion therefore suggests practical recommendations that may be implemented by restaurants in coping with power outages.

During some weather crises such as hurricanes, snowstorms, and possibly tsunamis, restaurants may be aware of what is heading towards them and have slightly more time to prepare. This advanced warning time should be used wisely to safely store perishable food products. If a restaurant has a large freezer or a large walk-in freezer, they may wish to move all perishable food into the colder temperatures of the freezer. Since the temperature of a freezer is much lower than the temperature in a refrigerator 40°F (4.4°C) food stored in the freezer will be able to hold temperature for longer when power is lost. In addition, they should cover foods with plastic wrap to prevent freezer burn. Wrapping the food will also ensure that the food maintains its quality in terms of texture and color for a longer time because it does not lose moisture and dry out.

The ideal temperature for a freezer is 0°F (–18°C). Research has shown that if a restaurant starts with this freezer temperature and loses power, a full freezer will be able hold temperature (if it is not opened) for about 48 h while a half-filled freezer will hold temperature for about 24 h (USDA, 2013). If a restaurant has multiple freezers, they may wish to consolidate their food into fewer freezers so that they are filled up as much as possible before the power outage begins in order to maintain the cold temperature of food as long as possible.

If there is advance warning, another way to maintain the temperature of the refrigerator (or freezer) might be to purchase dry ice and place this in cooling unit. Alternatively, frozen ice paddles (typically used for cooling of liquid foods such as soups prior to storage), or ice cubes (either made in the restaurant or purchased) might also be placed in the refrigerator or freezer.

From a practical standpoint, according to the FDA, 50 pounds of dry ice will keep an 18 cubic foot fully stocked freezer cold for about two days (USDA, 2013). But in order to get dry ice, a restaurant would need to make prior arrangements and to know possible suppliers who would be ready to deliver ice before or during the expected crisis. Alternatively, if a restaurant knows of a possible outage it could also fill empty food grade containers with potable water and freeze them. They can then be placed in the refrigerator when the power goes out to try and keep the refrigerator cold longer. An added advantage is that some potable water is available once the ice has melted, if water is also needed. Finally, if there is an anticipated power outage it's a smart idea for restaurants to lower the temperature of the refrigerator or freezer as much as possible till power returns. This will ensure the air inside the refrigerator or freezer is colder than normal, which may prolong the length of time that temperatures remain cold inside the refrigerator or freezer when the power does go out.

Keeping doors of the refrigerators and freezers closed is very important. To minimize the need to open doors (but still be able to monitor food), it is important to have a thermometer clearly visible from the outside of the unit that indicates the interior temperature, or have a built-in thermometer that indicates the internal refrigerator/freezer temperature on a dial or digital display located outside the refrigerator.

When the power goes out, the frozen food in the freezer starts to thaw starting from the outer layers of the food and moving towards the core. The core or center of the food product could still be frozen while the temperature of the outer surface of the food has already entered the temperature danger zone of 41°F (6°C) to 135°F (57°C). If this is the case, the food product needs to be thrown away if it has been in the temperature danger zone for more than four hours. On the other hand, when the power comes back on the food may still be safe to use if the food still has ice crystals on its surface and its temperature is below 41°F (6°C).

Additionally, if the outside air temperature is cold enough (below 41°F or 6°C), the restaurants might be able to temporarily store food outside if it is securely wrapped and otherwise protected from contamination from dust, dirt, insects, animals, or other potential contaminants. Alternatively, it might also be possible to bring snow or ice inside to keep food cold, but food would still need to be protected against all sources of contamination, including from the snow or ice.

When the power is out, there is heat transfer between the air in the refrigerator and food items placed in the refrigerator. This exchange continues until the air temperature and the temperature of food becomes the same. If a refrigerator is filled to capacity the transfer of heat takes longer to occur, hence food in a full refrigerator will go into the temperature danger zone much slower that food stored in the half-full refrigerator. Similar to the guideline given previously in regards to moving food into fewer (and therefore more full) freezers, restaurants may also wish to consider moving food to fewer (and therefore more full) refrigerators.

It is a common physics principle that hot air rises to the top. It rises because hot air is lighter than cold air as it molecules are lighter and farther apart. This phenomenon holds true even in the case of air in a refrigerator. Research has been conducted to understand this stratification in which the temperature of liquids (water) placed on the top, middle, and bottom shelves of a refrigerator were tracked after power was turned off. Results showed that the air temperature was much higher on the top shelf of a refrigerator as compared to the middle shelf or bottom shelf. Lowest air temperatures were found on the bottom shelf of the refrigerator. In fact, results showed that at the end of 60 min the temperature of the liquid in the top shelf was 5.2°F (2.9°C) warmer than that of the bottom shelf (Tyrewala, et al., 2009)

So what does this mean from a restaurant's point of view? It simply means that since the food on the top shelf of the refrigerator loses temperature fastest, highly perishable items such as dairy, meat, poultry, and fish products should never be placed on the top shelf of a refrigerator during a crisis when it is likely that the power will go out. Instead, food items, which are not potentially hazardous, such as some salads, fruits, or vegetables should be stored on the top shelf of the refrigerator, as these foods will not become unsafe or spoil as quickly when they enter the temperature danger zone.

Although proper placement of food in a refrigerator when the power goes out is important for temperature maintenance, another important concern is prevention of cross-contamination. Cross-contamination would be a very serious concern when foods with a higher cooking requirement splash, drip, or otherwise contaminate foods with a lower (or no) cooking requirement because possible pathogens would not be destroyed by heat. Typical guidelines for food placement in a refrigerator therefore, generally include a consideration of their cooking end point. Foods with the highest cooking end point should be placed at the bottom of the refrigerator and those with the lowest cooking end point (or ready to eat) at the top. As an example, chicken has the highest recommended cooking end point at 165°F (74°C) as compared to beef steaks or roasts at 145°F (63°C) or ready to eat foods (which do not require cooking). Among these three foods, chicken would be placed below the beef steaks or roasts and the ready to eat foods would be placed above the other two foods. In addition, raw foods must not be allowed to drip or contaminate other food products. For example, unsafe conditions are created if raw chicken drips onto lettuce or beef or other products. One final caution is that foods that are already at an unsafe temperature may potentially be a problem if they are placed in close proximity to other foods that may either insulate the "warm" food or themselves warm up.

The refrigerator study also conducted research to see if the position of the food on the shelf in the refrigerator impacted the amount of temperature it lost and heat it gained (Tyrewala, et al., 2009). Gallon size containers filled with water were placed in the front and the back of the refrigerator shelves. After time was allowed to stabilize the temperatures of the containers, the refrigerator was turned off and the refrigerator door was kept closed while temperature changes were measured. Results showed that the containers placed in the front of the refrigerator shelves (near the door of the refrigerator) were the warmest at the completion of the test period as compared to those placed in the back of the refrigerator. Why did this happen? The answer was hypothesized to be simply that among the four walls, the refrigerator door was the thinnest wall and hence offered the least resistance (and insulation) in terms of temperature change as compared to the other walls.

In addition, doors typically have a gasket or rubber strip along the inside perimeter of the refrigerator or freezer door. The gasket ensures that when the refrigerator door is closed the cold air from the refrigerator is trapped within the refrigerator and does not escape. The gasket also prevents hot air from entering the refrigerator, thus preventing a rise in the refrigerator temperature.

Gaskets play an important role in temperature maintenance and may cause temperature problems in the refrigerator if they are not functioning properly. If the gasket is loose or torn, it may not be able to seal in the cold air and warm air may move inside. This will mean the refrigerator will not stay as cold as it should and the refrigerator motor may have to work twice as hard to keep the refrigerator cool. This may also mean that food stored in the refrigerator will lose its freshness and will spoil faster.

Even worse, imagine if the refrigerator door has a defective gasket and the power goes out. Under usual circumstances the refrigerator might be able to maintain the temperature of food for a period of four hours. But with a defective gasket, the four hours could be reduced to a couple of hours and in some cases even a few minutes before the air temperature in the refrigerator goes above 41°F (6°C). Proper maintenance of the refrigerator (so that food starts at the coldest temperatures possible), and its gasket in particular (to best maintain those cold temperatures), is important in keeping food safe in a crisis.

If a gasket is found to be torn or loose when a crisis occurs, a few temporary solutions might be used, one of which is to use duct or other adhesive tape to seal the outside perimeter area of the door. This additional layer of tape might temporarily slow down the rate of heat gain and temperature loss. Another temporary solution might be to wrap the refrigerator door with cling film or plastic wrap. This process will be time consuming and labor intensive but if it is done using a larger cling film roll, this task may be easier to accomplish. Once the refrigerator is wrapped, one cannot open the refrigerator door, hence any needed food items should be removed before the refrigerator is wrapped with the cling film.

So from a restaurant's perspective, foods that are less perishable such as fruits, some salads, and vegetables should be placed near the door while highly perishable items such as dairy, meats, protein and fish should be stored at the back of the refrigerator shelf. During a crisis it is important to know in advance where to place different food items to ensure food stored in the refrigerator remains below the temperature danger zone for as long as possible.

The mass of food in a refrigerator is also important in determining how long food temperature is maintained when the power goes out. Research shows that food items with a larger mass take longer to go into the temperature danger zone and maintain temperature longer as compared to items with a smaller mass (Tyrewala, et al., 2009). This study found that gallon containers of water (as compared to half-gallon size containers of water) that were placed on the bottom shelf at the back of a refrigerator did not enter the temperature danger zone for six hours (Tyrewala, et al., 2009). Potentially, this might mean that food can be used when it is held properly in a refrigerator for a total of ten hours before it needs to be thrown away.

The ten hours period combines the six hours that the food is stored in the refrigerator during the power outage (if the temperature of the food is below 41°F or 6°C) and an additional four hours which is the maximum length of time the FDA recommends food be allowed in the temperature danger zone before it needs to be discarded. This is in contrast to the more conservative guidelines suggested by the FDA, which recommends that when the power goes out, food should be stored in a refrigerator for not more than four hours because refrigerator insulation may vary from one manufacturer to another.

Although the results of the research study by Tyrewala, et al. (2009) suggest an extended holding of food is possible, their research study also suggests that this is not recommended unless careful testing has been done to evaluate how long a specific operation's refrigerators can safely hold food. The cited research study was conducted with only one refrigerator make and model, the refrigerator gaskets were in good condition, and other conditions (such as not opening the door) were experimentally controlled. Therefore, an individual operation's refrigerator results are likely to vary. In addition, most important in an assessment of a refrigerator is the monitoring of the actual food temperature, as this provides the most accurate evaluation of food safety.

Some restaurants use heavy plastic curtains at the entrance of walk-ins, coolers, refrigerators and freezers to improve the safety of foods in storage areas. These heavy-duty plastic curtains have slits in them to enable employees to go in and out of the refrigerator or for employees to reach in and grab food products from the refrigerator. These heavy duty curtains have a dual role to play, first and foremost they stop any insects from flying into the refrigerator or the walk-in when the door is opened and second, they act as a barrier to prevent cold air from rushing out of the refrigerator and warm air from rushing in. At least one study however found that these curtains did not significantly reduce temperature loss and heat gain. In addition, there may be a down side to using these curtains, in that there is a high probability of them getting dirty and becoming unsanitary in which case they might cause more damage than good. Further research on the use of these plastic curtains in walk-in styles of refrigerators or freezers and with different styles of curtains is needed to determine their possible benefit under other conditions (Tyrewala, et al., 2009).

10.5 WATER EMERGENCIES

Like food, water is an essential requirement for human beings. In fact, potable water is required for almost all uses in restaurants and hotels. These include its use in: drinking as a beverage, cooking food, making ice to cool drinks, cleaning, washing dishes, hand washing and also for showers in hotels. Potable water is defined as water which is safe to consume and can be used for drinking and cooking purposes. Unfortunately, disruptions in the potable water supply can happen and have devastating impacts. With the growing concerns over global warming, the impacts on climate change have been devastating in some areas of the world. Severe droughts, floods, snowstorms, tornadoes, tsunamis, earthquakes and hurricanes can cut off the water supply to businesses as well as homes. These natural disasters can cause a significant loss of property and lives. In addition, they disrupt businesses. Restaurants in particular, cannot function without the availability of a potable water supply. In that case some restaurants may need to shut down. Other restaurants may be able to safely operate if they can still safely supply food and water.

"So What Can Restaurants Do During a Water Emergency?"

As mentioned above, potable water is an essential and very important component for restaurants to be in business. Though this may sound trivial, the extensive use of potable water in restaurants cannot be denied. So what can restaurants do during a water emergency? With natural disasters like floods, hurricanes, flash floods and tsunamis on the rise over the last decade restaurants need to have an action plan in place of what to do well in advance of a water emergency. In addition, local occurrences may temporarily disrupt the supply of potable water. Examples include broken water mains due to aging, maintenance needs, or weather conditions (such as freezing temperatures), construction accidents where a building's water pipes or water mains are accidentally damaged, or even broken pipes or plumbing within a building due to aging or weather conditions (such as freezing temperatures). In addition, drought conditions, flash floods resulting from extreme rainfall amounts, or flooding due to hurricanes are other examples, which may impact the safety of the water source for an entire area or region.

One of the first things restaurants need to do is keep themselves updated about weather conditions around them and weather systems heading towards them. This is very important as being forewarned is being forearmed. Every restaurant should maintain an emergency supply of water, which should be stored at an elevated height in the storage area of the restaurant. This is important in the case of a flood or tsunami where a restaurant might become flooded and storage of the potable emergency supply of water in an elevated area helps prevent this water from becoming contaminated by the floodwater. Bottled water is often used as an emergency supply because it can be used to drink as well as to cook. If needed, previously made ice (prior to the water emergency) might be allowed to melt and used as a source of potable water during the water emergency. Ice may also be purchased if it has been made from a potable water source.

During a water crisis, cleaning might also be minimized or delayed as long as food safety is not compromised (washing of floors might be one example). Using disposables during a water crisis is also a smart option in trying to minimize water use as washing of dishes is one of the largest uses of water in a foodservice.

But what should restaurants do if the water emergency lasts for a longer period of time? Sometimes after a flood or a tsunami water is available through faucets or taps in the restaurant. Unfortunately, in these situations it is difficult to know if the source from where the water is coming from is contaminated or not, hence it is always better to take necessary precautions. Any tap water being used after a water emergency should be boiled. Local regulatory authorities should also be consulted for water handling instructions. If boiling is recommended, the CDC recommends water should be boiled for one minute once the water starts boiling (CDC, 2013). This boiling of water will kill most harmful and disease causing bacteria in the water making it safe for consumption. This is also recommended for restaurants using water from private sources of water such as wells or lakes or rivers, which have become contaminated due to droughts or flooding. If the option of boiling is not available, bleach might be used to disinfect water. The CDC recommends using 1/8th teaspoon of household bleach per gallon of clear water, and once added and mixed, the water should be allowed to stand for 30 min before it can be used. For iodine, the CDC recommends using manufacturer's guidelines to disinfect water

(CDC, 2013). If the water is cloudy instead of clear, the water should be allowed to sit without disturbing it to allow sediments to settle in the bottom of the container. The clear water from the surface can then be disinfected and used for cooking. As a general rule a restaurant should not open after a water emergency until its sewage lines are clear, and there is potable water available in the rest rooms.

A real life example serves to illustrate how a water emergency can impact a hotel and restaurant. In 1995, extensive flooding of the Mississippi River occurred and affected numerous cities along the river including Alton, Illinois. During this time period, the 137-room Holiday Inn Hotel in Alton, Illinois was able to remain open due to their creative handling of the water emergency and national news reporters stayed there to report on the Mississippi flooding crisis. Initially from July 17 to August 1 there was a boil warning in Alton, Illinois. This meant it was possible to obtain water from the faucets, but that it needed to be boiled before being used. On August 1, floodwater entered the main pumping station, which then was unable to function for 10 days. At this point, no water could be obtained from faucets or was available for toilets or showers. To cope with the sudden water stoppage, the hotel management temporarily placed buckets of the chlorinated pool water outside the guest rooms for the purpose of refilling the toilets so that they could be flushed and during this first short time period no showers were possible. Bottled water was used for drinking and cooking purposes. The management then brought in 6000-gallon tankers, which were hooked up through the plumbing system to the main water line into the hotel. Roughly two tankers were needed per day for the hotel and about a half a tanker were used by the restaurant per day. When a tanker was empty, it would leave and a waiting tanker was then hooked up. The empty tanker would then drive to the nearest town where potable water was available for a refill. The longer the flood continued, the farther the empty tanker would have to drive. This continued until the crisis ended. The cost of re-plumbing the connections so that the tanker could deliver water to the main water line in the hotel was $10,000, which was paid for by the insurance. In summary, this hotel illustrates how quickly a water emergency can occur and the impacts this can have on a hotel and restaurant. This hotel was, in fact, able to survive the water crisis and continue functioning in spite of some very serious challenges.

Finally, restaurants should also move cooking utensils, flatware, and china, as well as food before a water emergency like flooding occurs to safe locations (perhaps elevating them) to prevent them from getting contaminated with the flood water. If cooking utensils, countertops, flatware and china do come in contact with floodwater, they should be thoroughly washed and sanitized. In the case of an extended contact time with floodwater where there is more damage to the building or equipment and they cannot be safely cleaned and sanitized for use, they must be discarded. The FDA recommends that any cans or food containers, which are damaged, dented or pierced due to flood water, or any other water emergency should be discarded. Only air tight cans which are not dented or damaged can be used after being washed with warm water and removal of the labels to ensure any dirt stuck in the labels is removed.

But what if restaurants are located in a region hit by a drought? Can restaurants do something to sustain themselves and keep the business going? A smart and simple long-term action plan is to install water tanks on restaurant rooftops. Water tanks in drought prone areas may help to provide a continuous supply of water when water use is restricted to one hour or in some cases for 30 min a day. Restaurants cannot conduct business and complete all tasks requiring the use of water in 30 min. Hence, restaurants in these areas may use the time when water is available to store water in overhead or rooftop water tanks. This ensures restaurants have water to use throughout the day even though they may still have to use it sparingly.

In some countries the utility companies cannot guarantee when water will be available throughout the day and the taps run dry. In such extreme circumstances, restaurants may have to depend on water tankers. Unfortunately, local and regional companies usually charge exorbitant prices for water during drought times. In addition, the water may appear clean but is not potable and will need to be boiled or disinfected before use. The benefit is that the water tankers may be able to fill overhead or rooftop water tanks of restaurants with water, which allows the restaurants to keep their businesses open. Another possible alternative may be to dig a well as a permanent source of water throughout the year. As easy and simple as it sounds, digging a well in a drought prone area has far-reaching consequences. One of the many consequences is it lowers the water level in the ground causing new wells to be dug deeper and deeper till they find water.

Even in this case, the water should be tested for safety and may need disinfection before it is used.

10.6 CONCLUSION

In this chapter, we have looked at various ways in which restaurants can store food in refrigerators and freezers safely during a power outage. This concept of crisis management and safe food storage during a power outage will only increase with importance as the world faces a growing number of both natural and man-made disasters. This chapter gives readers a glimpse of how important food safety is and how storage of foods is a major factor impacting food safety. After reading this chapter restaurateurs will know what to do during a power outage and how to store food in a refrigerator and freezer to ensure it remains safe for a longer period of time and is safe to consume. Important factors like the mass of food, the location of where the food is being placed in the refrigerator (front or back) and the shelf location (top, middle or bottom) have been discussed in this chapter. This chapter explains the concept of temperature danger zone and its importance in storing cold foods during an emergency. Several do's and don'ts have been mentioned in this chapter for restaurants to understand and implement during a power outage. This chapter also discusses what restaurants should do during a water emergency keeping in mind the scenario of a flood as well as of a drought.

The most important fact to remember about crises is that it is best to be prepared in advance. This chapter offers suggestions about how to handle power and water crises. Additional information is available on websites by the FDA and the CDC. Knowing how to handle a crisis before it happens allows for a rapid response and a rapid response may mean the difference between keeping food safe and having food that cannot be used because it might make people sick.

KEYWORDS

- **FATTOM**
- **food storages**
- **load shedding**
- **temperature danger zone**

REFERENCES

Centers for Disease Control and Prevention (2011) Cleaning and Sanitizing with Bleach after an Emergency, retrieved on August 21, 2013 from http://www.bt.cdc.gov/disasters/bleach.asp

Centers for Disease Control and Prevention (2013) Personal Preparation and Storage of Safe Water, retrieved on August 21, 2013 from http://www.cdc.gov/healthywater/emergency/safe_water/personal.html

Centers for Disease Control and Prevention (2013) Drink safe water, retrieved on August 21, 2013 from http://www.cdc.gov/healthywater/pdf/emergency/09_202278-A_Drink_Safe_Water_Flyer_508.pdf

Tyrewala, A., Almanza, B., Nelson, D. (2009) The effects of door opening and food placement on food temperature within the refrigerator when the power is lost during a disaster.

United Nations Department of Economic and Social Affairs, (2013) World population projected to reach 9.6 billion by 2050, retrieved on August 21, 2013 from http://www.un.org/en/development/desa/news/population/un-report-world-population-projected-to-reach-9-6-billion-by-2050.html

United States Census Bureau (2013) Countries and Areas Ranked by Population: 2013, retrieved on August 21, 2013 from http://www.census.gov/population/international/data/countryrank/rank.php

Gilbert, S., Whyte, R., Bayne, G., Lake, R., & Logt, P. (2007). Survey of internal temperatures of New Zealand domestic refrigerators. *British Food Journal, 109*(4), 323-329.

United States Food and Drug Administration (2013) Keep Listeria out of your kitchen, retrieved on August 21, 2013 from http://www.fda.gov/ForConsumers/ConsumerUpdates/ucm274114.htm#1

United States Department of Agriculture (2013) Keep Your Food Safe during Emergencies: Power Outages, Floods & Fires retrieved on August 21, 2013 from

http://www.fsis.usda.gov/wps/portal/fsis/topics/food-safety-education/get-answers/food-safety-fact-sheets/emergency-preparedness/keep-your-food-safe-during-emergencies/ct_index

CHAPTER 11

FRONT OF THE HOUSE SANITATION AND CONSUMERS' PERCEPTIONS OF SANITATION

HAEIK PARK, MS
Purdue University

CONTENTS

The number of Americans eating away from home has been increasing over the last few decades except during the time under recession (Kumcu and Kaufman, 2011). American consumers are now dining out more frequently than ever before (Variyam, 2005). This is one reason that the food industry has continued to expand over the last few decades. In fact, according to a report from the National Restaurant Association (NRA) (2013), restaurant sales increased rapidly during the 1970s, and have continued to increase so that it is projected to reach about $660.5 billion dollars in 2013. In 2013, $1.8 billion in sales are occurring in restaurants everyday (NRA, 2013).

There are many reasons for this increase including the fact that eating out has become a routine part of Americans' daily life as disposable income has increased (United States Census Bureau, 2012). That is not only because many people have a higher income, but because there also are many two-income households (Kant and Graubard, 2004). This also creates a stronger need for convenience when eating, according to Kant and Graubard (2004).

In addition, there are wide varieties of foods to try since America is a so-called cultural melting pot. Experiencing new tastes is an enjoyable activity for many people when traveling to different places. Since there are many cultures in the U.S., there are also many different kinds of ethnic restaurants in the U.S. According to the recent report by the market research group, Mintel, sales of ethnic food in the U.S. was $8,720 in 2012, and the ethnic food market is anticipated to grow about 20.3% (Mintel Oxygen reports, 2013). This report shows a diet trend in the U.S. People are eating at ethnic restaurants as never before.

In our daily lives, restaurants also play a significant role as a social activity. In modern society, restaurants have become an ideal place to meet with people and eat food at the same time. Eating together is an important activity in our society, which is a part of the relationships that we build with others.

At the same time, Ungku Fatimah, et al. (2011) have stated that consumers' concern about food safety has also increased. Regan (1991) has stated that more people are concerned about food and health than in the past, and 88% of American adults think that nutrition and diet are related to disease prevention. According to the Centers for Disease Control and Prevention (CDC) (2012), food has been the medium through which most common infections have been transmitted since 1996. Although the number of some kinds of foodborne illnesses has been decreasing, they still occur frequently. We often hear news about foodborne illness outbreaks occurring in restaurants through the media. In fact, about 1 in 6 (or 48 million) people get sick each year from eating contaminated food, with 128,000 hospitalizations, and 3,000 deaths annually (CDC, 2012).

The correlation between food and health is extensively recognized by restaurant customers (Regan, 1991). Consumers have numerous food safety and health concerns (Boo, Ghiselli, and Almanza, 2000) with sanitation as one of the most important factors for consumers when they consider where they want to eat. Worsfold (2006) mentioned that consumers' perceptions about food sanitation influence the consumer's consumption behaviors and sanitation dimensions are one of the significant keys in terms of restaurant selection.

As stated in Dulen's (1999) study, there are three major elements in consumers' restaurant dining experiences through which customers assess the quality of the restaurants. These three elements are food, atmospherics, and service. These factors are also suggested to be the main contributors to increasing revenue in the restaurant industry (Susskind and Chan, 2000).

Most of these three major elements are observable to the customers during their dining experiences. Sanitation aspects of the restaurant are also observable to consumers and should be included in each of the three major elements. In fact, previous research has always included sanitation in their measurements of the quality of restaurants.

In addition, consumers are often exposed to information about the restaurants before they decide to eat out. Information is obtained from a number of sources including the media as well as family members and friends. Henson et al. (2006) stated that sanitation information about restaurants influences one's assessment of the safety of food in restaurants and other foodservice establishments, and the results of restaurant assessments have an impact on restaurant selection.

Therefore, it is useful to better understand how consumers judge or assess the sanitation conditions of restaurants and how this affects their decision to visit a restaurant. This chapter identifies the important sanitation conditions in restaurants as they are perceived by consumers and summarizes how the information cues about restaurant sanitation influence consumers' behavioral intentions to revisit restaurants and their actual consumption behavior. In particular, this chapter will focus on consumers' perceptions regarding the front of house areas of foodservice establishments, such as the exterior of the restaurant, the dining room, the restrooms, and the behavior and appearance of wait staff and other restaurant employees.

11.1 CONSUMERS' PERCEPTIONS ABOUT FOOD SAFETY AND SANITATION IN RESTAURANTS

Before we talk about the public's perceptions of restaurant sanitation, we should define the term "food safety", since sanitation is a method to prevent from foodborne illness and ensure food safety. Food safety can be defined in two ways, broadly and narrowly (Ritson and Mai, 1998). From the narrow approach, food safety is the opposite of food risk, and would be the likelihood of not getting a disease as a result of consuming contaminated food or drink (Grunert, 2005). At the same time, from a broader perspective, food safety would include the nutritional qualities of food and wider concerns about the properties of unfamiliar foods, such as genetic modification (Grunert, 2005). Several terms are used in the literature when talking about food safety issues in the foodservice industry. Sanitation, hygiene, and cleanliness are the key terms that are most often used to describe the conditions of food safety in restaurants and foodservice establishments.

The concept of food safety for each individual will vary. Scientists and food experts, however, define food safety as the assessment of risk of consuming a contaminated food and drink. In addition, food safety should contain the concept of practices that preserve the quality of food to prevent contamination and foodborne illness. According to the CDC (2012), foodborne illness, also referred to as food poisoning, occurs as the result of eating harmful microorganisms present throughout the environment in soil, air, water, and in bodies of people and animals.

In the foodservice industry, there are five main factors that have been found to be associated with food safety issues; unsafe food sources, inadequate cooking, improper holding temperatures, contaminated equipment, and poor personal hygiene (FDA, 2010). The issue regarding these factors for restaurant consumers is that they cannot completely assess the level of food safety related to these five factors, as they are limited in their ability to watch all of the food preparation procedures.

On the other hand, consumers are able to observe almost all of the areas of the front of house in restaurants because these are open to the public (NRA, 2013). Front of house observable sanitation conditions in restaurants include the food, tableware, dishes, tables, chairs, walls, floors and other furnishings, wait staff, and the restroom. The back of the house in a restaurant is a restricted area that consumers cannot go enter and includes the receiving area, storage areas, dish washing areas, and the kitchen. Thus, consumers are generally only able to use the observable sanitation conditions (front of the house) to personally judge a restaurants' food safety level. These observations influence consumers' perceptions about the sanitation of a certain foodservice establishment. According to Grunert (2005), quality and safety perception is linked to consumers' food choices and consumer demand. Therefore, consumers' perception about sanitation conditions is thought to influence their consumption behavior in regards to restaurants.

Past experience may influence consumers' interest in food safety. Brunsø, et al. (2002) have stated that food safety may not be a major concern for consumers in food selection unless they have had an experience with food poisoning symptoms, such as nausea, vomiting, abdominal cramps, and diarrhea. Alternatively, consumers may be extremely aware of food poisoning because of media reports about an outbreak that has occurred as a result of eating contaminated food. As a result of such media reports, if a food health issue arises, such as *E. coli 0157:H7*, bird flu (although this has not yet been found to be transmitted to humans through food) or mad cow disease, people may in fact, become acutely concerned about food safety. Sales of all related food products, such as poultry and beef, have been shown to drop. Table 1 shows the CDC data on recent outbreaks in the U.S. and outside the U.S.

TABLE 1 Current Outbreak List (as of April 2013).

U.S.-Based Outbreaks	Outbreaks Affecting International Travelers
Cucumbers – *Salmonella* Saintpaul First announced April 2013	Avian Influenza A (H7N9) Virus in China First announced March 2013
Live Poultry – Human *Salmonella* Infections First announced April 2013	
Frozen Food Products – *E. coli* O121 First announced March 2013	
Chicken – *Salmonella* Heidelberg First announced February 2013	
Small Turtles – Human *Salmonella* Infections First announced March 2012	

Source: Centers for Disease Control and Prevention (CDC, 2013).

This shows that consumers' food safety perception is affected by the media and will greatly influence consumers' behavior in food selection even if they have not had a personal experience with food poisoning. Researchers agree that consumer's perceptions about food safety are an important influence on the food industry (Jordan and Elnagheeb, 1991). More specifically, McGuirk, Preston, and McCormick (2006) state that consumer's consumption behaviors are affected by the attributes related to food safety issues.

Grunert (2005) stated that there are two aspects to food safety, which play a significant role in consumers' perceptions. First, safety perceptions are influenced when major safety problems such as food scares occur. These are already mentioned above. The risk perception can control consumer's food choices and causes them to avoid certain foods or categories of food for some time until the risk disappears. Thus, safety perception is called a "sleeping giant" since once perceived it has a large ripple effect in times of crisis similar to what happens during a natural disaster such as a flood or earthquake.

Second, perception may be influenced by opinions about certain production technologies. Food irradiation and GMOs (genetically modified organisms) are representative examples. In these situations, production technologies might be considered by some as an unsafe production method. This may cause a negative perception to the use of these technologies.

This kind of negative perception may, in turn, directly influence the marketplace.

Consumers' concern about the safety of food products has risen recently (Ungku Fatimah et al., 2011). Jordan and Elnagheeb (1991) studied public perceptions about food safety, nutrition, taste, freshness, and antibiotics in animal feed. They found that the majority of the respondents in their study were concerned not only about safe, nutritious, and fresh food but also about the health effects of antibiotics in animal feed. This indicates that people are concerned about food safety in general when selecting food although it is not clear how these concerns affect consumers in terms of their experiences in foodservice establishments.

Alternative approaches to the study of perceptions about safety have also been suggested. Dosman, et al. (2002) emphasized the very complex and multidimensional nature in understanding individual and societal perceptions of environmental and health risks. Starr's (1969) approach is original in looking for a "revealed preference" with a focus on finding the characteristics of risk through the use of psychometric scaling. This approach was built upon by Slovic (1987) who explored the major attributes of risk and tried to find out the influence of the different attributes on risk perception. Later, et al. (1994) revealed that an individual's perception of health and food safety risks is affected by their socioeconomic characteristics and knowledge about the risks. Dosman, et al. (2002) further explored the relationship between socioeconomic determinants and perceptions of food safety related risk. They found that household income, number of children, gender, age, and voting preferences had a significant effect on personal risk perceptions.

Perceptions of risks associated with food in different countries was conducted by Frewer (2001) who evaluated different psychological factors that affect public perceptions of risk related to food, and the differences between and within countries about food. Some of Frewer's results included that women perceived greater food-related hazards as compared to men, and people who were Afro-Caribbean perceived greater food risks than did Asians.

Cullen (2005) evaluated the attributes that influence restaurant selection of Italian and Chinese restaurants in Dublin, Ireland. His study found that the perception of cleanliness of the restaurant is a key attribute for customers when selecting a Chinese or Italian restaurant. Cullen (2005) also found that the key attributes for restaurant selection were different for consumers based on their age, prior experience, mood, and the occasion involved. Cleanliness of the restaurant, however, was rated as one of the most important attributes in terms of restaurant choice for all consumers.

A study by Zopiatis and Pribic (2007) found similar results. Their study of college students' dining expectations and restaurant choices in Cyprus once again found that overall cleanliness of a restaurant was one of the three most important perceived attributes influencing college students' dining choices.

Extensive research on food safety perception of consumers has been conducted in the United States. Although the United States is thought to be one of the safest in the world (CDC, 2013), many foodborne illnesses still occur. In addition, people are reported to be more concerned about food safety and sanitation in restaurants than ever before (Boo et al., 2000; Ungku Fatimah et al., 2011). The following section summarizes many of the research studies on restaurant consumers' perceptions and consumption behavior in the United States.

Restaurant food safety information, such as inspection scores, has been suggested for many years to play a significant role in terms of restaurant choice (Fielding et al., 1999; Jin and Leslie, 2003, 2009; Jones et al., 2004; Simon et al., 2005). More recently, Knight, et al. (2007) explored the perceptions of food safety at restaurants and found that food safety is still one of the considerations in restaurant choice for a large number of consumers. An even more recent study by Lee et al. (2012) found similar results. Their study assessed consumer perceptions of food safety between Asian and Mexican restaurants and showed that Mexican restaurants had higher scores for trust in regards to food safety, and that food safety was an underlying factor in terms of restaurant selection.

In addition to the safety of food, consumers are also reported to consider foodservice hygiene or cleanliness when they dine out. Research by Barber et al. (2011) concluded that cleanliness attributes of physical environment play a significant role in terms of repeat patronage in consumer dining behavior. Another study by Barber and Scarcelli (2010) stated that cleanliness of the establishment (and particularly the restroom) was a factor in making a decision in selecting, staying, or returning to a restaurant.

Lee et al. (2012) assessed consumers' perceptions about ethnic restaurants. In the U.S., Chinese, Mexican, and Italian cuisines are ranked as the top three most popular ethnic restaurants among respondents (not counting American restaurants). Different perceptions regarding food safety however have been observed for consumers depending on the nationality of the ethnic restaurants. Customers had more trust in the safety of food in Mexican restaurants as compared to Asian restaurants such as Chinese, Japanese, or Indian restaurants. In this research, cleanliness of the kitchen was perceived as the most important factor influencing safety followed by restroom cleanliness and cooking temperature.

Moreover, Lee et al. (2012) found different perceptions depending on the consumers' age and gender. Their results suggested that women and seniors had more concerns about food safety while males in general, and young males in particular, worried less about food safety.

As shown above, people are concerned about food safety and sanitation conditions in relation to their dining experiences outside from home. Perception studies have been invaluable in exploring this relationship and are considered one of the most important areas in consumer behavior research, not only in the hospitality industry but in all other fields of consumer research. Marketers and researchers have been interested in the perception of products or businesses since consumers' perceptions of a product or business greatly helps establish its image. The established image is used to boost sales of products in any type of business.

There is no doubt about the importance of perception in relation to consumers' consumption behavior in the foodservice industry.

11.2 DIMENSIONS OF SANITATION CONDITIONS IN RESTAURANTS

In the United States, all foodservice establishments are generally inspected every six months to prevent foodborne illnesses from occurring as suggested in the *Food Code* published by U.S Food and Drug Administration (FDA). The *Food Code* is defined as "a model that assists food control jurisdictions at all levels of government by providing them with a scientifically sound technical and legal basis for regulating the retail and food service segment of the industry" (FDA, 2013). Foodservice health inspectors have a right to enter the back of the house of food establishments while they are inspecting. They check the conditions of foodservices in regards to numerous food safety factors, particularly in the five areas identified as the most common causes of foodborne illness: unsafe food sources, inadequate cooking, improper holding temperatures, contaminated equipment, and poor personal hygiene (FDA, 2010). Unlike consumers, health inspectors have access to observe both the front of the house and the back of the house to assess sanitation and safety conditions.

Consumers on the other hand, cannot go into the back of the house, and have to rely only on visible cues from the front of the house on food sanitation conditions. For example, consumers cannot assess the food storage conditions in the restaurant even if they visit the restaurant every day. According to Ungku Fatimah et al. (2011) there are limitations to the use front of the house conditions to assess food safety from the consumers' perspectives. Thus, using front of the house conditions to assess food safety factors may not be the best method of assessing sanitation conditions in restaurants, although it is commonly used.

Tangible objects are the factors that can be judged by customers in restaurants during their dining experiences. Then, what are the tangible objects in a full-service restaurant? According to Campbell (1967), the key elements that increase the appeal of a meal experience are food, atmospherics, and service. Customers use their experience of the three components, food, atmospherics, and service, for evaluating the quality of restaurants (Dulen, 1999). These three elements are thought to be the most important contributors that might assist with increasing sales in the restaurant industry (Susskind and Chan, 2000).

Many empirical studies have been conducted about customer's restaurant choice, satisfaction, and loyalty in the restaurant industry. Results show that food is one of the most important factors that influence consumers' satisfaction, loyalty, and consumption behavior in restaurants (Clark and Wood, 1999; Kivela 1997; Namkung and Jang, 2007; Sulek and Hensley, 2004; Tripp et al., 1995). Food attributes that are tangible factors in the restaurant setting, and therefore observable, include presentation, temperature, menu variety, healthy options, taste, and freshness (Kivela 1997; Raajpoot, 2002; Namkung and Jang, 2007). Food attributes also play a significant role in food safety, sanitation and hygiene research in the foodservice industry (Aksoydan, 2007; Leach et al., 2001; Ungku Fatimah et al., 2011; Worsfold, 2006). In these studies, tangible attributes of safe food include cleanliness, presentation of food, foreign objects in food, and freshness.

Food is the only product in a restaurant, and it is the most important thing for customers since the primary purpose in visiting a restaurant is to eat food. Thus, the food sanitation dimension is thought to be the most important factor that affects consumer perception in a restaurant setting. If there is an issue about food sanitation, consumers may react more quickly and strongly as compared to other factors.

Service is another factor that may affect consumers' food safety perceptions. Parasuraman et al. (1985, 1988) extensively studied service quality and developed a model of service quality that uses multidimensions to compare consumer expectations and actual performance in the service industry. The five factors that they used to measure service quality included: tangibles, reliability, responsiveness, assurance, and empathy. Among these five dimensions of service quality, tangible factors have been studied extensively by other researchers interested in service quality (Parasuraman et al, 1988; Perran, 1995; Turley and Milliman, 2000; Raajpoot, 2002). Tangible factors refer to the physical facilities, equipment, and the appearance of personnel (Parasuraman et al., 1988). The other four factors of service quality are considered intangible, and are associated with the social conditions of service between customers and the service providers.

Atmospherics can also be a tangible factor for restaurant consumers. The concept of atmospherics was introduced in 1970, and is regarded as a significant part of the service experience (Kotler, 1973). Kotler (1973) suggested that atmospherics or the physical environment is an important factor that influences consumers' consumption decision similar to the other factors of products and service. In fact, atmospherics contains both tangible and intangible environmental aspects. For instance, furniture would be a tangible feature and background music might be considered an intangible atmospheric feature in a restaurant.

Atmospherics is viewed as a complex concept in the restaurant industry. Because it is so complex, some researchers have categorized atmospherics into different dimensions. According to Bitner (1992), there are three dimensions of atmospherics, ambient conditions, spatial layout and functionality, and signs, which are symbols or artifacts. Spatial layout and the functionality dimension are part of the tangible dimensions and include interior and exterior design, equipment and furniture.

Barber and Scarcelli (2009) considered the restroom one of the most important dimensions of atmospherics. They evaluated restroom sanitation conditions using variables such as: evidence of insects, cleanliness of the toilet, soap availability, paper availability, cleanliness of the floor, odor, condition of the ceiling, and availability of hot water. Barber and Scarcelli (2009) found that restroom cleanliness and function positively influence consumers' perception of restaurants.

According to Brady and Cronin (2001) and Raajpoot (2002), customers consider atmospherics when evaluating the service qualities in service industries such as restaurants and hotels. They found that ambient conditions, facility design, and social factors were the three dimensions that influenced consumers' perceptions about the quality of atmospherics.

Over the last few decades, research studies have suggested several models for studying the many dimensions of the physical environment in the service industry (Baker, 1987; Berman and Evans, 1995; Bitner, 1992; Knutson et al., 1990; Parasuraman et al., 1988; Raajpoot, 2002; Ryu and Jang, 2008; Stevens et al., 1995; Turley and Milliman, 2000; Wakefield and Blodgett, 1996). Table 2 summarizes the many dimensions that have been used by researchers of the service industry.

TABLE 2 Dimensions of Physical Environment in Service Establishments.

Models used to study the physical environment	Dimensions Evaluated
Atmospherics	Ambient
(Baker, 1987)	Design (esthetics and functional)
	Social
SERVQUAL	
(Parasuraman et al., 1988)	
	Aesthetics
	Social
	Functional

TABLE 2 *(Continued)*

Models used to study the physical environment	Dimensions Evaluated
LODGSERV	Social
(Knutson et al., 1990)	Product/service
	Aesthetics
	Functional
SERVICESCAPE	Ambient
(Bitner, 1992)	Spatial layout and functionality
	Sign, symbol, and artifacts
Atmospherics	External
(Berman and Evans, 1995)	General interior
	Layout and design
	Point of purchase and decoration
DINESERV	Aesthetics
(Stevens et al., 1995)	Social
	Functional
SERVICESCAPE	Layout accessibility
(Wakefield and Blodgett, 1996)	Facility esthetics
	Seating comfort
	Electronic equipment/displays
	Facility cleanliness
TANGSERV	Ambient
(Raajpoot, 2002)	Design
	Product/service factors
DINSCAPE	Aesthetics
(Ryu and Jang, 2008)	Ambient
	Product/service
	Social

Source: Raajpoot (2002); Ryu and Jang (2008), Barber and Scarcelli (2010).

Because of these physical environment models, research studies have been able to suggest the most important dimensions in the service industry. These different dimensions have been widely used in many studies assessing consumers' perceptions about food safety, hygiene, and sanitation in the restaurant industry (Aksoydan, 2007; Barber and Scarcelli, 2010; Ungku Fatimah et al., 2011).

Leach et al. (2001), for example, studied sanitation conditions to find out how consumers assess food sanitation standards at public eating-places in U.K. Their assessed dimensions included the exterior of the premises, interior of the premises, staff, dining room tables, and food. Aksoydan (2007) also assessed the influence of sanitation conditions on restaurant choice, but used the four dimensions of hygiene, service, atmospherics, and food. Their hygiene dimension included cleanliness of food, servers' appearance, dishes, the kitchen, and the restroom. Barber and Scarcelli (2010) broadly examined the dimensions of restaurant cleanliness using five-point scaled responses to bi-polar questions. Five cleanliness dimensions were evaluated, such as exterior of a restaurant (e.g., cleanliness of building exterior), restroom appearance (e.g., cleanliness of floor), interior design of a restaurant (e.g., cleanliness of carpet and floors), restroom personal hygiene (e.g., availability of soap), and dining room appearance (e.g., cleanliness of table cloth and napkins). Finally, Ungku Fatimah et al. (2011) examined foodservice hygiene factors from the consumer perspective and suggested a model with four aspects to foodservice hygiene.

1. premise and practices,
2. staff and handling,
3. ambient scent, and
4. food and location.

To summarize, although individual studies have assessed slightly different dimensions, the combination of results from these studies suggest that there are at least six tangible "front of the house" dimensions that consumers are able to evaluate by observation and (according to these studies) use in forming their perceptions about food safety and sanitation in restaurants. They are as follows: the food, the service area (table and setting), dining room, exterior of the restaurant, the restroom, and the service staff. Based on a review of the literature, Table 3 elaborates the attributes as they have been assessed by researchers to measure the six "front of the house" dimensions that consumers' use in forming their perceptions of food safety and sanitation in restaurants.

TABLE 3 Tangible Food Safety Dimensions and Attributes Used to Measure These Dimensions in Restaurants.

Dimensions/attributes	Authors
Food	
Adequately cooked food	(Ungku Fatimah et al., 2011; Worsfold, 2006; Leach et al., 2001)
Freshness of food	(Leach et al., 2001; Ungku Fatimah et al., 2011)
Temperature of the food	(Leach et al., 2001)
Cleanliness of food	(Aksoydan, 2007)
No foreign object in food or drink served (insects, hair)	(Leach et al., 2001; Ungku Fatimah et al., 2011; Worsfold, 2006)
No flies are available around food	(Leach, et al., 2001; Ungku Fatimah et al., 2011)
Service area (table and setting)	
Cleanliness of cutlery (silverware, crockery)	(Aksoydan, 2007; Barber and Scarcelli, 2010; Cunningham et al., 2011; Henson et al., 2006; Leach et al., 2001; Ungku Fatimah et al., 2011; Satow, 2009; Worsfold, 2006)
Cleanliness of dishes	(Cunningham et al., 2011; Henson et al., 2006)
Cleanliness of eating area (table)	(Henson et al., 2006; Ungku Fatimah et al., 2011)
Cleanliness of cups	(Barber and Scarcelli, 2010; Leach et al., 2001; Ungku Fatimah et al., 2011)
Cleanliness of the service area and any surface in the restaurant	(Leach et al., 2001; Ungku Fatimah et al., 2011)
Clear and readable menu	(Aksoydan, 2007)
Cleanliness of any equipment	(Leach et al., 2001)
Cleanliness of tables, the surrounding walls, containers for seasonings, and cloths for wiping,	(Barber and Scarcelli, 2010; Cunningham et al., 2011; Jones et al., 2004; Leach et al., 2001; Satow, 2009),
Restroom	
Cleanliness of bathrooms (restroom)	(Barber and Scarcelli, 2009; Henson et al., 2006; Satow, 2009; Ungku Fatimah et al., 2011; Worsfold, 2006)
Cleanliness of floor	(Barber and Scarcelli, 2010)
Maintenance of wall, ceiling, and tiles	(Barber and Scarcelli, 2010)

TABLE 3 *(Continued)*

Dimensions/attributes	Authors
Odor in restroom	(Barber and Scarcelli, 2009, 2010)
Availability of toilet paper	(Barber and Scarcelli, 2009, 2010)
Availability of soap	(Barber and Scarcelli, 2009, 2010)
Availability of hot water	(Barber and Scarcelli, 2009, 2010)
Availability of paper towels/drying device	(Barber and Scarcelli, 2009, 2010)
Trash in toilet	(Barber and Scarcelli, 2009, 2010)
Cleanliness of sink	(Barber and Scarcelli, 2010; Ungku Fatimah et al., 2011)
General appearance (overall)	(Aksoydan, 2007; Henson et al., 2006; Worsfold, 2006)
Dining room	
Cleanliness of restaurant interior	(Barber and Scarcelli, 2010; Barber et al., 2011; Leach et al., 2001, Worsfold, 2006)
Cleanliness of floor	(Ungku Fatimah et al., 2011)
Cleanliness bar/lounge	(Barber et al., 2011)
Cleanliness windows	(Barber et al., 2011)
Cleanliness seat cushions	(Barber et al., 2011)
Cleanliness of counter surface	(Aksoydan, 2007; Henson et al., 2006; Leach et al., 2001; Ungku Fatimah et al., 2011; Worsfold, 2006)
Exterior	
The location of the restaurant	(Ungku Fatimah et al., 2011)
Cleanliness of the trash bin area	(Ungku Fatimah et al., 2011)
Cleanliness of landscaping and driveway	(Barber, 2011; Barber and Scarcelli, 2010)
Overall appearance of building exterior (paint, age, etc.)	(Barber and Scarcelli, 2010; Barber et al., 2011; Leach et al., 2001)
Cleanliness of parking lot	(Barber et al., 2011, Barber and Scarcelli, 2010)
Cleanliness of signage	(Barber et al., 2011)
Cleanliness of the neighborhood of a restaurant	(Barber and Scarcelli, 2010)

TABLE 3 *(Continued)*

Dimensions/attributes	Authors
Age of building	(Barber and Scarcelli, 2010)
Trash area or cigarettes, etc.	(Barber et al., 2011)
Service Staff (appearance and behavior)	
Appearance of staff (uniform, apron, hands, nails, hairstyle, accessories),	(Aksoydan, 2007; Al-Khatib and Al-Mitwalli, 2009; Ungku Fatimah et al., 2011; Worsfold, 2006)
The work staffs' short and clean fingernails	(Baş et al., 2006; Green, et al., 2007; Ungku Fatimah et al., 2011)
The work staffs in proper attire (e.g., wear apron, glove etc.)	(Al-Khatib and Al-Mitwalli, 2009; Baş et al., 2006; Satow, 2009)
Sanitation performance of the serving staff	(Leach et al., 2001; Ungku Fatimah et al., 2011)
Staffs' touching of food	(Satow, 2009)
Sanitation performance of the production staff	(Green, et al., 2007)
Touching body	(Worsfold, 2006)
Eating, drinking, tobacco use of kitchen staff	(Green, et al., 2007)
The work staffs' high standards of personal hygiene (no coughing or sneezing onto food or hands and then touching food, no biting nail, etc.)	(Aksoydan, 2007; Green, et al., 2007; Leach et al., 2001; Ungku Fatimah et al., 2011; Worsfold, 2006)

Consumers are generally able to observe almost all of these attributes though restaurants may differ greatly in the layout for their dining room and kitchen. One exception might be for kitchen staff and facilities, which are stated as tangible attributes in Table 3. Some restaurants have open or semiopen kitchens, which allow customers to view kitchen food preparation (the attributes of kitchen staff and facilities) whereas most other restaurants have enclosed kitchens where customers cannot view kitchen food preparation.

In the food dimension, freshness, cleanliness, no foreign objects (such as insects or hair), and appropriate temperature are the attributes most commonly used to measure consumers' perceptions about the sanitation level of food in restaurant settings (Barber and Scarcelli, 2009; Leach, et al., 2001; Ungku Fatimah et al., 2011; Worsfold, 2006). Because the main purpose in visiting a restaurant is to eat food, these attributes are critically important and play an important role in affecting consumers' perceptions about the sanitation conditions of a restaurant.

Atmospherics is a complex dimension since it is composed of many distinctive sections in restaurants. From a sanitation perspective, the restroom is one of the most important areas. This is because the restroom is generally regarded as the dirtiest place in a restaurant. If a restroom is clean, or the conditions of the restroom meet or exceed consumers' expectations, consumers are likely to believe that the overall sanitation conditions of the restaurant are also good. Thus, consumers may be concerned about the cleanliness of the restroom during their dining experience, and their perception about the cleanliness of the restroom can be a significant indicator in their assessment of the overall sanitation conditions in the restaurant (Barber and Scarcelli, 2009). Attributes used to assess the cleanliness of a restroom may be similar to those used by health inspectors during a restaurant inspection. These attributes can be categorized into cleanliness of the sink, toilets and urinals, overall environment, and drying equipment.

Other subcategories of atmospherics broadly include the interior of the building, exterior design, and ambience. The interior of the building consists of tables, chairs, table settings (dishes, cups, and silverware), floors, walls, windows, and decorations. Table settings are considered to be especially important indicators for consumers because these are personally used to consume the food. In addition, these are exposed to customers for the longest time and are located closest to the customers during their dining experience (as compared to the walls or ceiling for example). Exterior design of a restaurant can also be an indicator that influences consumers' perceptions about sanitation. Exterior design of a restaurant is similar to the first impression of a person. Thus, cleanliness of signage, windows, and the parking lot are important attributes in assessing the sanitation of a restaurant.

The last key dimension in restaurant cleanliness is service. In one sense, service is an intangible thing provided by servers in restaurants and is considered a social dimension. However, from a sanitation perspective, servers might also be considered tangible. Customers communicate with servers, and as a result, they observe servers' appearance and behavior. These are the two main attributes of the service dimension, which customers may assess. Cleanliness of personal appearance is regarded in terms of personal hygiene and includes the server's uniform, nails, hair style, and accessories. Certain behaviors of the servers may also be considered as attribute of service cleanliness, for example, touching food with bare hands.

These dimensions have been established throughout many studies in food safety and sanitation in restaurants. The results suggest that there are numerous attributes that can be used to measure each dimension to determine their impact on consumers' perceptions of the sanitation conditions in restaurants.

11.3 CONSUMERS' PERCEPTIONS AND CONSUMPTION BEHAVIORS

Barber et al. (2011) analyzed the relationship between physical service quality attributes and repeat patronage. The results of their study found that consumers decide to revisit based upon the perceived cleanliness attributes of restaurants. They emphasized the importance of consumer cleanliness perceptions in terms of consumers' repeat patronage.

Henson et al. (2006) found that perception of food safety plays a significant role in restaurant choice. For instance, if an outbreak occurs in a restaurant, the image will strongly influence consumers' mindset about the specific restaurant. This occurrence will also have a great impact for consumers in deciding where to eat. Aksoydan (2007) investigated what sanitation attributes influence consumers' choice of restaurant, and found that hygiene attributes such as cleanliness of the restaurant, food, china, service staffs, toilets, and the kitchen, influence consumers' restaurant choice. The most important attribute was found to be cleanliness of the food in restaurant selection. Barber (2009) conducted a study looking into a more specific area, the restroom, in regards to sanitation of restaurant and restaurant choice. The results show that restroom cleanliness and functioning had an impact on consumer choice of a restaurant and return to the restaurant. Many other research studies have also cited the importance of the restroom as one of the dimensions that consumers are concerned about when selecting a restaurant (Aksoydan, 2007; Henson et al., 2006; Leach et al., 2001). Furthermore, cleanliness of the restroom also affects consumers' revisit intention. Ungku Fatimah, et al. (2011) included a scent attribute (both inside and outside of the restaurant) to measure the perception of foodservice hygiene dimensions.

Gender differences have also been found in regards to perception of the importance of different attributes. Aksoydan (2007) found that males were more concerned about the cleanliness of china or cutlery as compared to other factors in terms of restaurant selection while these were not as important a factor for female consumers in relation to restaurant selection.

Lee et al. (2012) compared consumer perceptions about food safety in different ethnic restaurants, Asian and Mexican. Critical attributes were food quality, cleanliness of the kitchen, cooking and preparation of food, inspection scores, cleanliness of restrooms, and food storage.

Many studies have been conducted about the relationship between consumer perceptions about service quality (food, service, and atmospherics) and consumption behaviors, and it has been found that the food quality dimension of service quality is the most important dimension for customers in deciding where to eat (Clark and Wood, 1999; Kivela, 1997; Mattila, 2001; Susskind and Chan, 2000). The consumer perception of the food quality dimension includes food safety attributes, such as freshness, no foreign objects, and the temperature of food.

11.4 DISCUSSION

As people are now facing innumerable choices when dining out, it is important to understand how people make these choices. It is commonly reported that people are now more interested in food safety and sanitation conditions in restaurants. Many researchers have sought to explain what influences consumers' perceptions about food safety and sanitation conditions in restaurants. At the same time, the relationship between perceptions and consumer behavior is also being researched. The results of these studies suggest that consumers judge the degree or conditions of food safety and sanitation based upon tangible objects in and near the foodservice premises.

Much of the consumers' assessment in regards to food safety and sanitation conditions occurs as a result of their experience in the front of house because this is the place where consumers can observe tangible attributes such as the food, the dining room, restroom, and the staff and their behavior. Researchers segment consumers' assessment into the three elements of food, atmospherics, and service. These elements have been further categorized into at least six dimensions in research studies with many attributes suggested to measure the quality of each of these six dimensions in restaurants (Baker, 1987; Berman and Evans, 1995; Bitner, 1992; Knutson et al., 1990; Parasuraman et al., 1988; Raajpoot, 2002; Ryu and Jang, 2008; Stevens et al., 1995; Wakefield and Blodgett, 1996). Research in this area is complicated. In fact, some of these dimensions are mixed with tangible and intangible attributes. Others contain only tangible attributes. The evaluation of tangible attributes is most adaptable to studies that assess the influence of food safety and sanitation on consumers' perceptions. The dimensions of sanitation and food safety broadly include the food, table settings, the dining room, the exterior, the restroom, and the service staff. These tangible dimensions have been found to be indicators for consumers when they assess overall cleanliness or sanitation conditions in restaurants. The relationship between consumers' perception and consumption behaviors has been studied by many researchers (Aksoydan, 2007; Barber et al., 2011; Barber and Scarcelli, 2009, 2010; Cullen, 2005; Leach, et al., 2001; Lee et al., 2012; Ungku Fatimah, et al., 2011; Worsfold, 2006; Zopiatis and Pribic, 2007).

These researchers found that there are positive influences between these relationships. Some researchers have also studied sociodemographic characteristics, such as gender, education, and so on. They found that consumers may have different perceptions about food safety and sanitation conditions in restaurants depending on their gender and education level. In addition, perceptions may be different in regards to different types of ethnic restaurants, such as Mexican, Chinese, or Italian restaurants. However, even though consumers perceive sanitation conditions differently, it has been shown that consumer perceptions about food safety and sanitation have a great impact on their restaurant selection behavior. Additionally, the cleanliness of the restroom clearly influences consumers in their restaurant selection behavior. Based on consumers' experiences, they also decide whether they would come back or not. All of the research about food safety and sanitation demonstrate the importance of sanitation conditions for consumers' dining experiences in foodservice establishments.

Most people who dine out frequently think that sanitation conditions are very important in terms of restaurant choice (Wordsfold, 2006). Service providers can use the findings from studies to understand how customers perceive the quality of sanitation conditions. Restaurant owners or managers should recognize how important these sanitation conditions are as factors in changing consumers' consumption behaviors. The consequence of poor sanitation conditions might bring a decrease of sales in restaurants. Once a bad perception is imbedded in consumers' mindsets by a bad experience in a certain restaurant, it becomes very difficult to change a consumers' negative perception to a positive perception.

KEYWORDS

- **Food Code**
- **food safety**
- **front of the house**
- **revealed preference**

REFERENCES

Aksoydan, E. (2007). Hygiene factors influencing customers' choice of dining-out units: Findings from a study of university academic staff. *Journal of Food Safety, 27*(3), 300–316

Baker, J. (1987). The role of the environment in marketing services: The consumer perspective. In: Czeoeil, J., Congram, C.A., and Shanahan, J. (Eds.). The Services Challenge: Integrating for Competitive Advantage. American Marketing Association, Chicago, 55–80.

Barber, N., and Scarcelli, J. M. (2009). Clean restrooms: How important are they to restaurant consumers? *Journal of Foodservice, 20*(6), 309–320.

Barber, N., and Scarcelli, J. M. (2010). Enhancing the assessment of tangible service quality through the creation of a cleanliness measurement scale. *Managing Service Quality, 20*(1), 70–88.

Barber, N., Goodman, R. J., and Goh, B. K. (2011). Restaurant consumers repeat patronage: A service quality concern. *International Journal of Hospitality Management, 30*(2), 329–336.

Berman, B., and Evans, J.R. (1995). *Retail Management: A Strategic Approach.* Englewood Cliffs: Prentice-Hall, Inc.

Bitner, M. J. (1992). Servicescapes: The impact of physical surroundings on customers and employees. *Journal of Marketing, 56*(2), 57–71.

Boo, H. C., Ghiselli, R., and Almanza, B. (2000). Consumer perceptions and concerns about the healthfulness and safety of food served at fairs and festivals. *Event management, 6*(2), 85–92.

Brady, M. K., and Cronin Jr, J. J. (2001). Some new thoughts on conceptualizing perceived service quality: A hierarchical approach. *Journal of Marketing, 65*(3), 34–49.

Brunsø, K., Fjord, T. A., and Grunert, K. G. (2002). *Consumers' food choice and quality perception*: Aarhus School of Business, MAPP-Centre for Research on Customer Relations in the Food Sector. Retrieved from https://pure.au.dk/portal/files/32302886/wp77.pdf

Campbell-Smith, G. (1967). *Marketing of the meal experience: A fundamental approach.* Guildford: University of Surrey.

Centers for Disease Control and Prevention (CDC). (2012). Food Safety. Retrieved from http://www.cdc.gov/foodsafety/facts.html#howcontamination

Centers for Disease Control and Prevention (CDC). (2012). Trends in foodborne illness, 1996–2010. Retrieved from http://www.cdc.gov/foodborneburden/PDFs/FACTSHEET_B_TRENDS.PDF

Centers for Disease Control and Prevention (CDC). CDC Current Outbreak List (2013). Retrieved from http://www.cdc.gov/outbreaks/index.html

Clark, M. A., and Wood, R. C. (1999). Consumer loyalty in the restaurant industry: A preliminary exploration of the issues. *British Food Journal, 101*(4), 317–327.

Cullen, F. (2005). Factors influencing restaurant selection in Dublin. *Journal of Foodservice Business Research, 7*(2), 53–85.

Dosman, D. M., Adamowicz, W. L., and Hrudey, S. E. (2002). Socioeconomic determinants of health-and food safety-related risk perceptions. *Risk analysis,* 21(2), 307–318.

Dulen, J. (1999). Quality control. *Restaurants and Institutions, 109*(5), 38–41.

Fielding, J. E., Aguirre, A., Spear, M. C., and Frias, L. E. (1999). Making the grade: Changing the incentives in retail food establishment inspection. *American Journal of Preventive Medicine, 17*(3), 243.

Flynn, J., Slovic, P., and Mertz, C. K. (1994). Gender, race, and perception of environmental health risks. *Risk Analysis, 14*, 1101–1107.

Food and Drug Administration (FDA). (2010). Report on the Occurrence of Foodborne Illness Risk Factors in Selected Institutional Foodservice, Restaurant, and Retail Food Store Facility Types. US Public Health Service, FDA, Department of Health and Human Services, Washington, DC. Retrieved from http://www.fda.gov/Food/FoodSafety/RetailFoodProtection/FoodborneIllnessandRiskFactorReduction/RetailFoodRiskFactorStudies/ucm223293.htm

Food and Drug Administration (FDA). (2013). FDA Food Code. Retrieved from http://www.fda.gov/Food/GuidanceRegulation/RetailFoodProtection/FoodCode/default.htm

Frewer, L. (2001). Risk perception and risk communication about food safety issues. *Nutrition Bulletin, 25*(1), 31–33.

Green, L. R., Radke, V., Mason, R., Bushnell, L., Reimann, D. W., Mack, J. C.,... Selman, C. A. (2007). Factors related to food worker hand hygiene practices. *Journal of Food Protection, 70*(3), 661–666.

Grunert, K. G. (2005). Food quality and safety: Consumer perception and demand. *European Review of Agricultural Economics, 32*(3), 369–391.

Henson, S., Majowicz, S., Masakure, O., Sockett, P., Jones, A., Hart, R.,... Knowles, L. (2006). Consumer assessment of the safety of restaurants: The role of inspection notices and other information cues. *Journal of food safety, 26*(4), 275–301.

Jin, G. Z., and Leslie, P. (2003). The effect of information on product quality: Evidence from restaurant hygiene grade cards. *The Quarterly Journal of Economics, 118*(2), 409–451

Jin, G. Z., and Leslie, P. (2009). Reputational incentives for restaurant hygiene. *American Economic Journal: Microeconomics, 1*(1), 237–267.

Jones, T. F., Pavlin, B. I., LaFleur, B. J., Ingram, L. A., and Schaffner, W. (2004). Restaurant inspection scores and foodborne disease. *Emerging infectious Diseases,* 10(4), 688.

Jordan, J. L., and Elnagheeb, A. H. (1991). Public perceptions of food safety. *Journal of Food Distribution Research, 22*(3), 13–22.

Kant, A. K., and Graubard, B. I. (2004). Eating out in America, 1987–2000: trends and nutritional correlates. *Preventive Medicine, 38*(2), 243.

Kivela, J. J. (1997). Restaurant marketing: Selection and segmentation in Hong Kong. *International Journal of Contemporary Hospitality Management, 9*(3), 116–123.

Knight, A. J., Worosz, M. R., and Todd, E. (2007). Serving food safety: Consumer perceptions of food safety at restaurants. *International Journal of Contemporary Hospitality Management, 19*(6), 476–484.

Knutson, B., Stevens, P., Wullaert, C., Patton, M., and Yokoyama, F. (1990). LODGSERV: A service quality index for the lodging industry. *Journal of Hospitality and Tourism Research, 14*(2), 277–284.

Kotler, P. (1973). Atmospherics as a marketing tool. *Journal of Retailing, 49*, 48–64.

Kumcu, A., and Kaufman, P. (2011). Food spending adjustments during recessionary times. *Amber Waves, 9*(3), 10–17.

Leach, J., Mercer, H., Stew, G., and Denyer, S. (2001). Improving food hygiene standards– a customer focused approach. *British Food Journal, 103*(4), 238–252.

Lee, L. E., Niode, O., Simonne, A. H., and Bruhn, C. M. (2012). Consumer perceptions on food safety in Asian and Mexican restaurants. *Food Control, 26*(2), 531–538.

Mattila, A. S. (2001). Emotional bonding and restaurant loyalty. *The Cornell Hotel and Restaurant Administration Quarterly, 42*(6), 73–79.

McGuirk, A. M., Preston, W. P., and McCormick, A. (2006). Toward the development of marketing strategies for food safety attributes. *Agribusiness, 6*(4), 297–308.

Mintel Oxygen reports (2013). *Ethnic Food – US – January 2013*. Retrieved from www. http://academic.mintel.com.ezproxy.lib.purdue.edu/display/651052/?highlight=true#hit1

Namkung, Y., and Jang, S. (2007). Does food quality really matter in restaurant: Its impact of customer satisfaction and behavioral intentions? *Journal of Hospitality and Tourism Research, 31*(3), 387–410.

National Restaurant Association (NRA). (2013). A blooming trend: Gardens are sprouting up at restaurants across the country, Retrieved from http://www.restaurant.org/Manage-My-Restaurant/Marketing-Sales/Sustainability/Restaurant-Gardens

National Restaurant Association (NRA). (2013). Floor plans set the stage for success. Retrieved from http://www.restaurant.org/Manage-My-Restaurant/Operations/Openings-and-Closings/Floor-plans-set-the-stage-for-success

National Restaurant Association (NRA). (2013). Pocket factbook. Retrieved from http://www.restaurant.org/pdfs/research/Factbook2013_LetterSize.pdf

Parasuraman, A., Zeithaml, V. A., and Berry, L. L. (1985). A conceptual model of service quality and its implications for future research. *Journal of Marketing,* 49(4), 41–50.

Parasuraman, A., Zeithaml, V. A., and Berry, L. L. (1988). Servqual. *Journal of Retailing,* 64(1), 12–37.

Perran, A. (1995). Dimensions of service quality: A study in Istanbul. *Managing Service Quality, 5*(6), 39–43.

Raajpoot, N. A. (2002). TANGSERV: A multiple item scale for measuring tangible quality in the foodservice industry. *Journal of Foodservice Business Research, 5*(2), 109–127.

Regan, C. (1991). Nutrition awareness and the food service industry. *Annals of the New York Academy of Sciences,* 623(1), 392–399.

Ritson, C., and Mai, L. W. (1998). The economics of food safety. *Nutrition and Food Science, 98*(5), 253–259.

Ryu, K., and Jang, S. (2008). DINESCAPE: A scale for customers' perception of dining environments. *Journal of Foodservice Business Research, 11*(1), 2–22.

Simon, P.A., Leslie, P., Run, G., Jin, G.Z, Reporter, R., Aguirre, A., and Fielding, J.E. (2005). Impact of restaurant hygiene grade cards on foodborne-disease hospitalizations in Los Angeles County. *Journal of Environmental Health, 67*(7), 32–36.

Slovic, P. (1987). Perceptions of risk. *Science, 236,* 280–285.

Starr, C. (1969). Social benefit versus technological risk. *Science,165,* 1232–1238.

Stevens, P., Knutson, B., and Patton, M. (1995). DINESERV: A tool for measuring service quality in restaurants. *The Cornell Hotel and Restaurant Administration Quarterly, 36*(2), 5–60.

Sulek, J.M., and Hensley, R.L., (2004). The relative importance of food, atmosphere, and fairness of wait. *Cornell Hotel and Restaurant Administration Quarterly, 4*(3), 235–247.

Susskind, A. M., and Chan, E. K. (2000). How Restaurant Features Affect Check Aberages. *Cornell Hotel and Restaurant Administration Quarterly, 41*(6), 56–63.

Tripp, C., Greathouse, K. R., Shanklin, C. W., and Gregoire, M. B. (1995). Factors influencing restaurant selection by travelers who stop at visitor information centers. *Journal of Travel and Tourism Marketing, 4*(2), 41–50.

Turley, L.W. and Milliman, R.E. (2000). Atmospheric effects on shopping behavior: A review of experimental evidence. *Journal of Business Research, 49*(2), 193–211.

U.S. Census Bureau (2012). Disposable personal income per capita in current and constant (2005) dollars by state: 1980 to 2010. Retrieved from http://www.census.gov/compendia/statab/2012/tables/12s 0682.pdf

Ungku Fatimah, U. Z. A., Boo, H. C., Sambasivan, M., and Salleh, R. (2011). Foodservice hygiene factors—The consumer perspective. *International Journal of Hospitality Management, 30*(1), 38–45.

Variyam, J. (2005). Nutrition labeling in the food-away-from-home sector: An economic assessment. *USDA-ERS Economic Research Report* (4). Retrieved from http://www.ers.usda.gov/publications/err4/err4.pdf.

Wakefield, K. L., and Blodgett, J. G. (1996). The effect of the servicescape on customers' behavioral intentions in leisure service settings. *Journal of Services Marketing, 10*(6), 45–61.

Worsfold, D. (2006). Eating out: Consumer perceptions of food safety. *International Journal of Environmental Health Research, 16*(03), 219–229.

Zopiatis, A., and Pribic, J. (2007). College students' dining expectations in Cyprus. *British Food Journal, 109*(10), 765–776.

CHAPTER 12

SOCIAL MEDIA AND FOOD SAFETY RISK COMMUNICATION

SOOBIN SOO, PhD

Assistant Professor, The Ohio State University

CONTENTS

12.1 WHAT IS SOCIAL MEDIA?

Due to its increasing growth and popularity, social media has become the major communication channel used by both companies and consumers (Smith and Zook, 2011). The largest number of social media users is shown by Facebook which exceeded 1 billion in 2012, followed by Twitter (127 million), Google+ (90 million), Pinterest (21 million), LinkedIn (150 million), and Reddit (5.5 million) (Internet World Stats, 2012). Compared to 2008 when the number of Facebook users was only 34 million, the 1 billion users in 2012 reflect the tremendous growth and popularity of social media among consumers (Internet World Stats, 2012). Moreover, Skelton (2012) reported that about 66% of Americans were connected to one or more social media platforms in 2012.

12.1.1 DEFINITION OF SOCIAL MEDIA

With the growing interest in social media, several researchers have attempted to define exactly what social media is. Social media is generally defined as an online platform that allows users to share their experiences, insights, and perspectives in a variety of forms. Another definition of social media was suggested by Kaplan and Andreas (2010) as "social media is a group of internet-based applications that build on the ideological and technological foundations of Web 2.0, and that allow the creation and exchange of user generated content (Kaplan and Andreas, 2010, p. 3)." In addition, a study by Tinker et al. (2009) defined social media as "various electronic tools, technologies, and applications that facilitate interactive communication and content exchange, allowing the user to move back and forth easily between the roles of audience and author (Tinker et al., 2009, p. 17)."

12.1.2 TYPES OF SOCIAL MEDIA

Due to the large number of social media tools existing today, there has been no consistent systematic way of classifying the types of social media tools that are available. One of the clearest classification schemes was suggested by Tinker et al. (2009) and included seven types of tools that are available in social media today: social networking sites, blogs, microblogs, image/video sharing sites, virtual worlds, mobile text messaging, and social bookmarking. Social network sites (SNS) are online communities that enable users to stay connected and network with people who share common interests. The major social network sites include Facebook and MySpace. While blogs refer to individual websites containing regular commentaries and descriptions of events or materials, microblogs are another form of blogging managed by individuals, which allows users not only to update their status but also interact with others in their network. The major microblogs are Twitter and Flurk. Compared to traditional blogs which are more likely to be personal websites, the microblogs focus more on interaction with networks by updating messages, leaving comments, and being connected with others. Other social media tools include mobile text messaging, image/video sharing sites, and virtual worlds. Another classification of the types of available social media tools was attempted by Kaplan and Andreas (2010), categorizing social media tools into six groups by social presence/media richness and self-presentation/self-disclosure. The authors included six social media tools: blogs, social networking sites (e.g., Facebook), virtual social worlds (e.g., Second Life), collaborative projects (e.g., Wikipedia), content communities (e.g., YouTube), and virtual game worlds (e.g., World of Warcraft). For example, social networking sites are considered to exhibit a medium level of social presence and a high level of self-disclosure, while collaborative projects are considered to have a low level of social presence and a low level of self-disclosure. Table 1 presents their classification of social media tools depending on social presence/media richness and self-presentation/self-disclosure.

TABLE 1 Classification of social media tools (Kaplan and Andreas, 2010).

		Social presence/ Media richness		
		Low	**Medium**	**High**
Self-presentation/ Self-disclosure	High	Blogs	Social networking sites (e.g., Facebook)	Virtual social worlds (e.g., Second Life)
	Low	Collaborative projects (e.g., Wikipedia)	Content communities (e.g., YouTube)	Virtual game world (e.g., World of Warcraft)

12.1.3 SOCIAL MEDIA AND THE HOSPITALITY INDUSTRY

The hospitality industry has paid considerable attention to consumers' use of social media because the hospitality industry is highly dependent on internet users in a business-to-consumer context (Werthner and Ricci, 2004). Moreover, the experience-based nature of the hospitality industry increases the use of word-of-mouth (WOM) in the prepurchase evaluations of consumers (Litvin et al., 2008). The WOM communication stemming from the experiences of other people with hospitality products or services has become a preferred information source for consumers prior to making purchase decisions (Crotts, 1999). Dev et al. (2010) also emphasized the importance of social media-related hospitality research as a result of the eclipse of printed media and the rise of social media.

A large number of hospitality firms currently use social media to build relationships with consumers. For example, several restaurant companies have created and manage Facebook pages by updating pictures and comments (Table 2). McDonald's is the most popular restaurant brand Facebook page with 28,745,183 likes, 594,859 talking about this, and 3,747,333 visitors as of May 2013. Hotel brands such as Marriott and Hilton Hotels and Resorts also manage their Facebook pages and show more than 1 million likes. Travel search engine companies such as Expedia, TripAdvisor, and Priceline also manage their own Facebook pages.

TABLE 2 Examples of hospitality firms using Facebook pages (as of May 30, 2013).

Category	Hospitality firms	Number of likes	Number talking about this	Number of visitors
Food/Beverages	McDonald's	28,745,183	594,859	3,747,333
	Burger King	6,208,200	39,000	14,806
	In-N-Out burger	2,976,610	2,346	1,504,666
	Olive Garden Italian Restaurant	4,405,396	93,061	2,994,834
Travel/Leisure (Hotel)	Marriott Hotels and Resorts	1,181,616	57,354	—
	Hilton Hotels and Resorts	1,024,402	12,419	4,768,940
	Shangri-la Hotels and Resorts	684,781	35,868	—
	Renaissance Hotels	474,964	56,465	145,788

TABLE 2 *(Continued)*

Category	Hospitality firms	Number of likes	Number talking about this	Number of visitors
Travel/Leisure (tour-ism)	Expedia	1,801,892	18,407	—
	TripAdvisor	305,155	8,793	—
	Priceline.com	428,617	11,310	—
	Hotels.com	800,360	4,381	—

Academic research is also growing in the area of social media due to its growing popularity and importance in the hospitality industry. The topic of social media has been widely studied by hospitality researchers in all areas of the restaurant, tourism, and hotel industry.

12.1.4 SOCIAL MEDIA IN THE RESTAURANT INDUSTRY

The restaurant industry in the United States has generated $580 billion in sales, employed 12.7 million people throughout 945,000 locations across the nation as of 2010 (National Restaurant Association, 2010). In 2013, restaurant food sales are projected to reach $660 and 13.1 million people are being employed according to the forecasting report of the National Restaurant Association (NRA, 2013). Along with the growth of the restaurant industry, the effective use of technologies has been a major issue for restaurateurs in maximizing their profits. Understanding how to use social media in the restaurant industry is believed to bring numerous advantages to restaurateurs. Those benefits include enabling restaurateurs to perform effective restaurant management in terms of communication, marketing, recruiting, training, and overall operations (DiPietro et al., 2012).

Acknowledging the power of WOM communications on restaurant customers' behaviors, much attention has been paid to the role of online comments about restaurant experiences on future customers' visit intentions. In particular, Pantelidis (2010) examined 2,471 online restaurant comments in a restaurant review site in the United Kingdom. The author categorized comments regarding restaurant experiences into a variety of topics: food, service, atmosphere, price, menu, design, and décor. The study also analyzed positive and negative online reviews about customers' restaurant experiences, emphasizing the need for restaurants' strategic management of social media.

The usage pattern of social networking sites of large restaurant companies in the United States was investigated by DiPietro et al. (2012). The authors offered tips on how to effectively use social media for restaurant companies. The suggestions included that restaurateurs should use social media when seeking employees, getting feedback on new menu items, and spreading word-of-mouth about best practices of the restaurant system.

12.1.5 SOCIAL MEDIA IN THE TOURISM INDUSTRY

The internet has heavily influenced the tourism industry, particularly in the way users search for travel information (Buhalis and Law, 2008). In recent years, social media also has enabled users to share and exchange travel-related information and their travel experiences. This is thought to be a major trend in the tourism industry.

Researchers examined the use of social media in online travel information searches, noting the significant role of social media in the tourism industry. Xiang and Gretzel (2010) analyzed travelers' uses of a search engine for travel-related information in relation to nine U.S. destinations, finding that search engines are, in fact, very likely to direct travelers to the social networking sites (e.g., Facebook) and media sharing sites (e.g., YouTube). Once users visit the social networking sites, the sites may function as a marketing tool to disseminate favorable images and videos, which may capture future travelers' attention. Xiang, Wöber, and Fesenmaier (2008) asserted that a search engine such as Google or Bing plays an important role as the "gateway" to travel-related information as well as an effective marketing tool to reach and persuade future travelers.

12.1.6 SOCIAL MEDIA IN THE HOTEL INDUSTRY

Social media is at the center of a great deal of research in the hotel industry as well (Figure 1). The impact of online hotel reviews on consumers' online hotel booking intentions was examined by Sparks and Browning (2011), asserting that consumers are prone to use easy-access information when reading online hotel reviews. In addition, positively framed online hotel reviews together with numerical rating details were found to increase consumers' online booking intentions. Stringam and Gerdes (2010) analyzed 60,648 consumer ratings and comments from an online distribution site to identify significant factors influencing hotel ratings. The authors identified most frequently used words for hotels with either high or low guest ratings, implying the effect of online hotel reviews and ratings on future visitors' online hotel booking intentions.

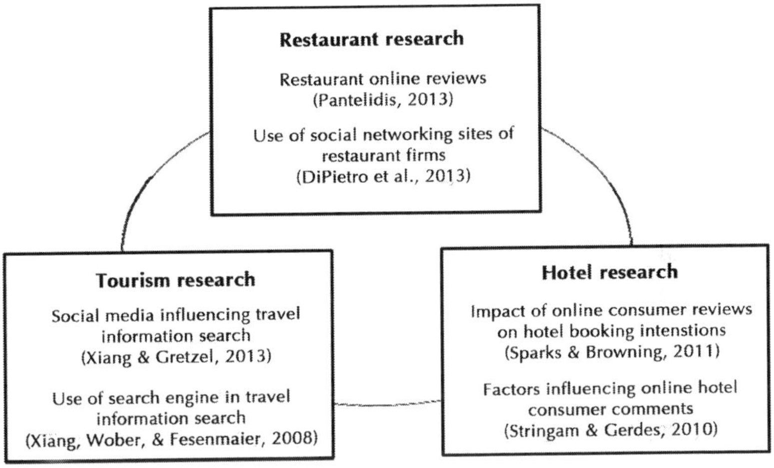

FIGURE 1 Examples of social media-related hospitality research.

12.2 FOOD SAFETY RISK COMMUNICATION THROUGH SOCIAL MEDIA

There are various types of risks threatening consumers' health and well-being. Fischer et al. (1991) classified risks into four categories according to the degree to which people are concerned about the most: health (22.9%), safety (22.4%), environment (44.1%), and society-related risks (10.6%). Health-related risks include cancer, heart disease, drugs, alcohol, and smoking, safety-related risks include motor vehicles, natural hazards, environment-related risks related to air/water pollution, hazardous chemicals, and society-related risks include war. Of those risks, food safety is of great concern not only to consumers in order to stay healthy and safe, but also to companies to maintain their brand loyalty and firm reputation.

12.2.1 DEFINITION OF RISK COMMUNICATION

Due to the numerous types of risks coupled with the unpredicted nature of these risks, risk communication has become crucial to companies and government agencies. Risk communication is defined as "communication intended to supply laypeople with the information they need to make informed, independent judgments about risks to health, safety, and the environment (Morgan, 2002, pg. 4)." Effective risk communication may result in reducing public outrage and minimizing future threats, while poor risk communication may cause public confusion and backfire. Thus, effective risk communication should focus on critical information that recipients need to understand to reduce the potential risks.

12.2.1.1 DISTINCTIVE FEATURES OF RISK AND CRISIS COMMUNICATION

Risk communication and crisis communication sound very similar at the same time that they show distinctive differences. Reynolds and Seeger (2005) suggested distinguished features of risk and crisis communication in terms of message content, tone, focus, frequency, message preparation, scope, and message type (Table 3). For example, risk communication contains messages that are about hazards and outrage, while crisis communication is more likely to deliver messages regarding current state and conditions of ongoing crises. While risk communication is shown frequently and prepared for prior to the outbreak of crises, crisis communication is more likely to take place after crises. In addition, the tone of risk communication is persuasive, while crisis communication tends to be informative.

TABLE 3 Distinctive features of risk and crisis communication (Reynolds and Seeger, 2005).

	Risk communication	Crisis communication
Message content	Hazards and outrage	Current state and conditions
Tone	Persuasive	Informative
Focus	Sender/message centered	Receiver/situation centered
Frequency	Frequent	Infrequent
Message preparation	Long-term, precrisis	Short-term, postcrisis
Scope	Personal	Personal, community, regional
Message type	Controlled and structured	Spontaneous and reactive

12.2.1.1.1 RISK AND CRISIS COMMUNICATION MODELS

Previous research has been dedicated to the development of an effective risk and crisis communication strategy as an approach to building relationships with consumers, responding to crises appropriately, and sustaining one's brand image and firm reputation. Strategy and communication models include image restoration discourse strategy, situational crisis communication strategy (SCCT), crisis and emergency risk communication (CERC) model, and food risk communication strategy.

12.2.1.1.2 IMAGE RESTORATION DISCOURSE STRATEGY

A foodborne illness outbreak can cause great harm to a corporate reputation as well as the image of a firm or an organization. Image is a central concept to public relations and the hospitality industry values its positive image because of its relatedness to the public. A food crisis threatening consumers' health may endanger the image of a hospitality firm, and even the survival of the firm.

Image restoration discourse strategy was developed by Benoit (1997) as an approach to understand crisis situations and a way to design an appropriate crisis communication strategy to protect the image of firms during a crisis. The author suggested four different image restoration strategies: denial, evasion of responsibility, a reduction in the offensiveness of the event, and corrective action. The decision on which strategy to be used depends on the nature and characteristics of ongoing crises.

12.2.1.1.3 SITUATIONAL CRISIS COMMUNICATION STRATEGY (SCCT)

Acknowledging the dynamic nature of crisis situations, Coombs (2004) introduced Situational crisis communication strategy (SCCT) as a mean of postcrisis strategy to protect a firm's reputation. The SCCT model incorporates how stakeholders will react to the crises, which in turn influences the type of appropriate crisis response strategies. The model identifies key variables influencing threats to the firm's reputation during a crisis: crisis responsibility, crisis history, and prior relational reputation. Based on these three variables, crisis managers are suggested to classify crises into three clusters: victim cluster (e.g., natural disaster, rumor), accidental cluster (technical-error accidents or product harm), and preventable cluster (e.g., human-error accidents or product harm).

Food safety events are considered either an accidental cluster or a preventable cluster due to the lack of control over the safety of foods. Based on the evaluation of crisis characteristics, the SCCT model suggests appropriate crisis response strategies such as compensation for victims, an apology which takes full responsibility, or a reminder of past good works to regain trust from the public.

12.2.1.2 CRISIS AND EMERGENCY RISK COMMUNICATION (CERC) MODEL

As an effort to combine the concept of risk and crisis communication, a study by Reynolds and Seeger (2005) developed the crisis and emergency risk communication (CERC) model. The CERC model identifies appropriate communication efforts depending on crisis stage: precrisis, initial event, maintenance, resolution, and evaluation.

According to the CERC model, the goal of precrisis communication is to educate the public and community by monitoring the emerging risks, increasing public awareness about possible risks, and promoting public preparation for possible risks. Once a crisis happens, the primary goal of communication is to reduce the uncertainty of risks. Reassuring the public about their safety is critical by partnering with government agencies and media. The next step is to inform and persuade the public about ongoing recovery and remediation efforts to assure the safety of consumers. Lastly, organizations may perform an evaluation of their crisis communication efforts during a crisis to determine specific actions that would improve or reduce the effectiveness of future crisis communication. The evaluation process after a crisis also helps design an effective precrisis communication strategy.

12.2.1.2.1 FOOD RISK COMMUNICATION STRATEGY

Considering the enormous impact of food safety events on hospitality firms, it is critical to understand key components of risk communication regarding food safety. Cope et al. (2010) suggested key elements of food risk communication and recommended the best way to deliver appropriate food safety information to the public. The identification of recipients is the first step in understanding who the interested consumers are and who are at particular risk. The second step is to determine which information is to be provided to the recipients. Depending on the needs of the recipients, customized and targeted information may be designed and disseminated by the communicators.

With regard to the food safety risks, communicators are advised to include such information: regulatory priorities, preventative measures, enforcement actions, and actions to improve future preparedness. The third step is to decide when to release information and how promptly the communication should take place. To prevent public outrage and unnecessary concern over potential risks, a timely and quick response is considered very important. Cope et al. (2010) noted that continuous efforts for engagement and rapid responses are the key to successful food risk communication. The last component of effective food risk communication is the way of communicating with the public, which includes prioritizing information recipients, providing balanced risk-benefit information, being transparent, and being consistent in food risk communication messages.

12.3 STRATEGIC USE OF SOCIAL MEDIA IN FOOD SAFETY RISK COMMUNICATION

As suggested by previous crisis research and these newly developed risk communication models, use of the appropriate risk communication strategies will enable crisis managers to effectively manage the food safety crises depending on the crisis stage: precrisis, during a crisis, or postcrisis.

12.3.1 PRECRISIS COMMUNICATION STRATEGY: USE SOCIAL MEDIA AS A TOOL TO EDUCATE THE PUBLIC

12.3.1.1 STAY CONNECTED WITH THE PUBLIC USING SOCIAL MEDIA

Risk communication should be performed even before a crisis occurs. This is termed precrisis communication. The purpose of precrisis communication is to build rapport with the public as well as stay connected with the public. Building strong trust and relationships with the public may protect a firm from extreme public outrage during a time of crisis.

12.3.1.2 PROVIDE FOOD SAFETY INFORMATION THROUGH SOCIAL MEDIA

Educating the public before a crisis is an effective precrisis communication strategy as well as a way to build relationships with the public. A survey by Tinker et al. (2009) investigated organizations' goals in using social media. The biggest reason for using social media was educating the public (37%), followed by encouraging public actions (25%), fostering two-way communication (17%), and networking among public health and emergency personnel (14%). Educating the public may be one of the most important keys to the success of risk communication before and during a crisis.

12.3.1.3 SELECTIVELY CHOOSE AND FOCUS ON A FEW SOCIAL MEDIA TOOLS

Despite the number of social media tools available, hospitality managers should not overuse those channels. The disadvantage of having too many social media tools was discussed by Tinker et al. (2009), and cited as being a reduction in the effectiveness and reliability of social media communication with the public. Rather than having too many social media tools which are not well-managed, selectively choosing a few social media tools and managing them well is thought to be a better precrisis communication strategy.

12.3.2 DURING A CRISIS COMMUNICATION STRATEGY: USE SOCIAL MEDIA AS A TOOL TO PROTECT THE PUBLIC FROM THE RISKS

12.3.2.1 DISSEMINATE URGENT FOOD SAFETY INFORMATION THROUGH SOCIAL MEDIA

Crisis managers acknowledge the advantage of social media in rapid information dissemination under urgent crisis situations. By showing a strong determination to protect the public during a crisis, a firm can regain trust and sustain relationships with the public. In contrast, the use of a denial strategy or a "being quiet" strategy during a crisis may increase public confusion and outrage. Active utilization of social media enables crisis managers to effectively and rapidly communicate with the public to protect the public health as well as a firm's reputation.

12.3.2.2 CORRECT WRONGFUL INFORMATION THROUGH SOCIAL MEDIA

Rumors and incorrect information may worsen the crisis situation. A well-known example was Wendy's finger incident in 2005 when a woman claimed that she found a finger-looking contaminant in her chili. Even though the event turned out to be a hoax, the rumor tarnished the brand image of Wendy's, resulting in a dramatic drop in their revenue (Braun-latour, Latour, and Loftus, 2006). Correcting wrongful information before it is too late is the key in preventing unnecessary public outrage and social media can be a communication tool in providing the correct information.

12.3.2.3 BE PROACTIVE, FAST, AND RESPONSIBLE IN SOCIAL MEDIA COMMUNICATION

The proactiveness of a crisis response strategy is also important in assuring the safety of food during a time of crises. In order to minimize possible harm to the public, a firm should perform proactive strategies such as recalling products or closing restaurants to eliminate possible risks to the public. Crisis managers should provide necessary information quickly and effectively to the general public. In addition, when any information is distributed by a firm through their social media tools, the firm should be responsible in the message.

12.3.3 POST-CRISIS COMMUNICATION STRATEGY: USE SOCIAL MEDIA AS A TOOL TO REBUILD AN IMPAIRED BRAND IMAGE AND REGAIN TRUST FROM THE PUBLIC

12.3.3.1 FOCUS ON REBUILDING RELATIONSHIPS WITH THE PUBLIC

The social media provides an interactive online platform to rebuild relationships with the public. Even after a crisis, firms should try to regain trust from consumers through constant efforts using social media communication. Although it is not easy to regain trust from the public after a crisis, constant and sincere recovery efforts may be a key to successful postcrisis management.

12.3.3.2 ACTIVELY PROMOTE A FOOD SAFETY CAMPAIGN TO SHOW GOOD WILL

Participating in food safety campaigns can be an effective strategy in showing one's willingness to protect the public health as well as one's efforts to monitor and ensure food safety. Conducting ecofriendly campaigns and participating in green movements are also strategies to repair a tarnished brand image after a crisis.

12.3.3.3 PARTNER WITH GOVERNMENT AGENCIES, HEALTH ORGANIZATIONS, AND COMMUNITIES

Being partners with government agencies, health-related organizations, and communities may also increase the effectiveness of postcrisis communication. For instance, hospitality firms may seek ways to prevent food safety events and educate the public in cooperation with health organizations such as the American Public Health Association (APHA) or government agencies such as the U.S. Department of Agriculture (USDA), the U.S. Food and Drug Administration (FDA), and the Centers for Disease Control and Prevention (CDC), or even local or state health departments (Figure 2).

Pre-Crisis	During a Crisis	Post-Crisis
• Provide food safety information through socail media. • Stay connected with the public using social media. • Selectively choose and focus on a few social media tools.	• Disseminate urgent food safety information throught social media. • Correct wrongful information throught social media. • Be proactive, fast, and responsible in social media	• Focus on rebuilding relationships with the public. • Actively promote food safety campaign showing good wills. • Partner with government agencies, health organizations, and communities.

FIGURE 2 Food safety risk communication using social media.

12.4 CONCLUSIONS

Due to the increasing popularity of social media as a communication channel, many restaurant companies manage their image through social media. Suggestions have been given in this chapter for how to effectively use social media prior to, during and after a food crisis. Crisis communication is being actively researched by all areas in hospitality and tourism and will continue to be an important area of research as consumers' use of social media continues to grow and companies expand their use of this newest marketing tool.

KEYWORDS

- **crisis and emergency risk communication model**
- **gateway**
- **social media**
- **word-of-mouth**

REFERENCES

Benoit, W. L. (1997). Image repair discourse and crisis communication. *Public relations review*, *23*(2), 177–186.

Braun-latour, K.A., Latour, M.S., Loftus, E.F., 2006. Is that a finger in my chili? Using affective advertizing for postcrisis brand repair. *Cornell Hospitality Quarterly*, 47(2), 106–120.

Buhalis, D., and Law, R. (2008). Progress in information technology and tourism management: 20 years on and 10 years after the Internet-The state of eTourism research. *Tourism Management*, 29(4), 609–623.

Coombs, W. T. (2004). Impact of past crises on current crisis communication insights from Situational Crisis Communication Theory. *Journal of Business Communication*, *41*(3), 265–289.

Cope, S., Frewer, L. J., Houghton, J., Rowe, G., Fischer, A. R. H., and De Jonge, J. (2010). Consumer perceptions of best practice in food risk communication and management: Implications for risk analysis policy. *Food Policy*, 35(4), 349–357.

Crotts, J. (1999). *Consumer decision making and prepurchase information search.* In Y. Mansfield and A. Pizam (Eds.), Consumer behavior in travel and tourism. New York, NY: Haworth Press.

Dev, C. S., Buschman, J. D., and Bowen, J. T. (2010). Hospitality marketing: A retrospective analysis (1960–2010) and predictions (2010–2020). *Cornell Hospitality Quarterly*, 51(4), 459–469.

DiPietro, R. B., Crews, T. B., Gustafson, C., and Strick, S. (2012). The Use of Social Networking Sites in the Restaurant Industry: Best Practices. *Journal of Foodservice Business Research,* 15(3), 265–284.

Fischer, G. W., Morgan, M. G., Fischhoff, B., Nair, I., and Lave, L. B. (1991). What risks are people concerned about. *Risk analysis*, 11(2), 303–314.

Internet World Stats. (2012). Facebook users in the world: Facebook usage and Facebook growth statistics by World geographic regions. Retrieved from http://www.internetworldstats.com/facebook.htm

Litvin, S. W., Goldsmith, R. E., and Pan, B. (2008). Electronic word-of-mouth in hospitality and tourism management. *Tourism management*, 29(3), 458–468.

Morgan, M. G. (2002). *Risk communication: A mental models approach.* Cambridge University Press.

National Restaurant Association. (2010). 2010 Restaurant industry overview. Retrieved March 17, 2010 from http://restaurant.org/research/facts/

National Restaurant Association. (2013). 2013 Restaurant industry forecast. Retrieved May 8, 2013 from http://www.restaurant.org/News-Research/Research/Facts-at-a-Glance

Pantelidis, I. S. (2010). Electronic meal experience: A content analysis of online restaurant comments. *Cornell Hospitality Quarterly*, 51(4), 483–491.

Reynolds, B., and Seeger, M.W. (2005). Crisis and emergency risk communication as an integrative model. *Journal of health communication*, 10(1), 43–55.

Skelton, A. (2012). Social Demographics: Who's Using Today's Biggest Networks. Retrieved from http://mashable.com/2012/03/09/social-media-demographics/

Smith, P. R., and Zook, Z. (2011). *Marketing communications: integrating offline and online with social media*. Kogan Page.

Sparks, B. A., and Browning, V. (2011). The impact of online reviews on hotel booking intentions and perception of trust. *Tourism Management*, 32(6), 1310–1323.

Stringam, B. B., and Gerdes Jr, J. (2010). An analysis of word-of-mouse ratings and guest comments of online hotel distribution sites. *Journal of Hospitality Marketing and Management*, 19(7), 773–796.

Tinker, T., Fouse, D., and Curre, D. (2009). Expert round table on social media and risk communication during times of crisis: strategic challenges and opportunities. Retrieved June, 13, 2011.

Werthner, H., and Ricci, F. (2004). E-commerce and tourism. *Communication of the ACM*, 17(12), 101–109.

Xiang, Z., and Gretzel, U. (2010). Role of social media in online travel information search. *Tourism Management*, 31(2), 179–188.

Xiang, Z., Wöber, K., and Fesenmaier, D. R. (2008). Representation of the online tourism domain in search engines. *Journal of Travel Research*, 47(2), 137–150.

PART 3

FOOD SAFETY AND CLEANLINESS IN SPECIAL ENVIRONMENTS

CHAPTER 13

FOOD SAFETY IN TEMPORARY FOODSERVICE ESTABLISHMENTS

CARL BEHNKE, PhD
Assistant Professor, Purdue University

CONTENTS

13.1 INTRODUCTION

Temporary foodservice establishments (TFE) are food production businesses that operate for no more than 14 consecutive days in conjunction with a single event (FDA, 2009b). These establishments are found in farmers' markets, fairs and festivals. They can be held on a street corner, in a ball park, on a dirt field or in a partially sheltered facility like a fairground or indoor market. Even a country club or hotel running an outdoor event, such as a banquet around the pool with cooking stations, while not specifically required to license as a TFE, would effectively be operating in a temporary facility.

Food prepared in a TFE for immediate consumption falls under the category of food handling; therefore, the food must be protected at every stage of production and appropriate temperatures must be maintained (Hofmann et al., 2007). TFEs represent a growing segment of the foodservice industry, and as such, pose unique challenges to foodservice operators. This chapter will examine safe food production in temporary foodservice establishments. Farmers' markets, as a growing trend in the U.S., will be used to illustrate many of the issues related to TFEs; however, most, if not all, concerns and suggestions discussed in this chapter can easily be related to TFE operations in fairs, festivals and roadside stands.

More and more, consumers are looking for ways to connect with their food; consequently, farmers' markets have experienced explosive growth in terms of popularity and economic influence. Research indicates that reasons for suppliers and consumers to engage in farmers' market activities are related to enjoyment, product quality and price, and social benefits (Brown, 2002). The U.S. Department of Agriculture (USDA) reported that in 1994 there were 1,755 markets, growing to 7,175 markets by the year 2011 (USDA, 2011). This growth trend has been beneficial for both producers and consumers. Small farmers and cottage industries have been able to gain direct exposure to consumers for a generally low cost. Consumers, in turn, have increasing access to fresh, regional products, while local communities have used farmers' markets as an economic revitalization tool. This trend of encouraging entry level consumer access for small producers, while meeting consumer and community demands, is certainly positive, yet also highlights the need for food safety in a growing industry.

A Michigan study of customer attitudes and behaviors towards locally grown foods and farmers' markets found that food quality and food safety were the most important factors for customers to visit farmers' markets, implying that they perceived these products to be better and safer than conventionally sourced products (Conner et al., 2010). Interviews of North Carolina market managers revealed that they did not think food safety was a significant concern, ostensibly because the increased freshness of the product and cleanliness due to decreased travel distances, increased washing and reduced levels of pesticides (Smathers, 2012), while another study found that vendors who packed their products in vacuum packaging, such as smoked fish, seriously underestimated the product risk because of a lack of knowledge of pathogenic anaerobes (Worsfold et al., 2004). Consumers, market managers and vendors seem convinced that their products are safer than conventionally produced products, and because of that, they don't need to be concerned about food safety; however, from a food safety standpoint, growth in farmers' market operations, especially those preparing and serving food in field conditions, comes with associated risks.

Traditionally, farmers' markets were focused around supplying local, fresh produce; the markets were simply a convenient location for farmers to bring their harvest and for customers to pick up their daily meal ingredients. Today, in addition to fresh produce, farmers' markets are much more diversified selling products such as baked goods, canned goods, candles, pottery, woven and knitted products, and ready-to-eat (RTE) foods. Raw produce sold at a farmers' market does present certain health risk issues; however, these issues are difficult to assess because while customers may consume the produce at the market, they will more likely take it home for further cooking or consumption. Since it can be expected that there will be substantial further raw produce handling, it is difficult to determine where food safety lapses may occur. The preparation and sale of ready-to-eat foods, on the other hand, is more complicated and problematic, and therefore, the primary focus of this chapter.

Ready-to-eat foods include those foods that are intended for immediate consumption. According to the Food and Drug Administration (FDA, 2009b), the category of ready-to-eat foods includes those foods that are in a form that is immediately edible without any subsequent preparation. RTE foods may include raw fruits and vegetables, potentially hazardous foods (PHF), and cooked or raw dishes (FDA, 2009b). Some RTE foods pose little risk to the consumer. For example, cookies baked in a remote bakery, packaged, sold and often consumed at a farmers' market are not likely to cause many foodborne illness outbreaks. Conversely, burgers that are cooked on a grill for immediate sale at the market offer multiple opportunities for cross contamination or time and temperature abuse contributing to foodborne illness. All one has to do is follow the flow of food to see challenging realities of onsite food production in a temporary foodservice establishment.

Hamburgers, hot dogs, and bratwursts are frequently served, not only in farmers' markets, but also in outdoor fairs and festivals, in venues that are considered TFEs. Since this type of vendor is fairly common, it offers a good example to use in following the flow of food. The flow begins at the start of day with the vendor leaving for the market. Hot dogs (precooked, held cold or frozen), hamburgers (raw, held cold or frozen), and bratwursts (raw or precooked, held cold or frozen) are transported by truck or van to the outdoor venue, usually held in a cooler. Buns, condiments, cheeses, and sides, such as potato chips, along with a grill, portable table or two, garbage can(s), cooking equipment and utensils, and promotional materials are also brought to the site with the proteins. Upon arrival, the vendor would probably proceed to set up tables and grills. He or she may, or may not, have an overhead awning or tent at his disposal, and could be setting up on blacktop, concrete, or even dirt. Once the production space is organized, the vendor will then start setting up the food – condiments by the cash register along with the plates, napkins and side options. Proteins arc set up by the grill, and then chafing dishes set up for assembling the various plates of food; in this hypothetical model, the food would go through a counterclockwise flow (see figure 1) from (1) protein >> (2) grill >> (3) chafing dish >> (4) assembly >> (5) plating >> (6) cash transaction, food transferred to consumer >> (7) consumer customizes product with condiments and leaves.

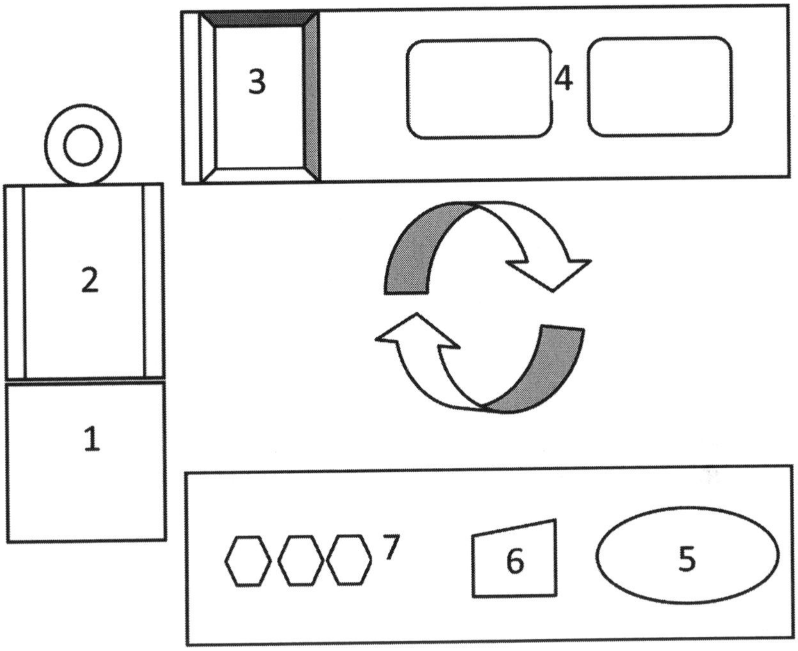

FIGURE 1 Typical TFE Food Flow Diagram.

This setup process would begin early in the morning with the vendor likely cooking food for sale by 10 a.m. and continuing through 2–3 p.m. Oftentimes, the vendor's vehicle is located in close proximity to the stall, sometimes right behind it, and it is not unusual for a vendor to work alone. Given these parameters, it is easy to imagine a vendor cooking the burgers (handling raw meat), immediately move to plating followed by a sales transaction. Points of concern in this flow of food are found in cold temperature control (keeping proteins properly chilled using a passive system) and hot temperature control (keeping food above 135°F after cooking), potential cross-contamination between handling raw food and subsequent cooked food, assembling food and handling cash, all by the same employee, not to mention the length of time food is exposed to high outdoor temperatures and contaminants.

Other food production considerations unique to TFEs are environmental in nature. For example, farmers' markets can be held indoors, in semipermanent facilities, or more often, outdoors, under tents. A survey of market managers found that two thirds of markets operate in temporary facilities (Ragland and Tropp, 2009). Food produced onsite is often exposed to significantly more dirt, dust and flies in a temporary outdoor facility than in a permanent indoor facility. Potable water and electricity are often unavailable, especially in outdoor markets – critical components for safe food production. Without modern air conditioning, temperatures in the market, especially near the grill where food is being stored and cooked, can get excessively hot, especially given that most markets are seasonal, running from late April through mid-late September, the hottest months of the year in many places. Given the temporary nature of the establishment, there are often few places to store food, so it is not uncommon to see food sitting on the floor of a van, which may or may not be regularly cleaned. It is easy to see how producing food safely in farmers' markets, fairs, and festivals can be more challenging than in a permanent restaurant facility. Moving forward, this chapter will systematically examine the issues related to safe food production in a temporary foodservice establishment, concluding with practical suggestions for managers producing food in this challenging setting.

13.2 REGULATING TEMPORARY FOODSERVICE ESTABLISHMENTS

Temporary foodservice establishments fall under the purview of local health departments. Health inspectors generally try to inspect vendors at least once or twice during the season. However, given the temporary, and somewhat fluid, nature of these operations, inspectors may not be able to address the issues as promptly or effectively as they might with a more permanent operation, such as a restaurant. Using fairs as an example, a fair might be held in a community for a week or less. During this time, there may be several vendors producing food onsite. Depending upon state regulation, vendors may or may not be required to be food safety certified. For example, state law in Indiana permits nonprofit operators to produce and sell food without any food safety certification (Indiana State Department of Health, 2006). Burt et al. (2003), in a study of New York City mobile food cart vendors, noted that vending requirements were minimal, including a vendor license, cart permit and either a valid food safety certificate, or just a current receipt showing registration for a food safety course; evidence of completing the course was not required. It is easy to see the risks should an untrained operator be producing food for a fair that may only last a weekend. A study of a large regional festival in Indiana found that 37% and 52% of vendors, in 2006 and 2007 respectively, had no prior foodservice work experience; many were untrained volunteers assisting nonprofit organizations (Lee et al., 2010). Given the incubation rates of some foodborne illnesses, the fair may be over and vendors departed before anyone experiences symptoms of distress. And, with the transitory nature of fairgoers and vendors, it can be very difficult for inspectors to investigate possible outbreaks; customers have returned home, which may be hundreds of miles away, and vendors may have moved their establishments to another setting.

Other issues facing health inspectors are resource limitations. Health inspectors are responsible for a large number of food production facilities, both temporary and permanent. Given that restaurants generally have more complicated menus and operations, and serve significantly more meals over the course of a year than temporary foodservice vendors, they tend to be the health department's focus of attention. Still, the FDA does specify that a TFE operation that "prepares, sells, or serves unpackaged potentially hazardous food (Time/Temperature Control for Safety food) and that (1) has improvised rather than permanent facilities or equipment for accomplishing functions such as handwashing, food preparation and protection, food temperature control, warewashing, providing drinking water, waste retention and disposal, and insect or rodent control" should be periodically inspected throughout its permit period (FDA, 2009b, Chapter 8, p. 203). However, given the popularity and growth of temporary foodservice establishments, such as those found in farmers' markets, health department resources are stressed.

A report by the Trust for America's Health (Levi et al., 2012) identified significant differences in state and federal funding for public health, noting significant cuts at the state and local levels; 40 states decreased their public health budgets from 2009 to 2010, another 30 states decreased their budgets two years in a row, while 15 states decreased them for 3 consecutive years. An examination of variation in public health spending on a geographic basis found that local public health spending has remained the same, at $29.57 per capita, since 1996 and is only slightly higher than it was in 1993 when it reached a median rate of $26.26 per capita (Mays and Smith, 2009). Lastly, the Association of State and Territorial Health Officials (2012) noted that approximately 17,800 jobs in state and territorial health offices were lost through attrition and layoffs since July 2008.

A question that one might ask is, "How are Temporary Foodservice Establishments regulated?" The answer is complicated in that it differs from state to state and depends upon what is being sold. If a vendor is selling processed meats, such as presmoked pork chops packaged for later cooking and service, that vendor would fall under USDA jurisdiction. If food is being prepared on site for immediate consumption, then it clearly falls under the health department responsibility, which generally will follow guidelines grounded in the FDA's food code; however, in some states, certain foods, generally those considered nonpotentially hazardous, are allowed to be produced at home, so long as they meet specific labeling requirements. For example, Hamilton (2002) noted that Maine's Department of Agriculture has specific rules covering Home Manufactured Foods, while Indiana also permits home-based vendor sales so long as the food is labeled with 10 point font stating "This product is home produced and processed and the production area has not been inspected by the state department of health" (Indiana Code, 2012). Also, each market develops its own rules governing what types of food may be sold along with regional restrictions and rules detailing how often a vendor needs to be present at the market. These rules are enforced by the market manager. So, to make it a little more confusing, while a state may permit home-based vendor (HBV) sales, individual markets may have rules forbidding them.

The risks are clear — vendors who may or may not be trained in safe food production, the short-term nature of fairs and farmers' markets (usually once or twice a week), health inspectors who attempt to ensure public safety in an environment of limited resources, and a regulatory environment that is confusing at best. Add in the unique challenges associated with temporary foodservice establishments, such as environmental exposure, lack of power or potable water, lack of hand washing and restroom facilities, and difficulties in temperature control, and it becomes easy to see how safe food production may be a concern. So, how safe is the food produced in a temporary foodservice environment?

13.3 FOODBORNE ILLNESS OUTBREAKS ASSOCIATED WITH TEMPORARY FOODSERVICE ESTABLISHMENTS

As previously mentioned, the short-term nature of a temporary foodservice establishment, with vendors moving from market to market and each with a different customer base, makes it difficult for inspectors to follow-up on reports of food poisoning; consequently, reports of foodborne illness outbreaks associated with TFE production are limited. That is not to say they don't exist.

In 2011, the FDA issued a Safety Recall for fresh strawberries from Washington County, Oregon (FDA, 2011). These berries were sold in farmers' markets and roadside stands and were implicated as the source of *E. coli* O157:H7 that sickened at least 15 people including one fatality. Iowa experienced an outbreak of Salmonella Newport associated with guacamole sold at farmers' markets and fairs in 2010. Thirty-nine illnesses were reported, five serious enough to warrant hospitalization (CDC).

Perhaps one of the most serious confirmed foodborne illness outbreak associated with TFE production occurred in July 2007 during the Taste of Chicago food festival. The CDC confirmed that this outbreak of Salmonella Heidelberg was linked to hummus and sickened 802 patrons, 29 serious enough to require hospitalization (CDC). These cases provide examples as to the health concerns associated with food prepared and served in TFE operations. Further investigation of the CDC Foodborne Outbreak Online Database reveals 3,518 confirmed and 844 suspected illnesses in the U.S. from 1998 through 2011 associated with fairs, festivals and temporary mobile services.

In its 2009 "Report on the Occurrence of Foodborne Illness Risk Factors," the FDA identified the operational areas most in need of improvement as employee hand washing; time and temperature control, specifically cold holding; date marking of RTE foods; and the cleaning and sanitization of food contact surfaces. However, the unique conditions associated with TFE operations, outdoor establishments with limited utility access, require conscientious efforts on the part of vendors and employees, something that research demonstrates tends to be lacking. The FDA emphasis coupled with research findings documenting related deficiencies is useful, since many of the issues associated with foodborne illness outbreaks can be mitigated by taking fundamental precautions, such as appropriate hand washing, preventing bare hand contact with RTE foods, preventing time and temperature abuse, and maintaining sanitary food contact surfaces.

13.4 HAND WASHING IN TFE OPERATIONS

Hand washing is the first line of defense in preventing foodborne illness outbreaks for both employees and customers. According to FDA Food Code (2009) employees need to wash hands under clean running warm water with soap, vigorously rubbing their hands together for at least 10–15 sec. After rinsing, employees are to dry their hands with disposable paper towels, and forced or heated air dryers. Health departments realize that outdoor venues are often lacking in terms of physical infrastructure, so a compromise hand washing station for TFE employees often consists of a cooler of hot tap water, which will stay warm for several hours, placed above a catch basin, typically a bucket, with a roll of paper towels and pump dispenser of soap. This equipment needs to be segregated from the food production, meaning it shouldn't share a table with food preparation, but still should be in close enough proximity to facilitate regular usage. Unfortunately, even with a compromise setup as described, hand washing stations are not always found in vendor stalls at fairs, festivals and farmers' markets.

The FDA (2009a) report concluded that lack of convenient, accessible and properly stocked hand washing stations in fast food restaurants (most analogous to TFE operations) contributed to inadequate hand washing. Hand washing in these operations was inadequately performed 38.8% of the time. If the violation rate is this high in permanent, fast food restaurants where hand washing facilities are designed into the production space; how bad are the violations in TFE operations, where hand washing facilities are minimal at best? Other studies have also noted serious hand washing deficiencies associated with farmers' market vendors.

An observational study of farmers' market TFE vendors in NW Indiana found that employees performed adequate hand washing 6.3% of the time, meaning they failed to wash their hands in accordance with the Indiana Health Code 93.7% of the time. Twenty six (35.6%) of the 73 observed employees did not have hand-washing stations in their work area. Most employees (64.4%) had access to hand washing stations, but clearly did not wash their hands given the 6.3% compliance rate (Behnke and Seo, in press). It seems that if employees wanted to wash their hands, in many cases they did not have the facilities to do so, and conversely, when they did have the facilities, they generally chose not to do so. Inadequate facilities and lack of training seem to minimize hand washing in TFE environments. The implications are that employees, many of whom are working multiple positions handling cooking duties, assembling food plates and handling cash transactions, represented a potential source of cross-contamination, which could lead to foodborne illness outbreaks.

The King County (Oregon) Board of Health (2011) noted that hand washing violations were recorded 73 times during farmers' market inspections from January 1 through August 24, 2011. Another study found that 67% of vendors were observed serving food with bare hands. Significantly, some vendors were observed handling food with visibly dirty hands, while none washed their hands or changed their gloves; cross contamination was clearly noted (Burt et al., 2003). A study of cheese vendors at 19 farmers markets in Ontario, Canada found them to be generally deficient in the practice of hand washing; 24% of the vendors did not have hand washing sinks available and 88% of food handlers did not attempt to wash their hands before or after serving patrons (Teng et al., 2004).

From the customer perspective, in addition to being located outdoors with all of the attendant cleanliness issues, many fairs and festivals have activities involving animals, while pets may or may not be allowed in farmers' markets. Customers could be involved in handling these animals, and then subsequently move on to eating some fair food prepared by a TFE vendor. Colorado State University's extension department identified cases in Australia, Colorado, Washington and Minnesota where foodborne outbreaks appeared to be linked to live animal contact (Safefood News, 1998). Smathers' (2012) study of 168 vendors in North Carolina farmers' markets revealed that 37% allowed live animals to be present, and that in some cases dogs were observed in the vendor stalls. Without access to public hand washing stations, customers have few choices at their disposal for washing their hands. Unfortunately, should this scenario lead to a case of foodborne illness, it is quite likely to be ascribed to food prepared and consumed at the fair or market, rather than the customers' behavior in interacting with animals. For this reason, it is especially important for outdoor venues to have accessible and properly maintained hand washing stations, which is not always the case. Hand washing stations are not always required in outdoor public areas, and when they are responsibility for maintaining them is not always clear; therefore, in the rush of activities, these stations may not be regularly supplied with soap or paper towels. Hofmann et al.'s (2009) study of Indiana farmers' markets found that just over 60% had public restroom facilities, and only 53.1% had hand-washing facilities. Smathers' (2012) study found that only 54% of farmers' markets provided restroom facilities in close proximity to the market; furthermore, only 49% of the markets provided hand-washing facilities. Without facilities there is no way for customers to practice reasonable hand sanitation.

Hand washing is the first line of defense in preventing foodborne illness outbreaks; however, it is clear that providing and encouraging adequate hand washing for both employees and customers in outdoor venues is problematic at best.

13.5 HANDLING RTE FOODS

Food service workers are not supposed to handle RTE foods with their bare hands. Because of this, employees use utensils, such as spoons, tongs and spatulas, or wear food-approved gloves while manipulating food through the final assembly processes. In an outdoor environment, where there are no three compartment sinks for dishwashing, if these tools get contaminated, for example they are dropped on the ground, then they need to be removed from use until cleaned and sanitized. This means that TFE vendors should have backup tools at their disposal. With regards to glove usage, employees are supposed to wash their hands before putting on gloves, which are meant to be single use, and to change the gloves between distinctly different tasks. Unfortunately, this is not always the case.

A study of 73 TFE employees at farmers' markets identified 11 instances of gloves being changed between discrete tasks over the course of 28 h of observation (Behnke and Seo, in press). Hofmann et al. (2007) noted that RTE foods need closer regulation due to their increased risk; unfortunately, the FDA's report (2009a) found prevention of bare hand contact with RTE foods in permanent fast food operations was in violation 26.3% of the time; in temporary operations, without trained staff and limited infrastructure, this violation rate is likely much higher.

Another challenge with regards to food production and RTE foods lies with the nature of TFE operations. Permanent restaurants tend to have staff and facilities that encourage segregation of duties, for example the grill cook grills the food and passes it over to the middleman who then plates it up. In this scenario, the employee handling the raw proteins on the grill is different from the employee who assembles the final plates. Most TFE operations tend to be very small and have a limited number of employees, oftentimes just one. This limitation forces TFE employees to multitask, thus increasing the chances of food contamination. Behnke and Seo's (in press) study found a significant difference in the frequency of hand washing violations based upon the number of employees; TFE operations with less employees had more hand washing violations than those operations who had more employees, consequently, more segregation of duties.

The use of food samples as a marketing tool in farmers' markets is common; however, this, too, is not without risk. Unpasteurized cheese sold and sampled at a farmers' market in Alberta, Canada was contaminated with *E. coli* O157:H7 gastroenteritis. While the contamination seemed to have occurred in the processing plant, samples passed out at the market were also implicated in the outbreak (Honish et al., 2005). Samples need to be protected from bare hand contact by both the employees and the customers, and ideally, should be covered to prevent any incidental contamination, such as might occur with coughing or sneezing.

13.6 TIME AND TEMPERATURE CONTROL

Time and temperature abuse occurs when food is allowed to rest in the danger zone. By keeping food below 41°F or above 135°F, cooks are able to avoid the danger zone, where pathogenic microorganisms thrive. In permanent facilities, this is achieved through the use of refrigeration or cooking and holding units, such as warmers, ovens or steam tables. In outdoor venues, this is much more challenging. Some markets and fairs have access to electricity, allowing the use of refrigeration; however, this requires the purchase of expensive equipment on the part of the vendors. Since TFE operations are generally operated by entrepreneurs on a very tight budget, the purchase of a mobile refrigerator could easily exceed their budgetary limits. For this reason, many TFE operations use passive cold holding systems, such as coolers, ice buckets, and refrigerated gel packs. The challenge with these methods is that they are inconsistent; how well they keep food chilled is dependent upon the exterior temperature, the frequency with which employees open and close the coolers, how much product is being held cold and how long the food is held throughout preparation and service.

Refrigeration issues have been observed in a number of studies focusing on TFE operators. Forty seven percent of cheese vendors in 19 farmers markets near Ontario were found to have refrigeration challenges (Teng et al., 2004). The Board of Health in King County, Oregon, identified 51 refrigeration violations in farmers' market inspections over an eight-month period (2011). A study examining the food hygiene and safety at farmers' markets specifically noted that the lack of temperature control during transport and display of food was of concern (Worsfold et al., 2004), while Smathers' (2012) study of farmers' market vendors observed time-temperature controlled food violations, most commonly with food samples. The author also noted that throughout the study only one thermometer was observed in use (Smathers, 2012); without thermometers, employees have no way of monitoring food temperatures. Anecdotally, a news report from 2011 noted that uninspected raw chicken contaminated with salmonella (while not illegal, this points to the responsibility of the consumer to properly cook meat and poultry) and unrefrigerated raw eggs sitting out in 90°F + temperatures were found in farmers' markets in Washington, DC, sourced from small farms exempt from federal food safety inspection (French et al., 2011).

At the other end of the danger zone, electricity would permit vendors to use portable electric ovens, but again this is an expensive proposition.

There are ovens and stoves that work off propane, which may be acceptable based upon market rules; however, these are large, expensive and more complicated to transport, set up and take down at the end of the day. Hot holding can also be achieved through the use of canned fuel and chafing dishes, as long as they are not exposed to windy conditions, which minimize the effectiveness of canned fuel and can also present a fire hazard. Grills, charcoal or propane, are often used to cook food, and if the grill is large enough, can also be a means of holding food above 135°F. If open flame heat sources are used, a properly rated fire extinguisher should be easily accessible.

Keeping cold food cold and hot food hot is a key principle of professional food preparation; however, this is easier said than done in a temporary foodservice establishment. Since vendors are not always certified for food safety, they need to be made aware of the importance of this principle. Thermometers must be made available to employees and should be used to verify temperatures throughout the day, while temperature logs should be maintained starting with the point when the food is first removed from refrigeration for transport to the market. Potentially hazardous food that is in the danger zone where ambient temperatures may exceed 70°F, a situation common in outdoor venues, for more than four hours needs to be discarded. Maintaining temperature logs will help vendors make this determination, and is a requirement in the FDA Food Code for food when time without temperature control methods are applied (FDA, 2009b).

13.7 MAINTAINING CLEAN AND SANITARY FOOD CONTACT SURFACES

Hand washing is a means of reducing pathogenic microorganisms levels on the surface of an employee's hands and arms. Preventing bare hand contact with RTE foods is meant to minimize the transfer of any remaining pathogens from the employee's hands to food that will not receive any additional cooking. Time and temperature control is intended to minimize the length of time those pathogens that do manage to come into contact with RTE foods stay in ideal incubation temperature of the danger zone, thus minimizing the amount of replication. The last FDA operational area identified as most in need of improvement was the cleaning and sanitizing of food contact surfaces (FDA, 2009a), another means of minimizing the transfer of pathogens from one product to another – cross contamination.

Cross contamination occurs when pathogens are transferred from one surface or food to another (NRAEF, 2012). In permanent facilities, cooks tend to have adequate space to work and multiple cutting boards designated for different food production tasks. In temporary foodservice establishments, however, space is at a minimum, and with few employees, generally only one or two cutting boards. In a permanent establishment, the production surface is likely to be stainless steel; however, in a TFE operation, it may well be a wooden portable. These limitations represent an increased risk of cross contamination; without segregated duties and managerial diligence, food produced and served in a TFE operation could easily lead to a foodborne illness outbreak. Burt et al.'s (2003) study of mobile food cart vendors found that four out of ten vendors clearly cross contaminated bread products served to customers with bacteria from raw poultry or beef. The FDA found a noncompliance rate of 35% for contaminated equipment in fast food restaurants, the closest analogy to most temporary foodservice establishments (FDA, 2009a), while 29% of cheese vendors at Ontario farmers' markets were observed with work surfaces that were unclean (Teng et al., 2004).

The four operational areas of focus: (1) hand washing, (2) handling RTE foods, (3) time and temperature control, and (4) maintaining clean and sanitary food contact surfaces, represent steps in either minimizing pathogenic contamination or microbial growth. Simply addressing these four areas would go a long way towards ensuring safe food production and service in temporary foodservice establishments. This will benefit customers by reducing their exposure to foodborne illness, vendors by reducing liability, and the market by protecting its reputation (Beecher, 2013); a foodborne illness outbreak associated with a specific market and vendor could have a significant impact upon patronage and commercial viability. Hofmann et al. (2009) found that smaller markets were more at risk than larger markets in that they have less insurance coverage in case of a foodborne illness outbreak. Such an outbreak in a small market could easily lead to the market's closure.

13.8 RECOMMENDATIONS

Producing and serving food in a temporary foodservice establishment clearly comes with a set of unique challenges. Yet, this is a growing and popular business, and vendors have an obligation to protect the integrity of the food and the health of the population they serve. What follows is a series of recommendations designed to help TFE vendors protect their customers and their business.

Hand washing must become second nature to anyone who works with food, and TFE vendors are not exempt. Figure 2 provides an example of an appropriate hand washing station.

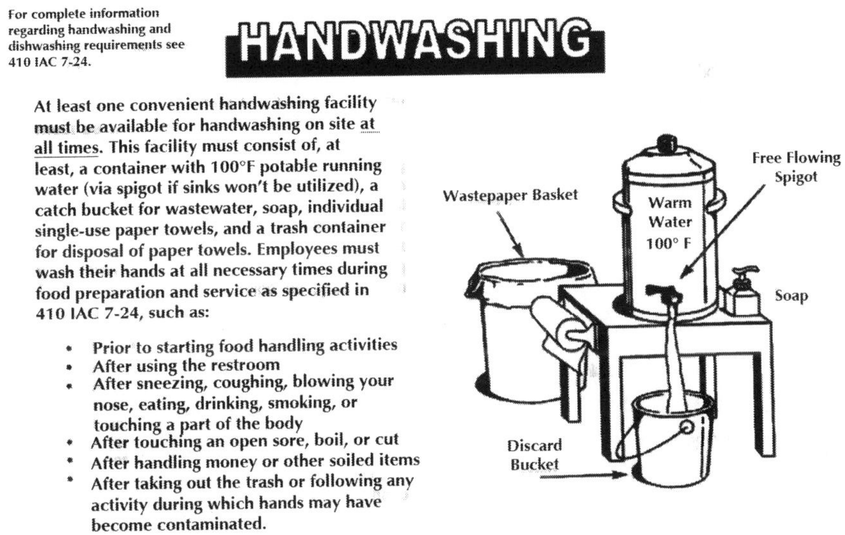

FIGURE 2 Example of a Temporary Hand Washing Station (reprinted with permission of Dr. Jennifer Dennis from Purdue Extension Knowledge to Go Brochure EC-740).

It is important to note that this arrangement needs to be in close proximity to the food production space, and that the manager needs to mandate and enforce its use. Given the seasonal, part time nature of TFE operations and the critical role that food workers play in preventing the transmission of foodborne illnesses, the FDA notes that it is the responsibility of the supervisor to communicate and demonstrate proper hygiene practices (1998).

Observational studies have identified employees violating basic food handling protocols, such as handling money, touching personal belongings or coughing and sneezing (Behnke and Seo, in press; Smathers, 2012) or handling food with visibly dirty hands without subsequent hand washing (Burt et al., 2003). These behaviors can be traced to a lack of training and managerial reinforcement. The FDA's report (2009a) examining foodborne risk factors concluded that the presence of sanitation-certified food managers correlated significantly with improved compliance rates; while this particular conclusion pertained to full-service restaurants, the concept easily translates to temporary establishments; managers need to (1) know the rules, (2) teach the rules, and (3) enforce the rules. However, as Brown noted, farmers' market vendors come from many different disciplines and occupations (2002), so it is conceivable that their knowledge levels with regards to food safety are widely divergent. Perhaps, since TFE operations are limited in scope as compared to permanent, full service restaurants, they may not need to go through a full food safety certification course; a course based upon the fundamentals as experienced in temporary operations may be easier to learn and follow, and therefore, more effective.

At the same time, the market managers represent another defense against foodborne illnesses. The FDA (2009a) holds that market managers should be responsible for ensuring appropriate operational procedures, such as the proper storage of food, prevention of cross contamination and proper cleaning and sanitizing, are followed. At the very least, market managers should be trained in food safety and regularly conduct training, monitoring and verification. The challenge here, as Ragland and Tropp (2009) noted, is that many market managers are volunteers or part time employees, making it difficult to establish and maintain a basic level of sanitation.

From a customer standpoint, there is a clear need for properly stocked and maintained public hand washing facilities (Worsfold et al., 2004) that are accessible to all and in close proximity so as to be convenient and more likely used (FDA, 1998). These facilities need to be cleaned on a regular basis, and basins and wastebaskets should routinely be cleaned and sanitized. Also, any signage that is placed in public view should consider the languages of the majority of patrons in order to be most effective (Smathers, 2012).

Handling ready-to-eat foods is a principle concern, both in permanent and temporary foodservice establishments. The basic goal of not allowing bare hand contact with food that will receive no further cooking is rooted in preventing cross contamination. In a TFE, this means that there needs to be a barrier between bare hands and RTE foods, for example, gloves or utensils. If gloves are to be the principle barrier used, hands still must be washed prior to putting them on and then they must be changed between tasks. If an employee is solely cooking burgers on a grill, then that employee need not change gloves between every hamburger he or she handles; however, as soon as that employee moves to another task, such as adding cheese or plating up the burger, then a glove change with appropriate hand washing is warranted. This is obviously a challenge in TFE operations, where oftentimes there is only one employee, so it may be more realistic to use utensils dedicated to specific tasks. These utensils need to be changed a minimum of every four hours, or as needed if soiled beyond normal operations, for example if a pair of tongs falls to the ground. Ideally, vendors should supply extra utensils to cover the day's work requirements (Hofmann et al., 2007). Should this not be feasible, then a proper dishwashing station, consisting of a three compartments (bus tubs are commonly used) set up in the following sequence: (1) wash, (2) rinse and (3) sanitize needs to be provided (see Figure 3). Since hot water sanitizing is not feasible in a TFE, cold-water sanitizing processes using chlorine (25–200 ppm), quaternary ammonia (200 ppm) or iodine (12.5–25 ppm) need to be used, and equipment air-dried. In this situation, chemical test kits must also be provided and properly used at each location (Hofmann, Dennis, Gilliam, and Vargas, 2007).

With regards to food samples, vendors should use toothpicks or tongs, or alternatively place the samples on small plates, napkins, or similar small serving containers so that there is no bare hand contact with the sample. The samples should be covered, either by a hard surface, such as a clear plastic display, or by plastic wrap when not actively being passed out. Samples should be frequently changed so that they present better, and also don't run the risk of time and temperature abuse.

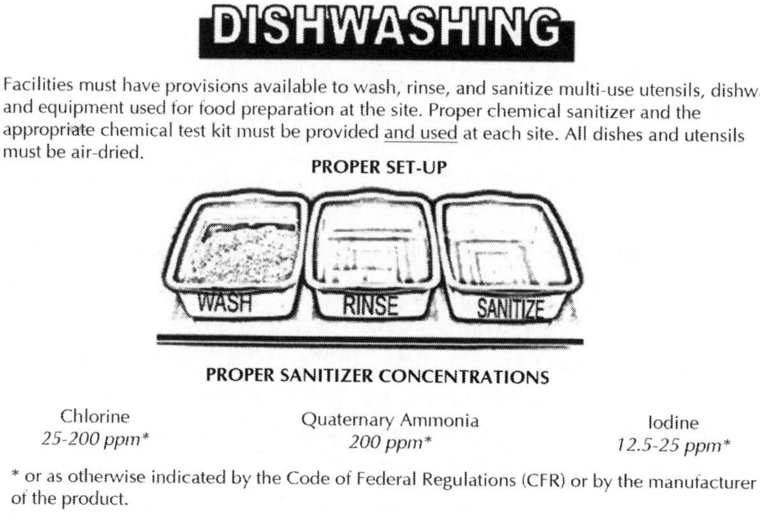

FIGURE 3 Example of a Temporary Dish Washing Station (reprinted with permission of Dr. Jennifer Dennis from Purdue Extension Knowledge to Go Brochure EC-740).

Time and temperature control practices are designed to inhibit the growth of pathogens. Given the outdoor, exposed nature of TFE operations, temperature control is especially important, and correspondingly difficult. Potentially hazardous foods, which include animal foods that are raw or heat treated; plant foods that are heat treated; specific raw plant foods, such as raw seed sprouts, cut raw melons, leafy greens or tomatoes; or untreated garlic-oil mixtures are examples of potentially hazardous foods (FDA, 2009a) that need to be controlled for time and temperature exposure.

In an ideal situation, TFE operations should have access to electric power and refrigeration in order to maintain a constant temperature below 41°F; however, that is not commonly found in U.S. temporary venues. In the absence of refrigeration, other means must be used. If cooked prior to arrival at the market, food destined for service in a TFE needs to be properly handled, which in this case means quickly chilled within two hours from 135°F down to 70°F and then within four hours from 70°F to 41°F or below. This can be achieved by cooling food in shallow pans under refrigeration, dividing the food into smaller portions, and placing the food into an ice water batch and stirring frequently (FDA, 2009a).

In preparation for, during transport and while on site, potentially hazardous foods must remain in a refrigerated state, most commonly achieved by using insulated coolers. Hofmann et al. (2007) stated that if ice is to be used, the food cannot come into direct contact with the ice, and containers must be sealed to prevent water entry. Freezer gel packs can be used so long as they have been adequately prefrozen. Dry ice is also an acceptable means of chiling food; however, caution is warranted so that employees do not come into direct contact with the dry ice, which can be dangerous. Ice that is used to refrigerate food products cannot be used for beverage service; ice meant for consumption must come from an approved source, must be held in closed containers, and handled with approved utensils. Ice intended for consumption cannot be used for food or beverage storage (Hofmann et al., 2007).

Upon commencing food preparations at TFE site, food destined for hot service must be rapidly reheated above 165°F for a minimum of 15 sec in accordance with the Food Code (FDA, 2009a). Once reheated, the food should be immediately served, or alternatively, held hot (above 135°F). This can be achieved through the use of crock pots or slow cookers (assuming that power is available), or in chafing dishes heated with canned fuel. If chafing dishes are used, consider wrapping foil around the bottom of the chafer as a wind break; a breeze blowing over the open flame reduces its heating effectiveness, and also represents a fire hazard. Anytime open flames, such as chafing dishes and charcoal grills, are present in a TFE, an appropriately rated fire extinguisher needs to be present.

Food that has been reheated to 165°F for more than 15 sec may then be held as described above 135°F for service for a maximum of four hours before needing to be discarded. Food that is intended to be served in a chilled state, for example, prepared salads or cut fruit, may be held for a maximum of six hours once they exceed a refrigerated (below 41°F) state, after which it must be discarded. These procedures follow the use of time as a public health control and are found in Section 3–501.19 of the Food Code (FDA, 2009a). Given the operational constraints associated with temporary foodservice establishments, many operators will find themselves working under this section of the code. If this is the case, then vendors and employees need to have in place written procedures available for health inspectors detailing compliance, and also need to clearly label the food to indicate when it will exceed these time parameters. One point for emphasis, it is critical to discard the food once time has expired. Given that many markets run for 4–6 h, this means that vendors and employees must not retain the food for future use on another day, or they will put the consumer, and their business, at risk. Lastly, temperature control is impossible if thermometers are not available. Vendors need to provide properly calibrated thermometers and instruct employees in their use.

Maintaining clean and sanitary food contact surfaces is an important step in minimizing cross contamination; however, this is much more challenging in a temporary foodservice establishment than in a permanent facility. Addressing this challenge begins with the menu and recipes. Vendors need to carefully consider what they intend to serve and should limit the number of raw ingredients, or at least, minimize the amount of food preparation that actually occurs on site, so that employees can focus more on assembly rather than from scratch-preparation. More than one cutting board is needed; should one become soiled or an employee switch to a different task, then a clean board would be available. Sanitizing wipes should be available in the vendor stall, and used to clean the work surfaces regularly throughout service. Extra utensils and production equipment, such as bowls or pans, need to be brought to the site so that utensils can be rotated out of use as needed throughout the day. Garbage needs to be located away from the production space, and if possible, sealed so that it does not attract flies. Vehicles tend to be dirty, so employees need to be aware that if they have to get anything from the vehicle, they need to wash their hands before returning to food preparation or risk cross contaminating the food that they handle.

At the vendor's stall, food needs to be covered, starting with an awning or tent as a basic precaution, and also a means of keeping the sun off food products. Within the stall, in addition to temperature control, food needs to be covered with barriers, such as plastic wrap, foil or held in containers with sealable lids in order to protect it from environmental contaminants, such as dirt or dust. Food that will not receive any further preparation, for example, breads or pies, should be wrapped prior to bringing them to the outdoor site. Pets should not be allowed in the food production areas. Since many vendors are family operations, children are commonly observed in stalls, however, unless the children are properly trained, supervised and legally allowed to work, they should not be involved in food production. Food needs to be stored off of the ground and when at all possible, barriers should be placed between the customer and the food (Teng et al., 2004).

Employee behaviors must also be addressed. Employees involved in food production should wear hats or hairnets, and need to understand that every time they touch a delivery vehicle or other unclean surface, they must wash their hands prior to returning to food preparation. Employees who are ill should not be allowed to work around food, as coughing and sneezing represent a distinct opportunity for cross contamination. Employees need to avoid shaking hands with customers, touching pets or handling money prior to preparing food; if they do, then they need to wash their hands prior to returning to food preparation. Employees should not eat or smoke while working at in the booth and must be trained in proper glove usage (Beecher, 2013). These employee behaviors highlight the need for properly training TFE employees.

Experiencing a foodborne illness is not pleasant, in fact, it can be downright lethal. Markets, vendors and employees associated with food production are all obligated to take preventative measures. The risks associated with foodborne illnesses are too great and the consequences too severe, so it is in the interest of all to be aware and diligent in the prevention of foodborne illnesses.

13.9 REFLECTION

1) Permanent restaurants and temporary foodservice establishments all produce food; yet, are uniquely different. Compare and contrast the two types of operations identifying three unique differences between them.
2) Design a menu specifically for a temporary foodservice establishment that minimizes the risk of foodborne illness. Identify three specific steps related to your menu that achieve this goal.
3) Using the menu developed above, identify all of the points at which:
4) Hand washing should occur
   a. Food may be at risk of time and temperature abuse
   b. Cross contamination could happen

5) Using the information learned in this chapter, develop a food safety brochure designed around the unique requirements of temporary foodservice establishments.

13.10 ADDITIONAL RESOURCES

- FDA Food Code http://www.fda.gov/Food/GuidanceRegulation/RetailFoodProtection/FoodCode/default.htm
- CDC Foodborne Outbreak Online Database (FOOD) http://wwwn.cdc.gov/foodborneoutbreaks/Default.aspx
- USDA Agricultural Marketing Service http://www.ams.usda.gov/AMSv1.0/FARMERSMARKETS
- Farmer's Market Food Safety http://www.foodsafety.gov/blog/farmers_market.html
- Purdue University Extension http://www3.ag.purdue.edu/extension/Pages/default.aspx
- North Carolina's Good Farmers Market Practices http://www.ncgoodfarmersmarketpractices.com/

KEYWORDS

- **foodborne illness**
- **potentially hazardous foods**
- **ready-to-eat foods**
- **temporary foodservice establishments**

REFERENCES

Association of State and Territorial Health Officials (ASTHO). (2012, March). *Budget cuts continue to affect the health of Americans: Update March 2012*. (Research Brief). Washington, DC: ASTHO.

Beecher, C. (2013, April). Farmers market vendors, managers get up to speed on food safety ABCs. *Food Safety News*. Retrieved from http://www.foodsafetynews.com/2013/04/farmers-market-vendors-managers-get-up-to-speed-on-food-safety-abcs/#.UflKx0rpxXg

Behnke, C., and Seo, S. (in press). Using smartphone technology to assess the food safety practices of farmers' markets foodservice employees. *Journal of Foodservice Business Research.*

Brown, A. (2002). Farmers' market research 1940–2000: An inventory and review. *American Journal of Alternative Agriculture, 17(4),* 167–176. doi: 10.1079/AJAA200218

Burt, B., Volel, C., and Finkel, M. (2003). Safety of vendor-prepared foods: Evaluation of 10 processing mobile food vendors in Manhattan. *Public Health Reports (1974-),* 118(5), 470–476

CDC. (2013). Foodborne Outbreak Online Database. Retrieved from http://wwwn.cdc.gov/foodborneoutbreaks/Default.aspx on 7/26/13.

Conner, D., Colasanti, K., Ross, R. B., and Smalley, S. C. (2010). Locally grown foods and farmers markets: Consumer attitudes and behaviors. *Sustainability,* 2(3), 742–756. doi: 10.3390/su2030742.

Food and Drug Administration (FDA). (1998). Guide to minimize microbial food safety hazards for fresh fruits and vegetables. Washington, DC.: Center for Food Safety and Applied Nutrition.

Food and Drug Administration (FDA). (2009a). FDA Report on the occurrence of foodborne illness risk factors in selected institutional foodservice, restaurant, and retail food store facility types (Report No. 3). Silver Spring, MD.: FDA.

Food and Drug Administration (FDA), U.S. (2009b). FDA Food Code. Retrieved from http://www.fda.gov/Food/GuidanceRegulation/RetailFoodProtection/FoodCode/default.htm.

Food and Drug Administration (FDA), U.S. (2011). Fresh strawberries from Washington county farm implicated in E. coli O157 outbreak in NW Oregon. Safety Recall. Retrieved from www.fda.gov/Safety/Recalls/ucm267667.htm.

French, E., Kramer, M., and Clark, M. (2011). Food safety issues flare in the shadow of U.S. capitol. *News 21.* Retrieved from http://foodsafety.news21.com/2011/local/capitol-poultry/index.html.

Hamilton, N. D. (2002). *Farmers markets rules, regulations and opportunities.* National Center for Agricultural Law Research and Information.

Hofmann, C., Dennis, J. H., and Marshall, M. (2009). Factors influencing the growth of farmers' markets in Indiana. *Hort Science, 44(3),* 712–716.

Hofmann, C., Dennis, J., Gilliam, A. S., and Vargas, S. (2007, March). *Food safety regulations for farmers' markets (EC-740).* West Lafayette, IN.: Purdue Extension.

Honish, L., Predy, G., Hislop. N., Chui, L., Kowalewska-Grochowska, K., Trottier, L., Kreplin, C., and Zazulak, I. (2005). An outbreak of *E. coli* O157:H7 hemorrhagic colitis associated with unpasteurized gouda cheese. *Revue Canadienne de Santé Publique.* 96(3). 182–184.

Indiana Code § 16–42–5-29 (2012). Retrieved from http://www.in.gov/legislative/ic/code/title16/ar42/ch5.html.

Indiana State Department of Health (I.S.D.H.). (2006). Certificate of food handler requirements: Title 410 IAC 7–22. Retrieved from http://www.in.gov/isdh/files/FoodHandlerFinal.pdf.

King County Board of Health. (2011, September 15). *Meeting minutes.* Retrieved from http://blog.seattlepi.com/thebigblog/files/2011/09/BOH_9_15_11_meeting_packet.pdf

Lee, J., Almanza, B., and Nelson, D. (2010). Food safety at fairs and festivals: Vendor knowledge and violations at a regional festival. Event Management 14: 215–223.

Levi, J., Segal, L., St. Laurent, R., and Lang, A. (2012). Investing in America's health: A state-by-state look at public health funding and key health facts. Issue Report. Washington D.C.: Trust for America's Health.

Mays, G. P., and Smith, S. A. (2009). Geographic variation in public health spending: Correlates and consequences. *Health services research*, 44(5p2). 1796–1817.

Ragland, E., and Tropp, D. (2009). USDA national farmers market manager survey 2006. Retrieved from http://www.ams.usda.gov/AMSv1.0/getfile?dDocName=STELPRDC5077203.

Safefood News. (1998). Fairs, farms, and petting zoos…A source of foodborne illness? 3(1). Colorado State University Extension Newsletter. Retrieved from http://www.ext.colostate.edu/safefood/newsltr/v3n1s 09.html.

National Restaurant Association Educational Foundation (NRAEF). (2012). ServSafe 6th edition Coursebook. Chicago, IL.

Smathers, S. A. (2012). *Evaluation, development, and implementation of an education curriculum to enhance food safety practices at North Carolina farmers' markets* (Master's thesis). Retrieved from http://repository.lib.ncsu.edu/ir/handle/1840.16/8094.

Teng, D., Wilcock, A., and Aung, M. (2004). Cheese quality at farmers markets: Observation of vendor practices and survey of consumer perceptions. *Food Control*, 15(7), 579–587. doi:10.1016/j.foodcont.2003.09.005.

USDA. (2011). Farmers Market Growth: 1994–2011. Agricultural Marketing Service. Retrieved from http://www.ams.usda.gov/AMSv1.0/ams.fetchTemplateData.do?template=TemplateS&navID=WholesaleandFarmersMarkets&leftNav=WholesaleandFarmersMarkets&page=WFMFarmersMarketGrowth&description=Farmers%20M arket%20Growth&acct=frmrdirmkt.

Worsfold, D., P. Worsfold, and C. Griffith. 2004. An assessment of food hygiene and safety at farmers markets. International Journal of Environmental Health Research 14: 109–119.

CHAPTER 14

HOTEL GUEST ROOM CLEANING: A SYSTEMATIC APPROACH

SHERYL F. KLINE, PhD

Professor, University of Delaware

BARBARA A. ALMANZA, PhD, RD

Professor, Purdue University

JACK "JAY" NEAL, PhD

Assistant Professor, University of Houston

CONTENTS

The food service industry has had a long-standing awareness and use of food sanitation practices. The practice of food safety has a basis in scientific research and includes all aspects of food handling from the farm to the table. The hotel industry, on the other hand, has had a very different approach to guest room cleaning and sanitizing. Guest room cleaning practices range from the notion that a room that looks clean is clean, to one that resembles the procedures used by hospitals to clean and sanitize patient rooms. The need to prevent the spread of disease is a primary motivator for cleaning practices used in hospital patient rooms. In large part, prevention of foodborne illness is also a primary motivator in restaurants' cleaning practices. In general however, the need for disease prevention has not been as clearly recognized for hotel rooms.

Similar to consumers in restaurants, there are cleaning issues that are clearly important to hotel guests. Hotel rooms, as well as restaurants, must not only look clean, they must be sanitary for their guests. Consumer perception of the clean (or unclean) condition of both restaurants and hotel rooms affects consumers' opinions of that property. So the question remains, what is a best or better practice for hotel guest room cleaning?

14.1 LITERATURE REVIEW

14.1.1 THE ISSUE BOTH REAL AND PERCEIVED

One of the biggest challenges for human resource departments is ensuring that employees are properly trained and have the knowledge and tools necessary to perform their jobs efficiently and effectively. Most room attendants are trained on how to clean a room and the focus is on the esthetics of the room but a room that looks clean does not ensure the sanitary conditions of the room. Improperly cleaned guestrooms may lead not only to poor hotel reviews and a lack of guest satisfaction, but also to unhealthy conditions and the possible spread of disease.

Some common diseases that guests can be exposed to in hotel rooms include the common cold, athletes' foot, influenza, herpes, staphylococcus infections, streptococcus infections, hepatitis, salmonellosis, trench mouth, intestinal flu, mononucleosis, tuberculosis and Legionnaires' disease (Dykstra, 1990). Disease outbreaks in hotels have been reported widely in news stories and are therefore salient issues for travelers.

Legionnaires' disease, an uncommon pneumonia with serious consequences especially for older age groups (CDC, 2012), was identified in 1976 at a hotel in Philadelphia (Freije, 2006), but an outbreak has also been reported in a 4000-room hotel on the Las Vegas Strip in July 2011 (Ritter, 2011). In August 2012, eight people became infected, two of whom died from an outbreak of Legionnaires' disease linked to a hotel in downtown Chicago (Smith, 2012).

Another prominent disease linked to hotel outbreaks is the Severe Acute Repertory Syndrome (SARS) virus. The origin of the spread of SARS was sourced to a single guest and a doctor who spent the night in a Hong Kong hotel in February 2003 (Bell, 2004; WHO, 2004). This resulted in the spread of a global flu epidemic that devastated the tourism industry in Asia. Another noteworthy disease publicized in the media is Norovirus also called Norwalk-like virus. This virus which has been linked to many prominent cases in the cruise ship industry has also been found in hotels. This virus is highly contagious and resilient (AP, 2004; CDC; 2013). Last, but not least, bed bugs (*Cimex lectularius)*, parasites of warm-blooded animals traveling around the world in luggage and on clothing (Shoemaker, 2011) are generally believed not to spread disease but have been considered to possibly harbor diseases such hepatitis (James, 2003) and have been found to be vectors for methicillin resistant *Staphylococcus aureus* (MRSA) (Lowe and Romney, 2013). The threat of catching a disease in a hotel room is certainly real but may be quite rare. However, media attention to hotel disease outbreaks have created acute awareness of these perceived and real threats among many hotel guests who previously would not have thought twice about cleaning in hotel rooms.

14.1.1.1 CLEANING, SANITIZATION, AND THE POTENTIAL FOR THE SPREAD OF DISEASES IN HOTEL ROOMS

Cleaning refers to the removal of visible dirt while sanitizing describes the methods needed to reduce microorganisms to safe levels so individuals will not become sick. Improper cleaning practices and the lack of a systemized procedure used in the hotel industry can also result in the spread of infectious diseases. The importance of a clean and healthy guestroom environment is clear to the industry and is well described in housekeeping textbooks (Jones, 2007; Nitschke and Frye, 2008; Casado, 2011). Although the act of cleaning a hotel room is repeated by room attendants thousands of times a day in hotels around the world, the systems and techniques for guestroom cleaning procedures still differ greatly from one hotel to another. Also changing is the time it takes to clean a standard hotel room. The number of minutes allowed has been shrinking as a result of economic pressures and changes in room design. In some cases, some hotel companies have placed more emphasis on quantity and efficiency, rather than quality.

Cleanliness is an important factor for a hotel customer's satisfaction. A study by Dolnicar (2002) showed that cleanliness was ranked as the most important factor considered by business travelers when selecting a hotel. Unfortunately, this study also pointed out that hotel cleanliness and hygiene have been identified by the same population as the most disappointing factor when staying in one, two, three and four star hotels (Dolnicar, 2002). A 2011 study conducted on both full-service and limited-service hotel reviews posted on the travel blog TripAdvisor.com showed that in 15.5% of the cases, cleanliness was the overall determinant of customer delight, second only to a general concept of customer service (Magnini et al., 2011).

At the same time, studies have shown that it is possible for guests to be ill or harmed by diseases and viruses while staying in a hotel. In 2007, a study conducted in hotel rooms on the transmission of rhinovirus, a common cold virus, concluded that 35% of the examined sites were contaminated, particularly places such as door handles and pens provided by the hotel, and that the virus could be transmitted to the next guest in 33% of the cases (Winther, et al., 2007). While this study did not touch upon the issue of cleaning and sanitizing such contaminated rooms, it can be concluded that without proper procedures, contaminated hotel rooms could be the transmission environments for such viruses. Fairly or unfairly, hotels have been linked to illness and the spread of disease.

Several studies have shown that diseases can be transmitted through contact with contaminated surfaces in hotel rooms. Winther et al.'s (2007) study demonstrated that it is possible for the common cold virus to survive on guestroom surfaces for as long as 18 h. In addition to the common cold virus, there are eleven11 known microorganism that guests and housekeeping employees may come in contact within guestrooms (Jones, p. 283) including norovirus. This infectious disease causes diarrhea and vomiting and can survive on surfaces such as door knobs and other high touch areas for a long period of time (CDC, 2012). One of the more well-known outbreaks of norovirus associated with hotels occurred in November 2004 at the Flamingo and Monte Carlo hotels in Las Vegas where over 1,200 employees and guests were afflicted with this virus (Jones, 2006).

Although cleaning models have been suggested, and possible disease transmission has been demonstrated, little research has been done on how to empirically test and measure cleanliness in hotel rooms. In 2011 one study attempted to identify the most contaminated surfaces in a hotel room (Almanza et al., 2011).

The selection of surfaces tested was based on previous studies of environmental contamination of high touch areas (Shoemaker, 2007; Winther et al., 2007, 2011). These surfaces included: internal door handle of the room, light switch closest to the entrance, carpet at the entryway of the hotel room, headboard, bedside lamp switch, telephone keypad, TV remote keypad, bathroom internal door handle, bathroom floor, bathroom faucet, bathroom sink, shower floor, toilet paper holder, toilet basin, glove from maid cart, towel from maid cart, mop from a maid cart, and lip of a nondisposable coffee mug (if available). The results are reported in Figure 1. This figure shows the number of coliforms detected on each surface. Coliform bacteria are present in the environment and are also found in the feces of warm-blooded animals and humans. Although the presence of coliform bacteria does not necessarily indicate disease causing bacteria are present, they indicate that there is the potential for disease causing bacteria to be present through lack of cleaning. The results of this study clearly showed that bacterial contamination does occur in hotel rooms and suggested which surfaces should be assessed in an evaluation of best cleaning practices.

Although this scientific approach is rare in the guest room cleaning literature it is interesting to see that it is a valid approach to determining the levels of clean in a guest room and on surfaces of items in the guest room. In summary, the review of the literature has suggested diseases may be transmitted through contaminated surfaces in hotel rooms. In addition, not all areas of a guest room are equally dirty and cleaning procedures are needed to address this reality. The public has an expectation that having a clean guest room is a basic amenity and clean rooms prevent customer dissatisfaction (Torres and Kline, 2006). Hotel guests care about the cleanliness of a room and clean hotels are a given minimum requirement that guests desire when selecting a hotel and when staying at a hotel.

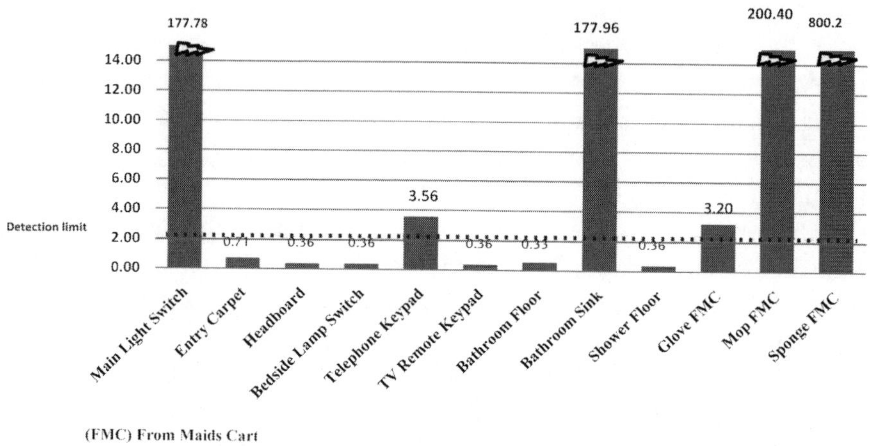

FIGURE 1 Enumeration of Coliforms on Hotel Surfaces.

14.1.2 HOW TO CLEAN A GUEST ROOM

While articles do not fully address the issue of cleaning contaminated rooms, it can be concluded that without proper cleaning and sanitizing procedures, contaminated hotel room surfaces could very well provide the transmission environments for viruses. Proper training and cleaning procedures could limit or prevent the spread of disease. There are also no agreed upon procedures for cleaning and sanitizing guestrooms. A variety of cleaning practices are detailed however in several well-written housekeeping textbooks used by hospitality and tourism management programs and by hotel companies in the U.S.

The most prominent housekeeping textbooks are written by Jones (2007), Nitschke and Frye (2008), and Casado (2011); however, they offer a varied list of methods used to clean rooms. For example, Jones (2007) describes how to disinfect areas of a guestroom and provides a list of steps used to clean these rooms. Nitschke and Frye (2008) have perhaps the most detailed list of steps to clean a guestroom and include a "how to" section for each of the steps, however, they do not go into detail with regards to sanitizing these rooms. Only Casado's (2011) textbook has a detailed checklist that can be used to inspect a room after it is clean.

These textbooks include the cleaning procedures for bathrooms and bedrooms in the hotel guestroom. The books describe the time it takes to clean an average guest room and in general state that housekeepers clean 14–16 rooms per eight hour shift, spending approximately 30 min on each room (Cassado, 2011; Jones, 2007). They also mention specific items within these rooms; a common feature observed was that they are very general and that they leave room for interpretation. Even the one published by the American Hotel and Lodging Association states that the task descriptions are for "illustrative purposes only and should not be construed as recommendations or standards" (Nitschke and Frye, 2008, p. 405).

If mentioned, the suggested cleaning supplies are mostly referred to in very generic terms such as "cleaning solution" or "all-purpose cleaner/sanitizer", "damp cloth" or sometimes "microfiber cloth." Also, it seems to be very common in cleaning guidelines to combine the actual cleaning procedures with housekeeping procedures. Housekeeping procedures have less to do with cleanliness and more to do with neatness and function of the rooms. This would include tasks such as replacing guestrooms supplies or reporting of lost items, malfunctioning equipment or damaged furniture. Housekeeping books typically focus on both cleaning and housekeeping procedures.

Some steps are more commonly mentioned in textbooks in regards to cleaning guestrooms. They are: to start the cleanup with removing trash from the room and stripping the bed; check the mattress for stains (without going into much detail about what these stains may mean and that they could be connected with serious infestations, such as bed bugs); wear gloves and even goggles when cleaning the bathroom or toilet; use specific brushes and cloths for toilet cleaning only, finish the cleanup of a room with vacuuming. It is important to emphasize that these textbooks do not provide a singular approach to the cleaning and sanitizing of hotel guestrooms. In addition, no research has been done to compile the best practices used by hotel companies in cleaning and sanitizing hotel rooms. Therefore actual cleaning methods and ideal cleaning methods to assure safe and sanitary hotel rooms are not clearly understood.

Hotel companies in the U. S. also have defined cleaning procedures and the following is a summary of two major hotel chains' manuals, representing over 43% of the hotel guestrooms in the largest ten hotel companies in the USA ("2011 US Hotel", 2011). For this summary, Kline et al. (2012) performed a content analysis to find similarities in cleaning procedures actually used by the industry. It is interesting to observe that these procedures are much more detailed than those found in the textbooks.

The similarities in the order of the overall guestroom cleaning procedures found in the manuals of the two major hotel chains are as following:

- Positioning the cart in front of the open door of the room;
- Turning on the lights and opening the drapes for maximum light exposure;
- Removing "requested" items (room service trays, etc.) and trash;
- Stripping the bed of used linens;
- Cleaning the bathroom after cleaning the room, but applying the cleaner before cleaning the room to allow the chemical to begin to work;
- Vacuuming from the back of the room towards the door as the final step in the cleaning procedure;
- Adjusting the room temperature as one of the last procedures before leaving the room.

The differences in the order of the overall guestroom cleaning procedures found in the manuals of the two major hotel chains are as following:

- The timing of spraying the bathroom – before or after stripping the bed;
- The timing of dusting the room – before or after making the bed;
- The timing of the last visual check-up/inspection – before or after vacuuming.
- Also, only one of the housekeeping manuals mentioned the following:
- Checking the bed for stains – nevertheless, no further procedure is mentioned if such stains are identified;
- Removing dirty glasses and replacing them with new ones – to be noted here is that the room attendant is not cleaning the glasses and this change in procedure could be the result of the recent negative news expose in an Atlanta hotel (Marco, 2007);

- Cleaning the entrance door, threshold and light – the main light switch was found to have a high bacteria count (Stroia et al., 2012).

Based on the content analysis, the following procedural model (Figure 2) for hotel guestroom cleaning was developed. Overall the cleaning sequence was found relatively consistent within the two housekeeping manuals, the differences being indicated by dashed-outline boxes.:

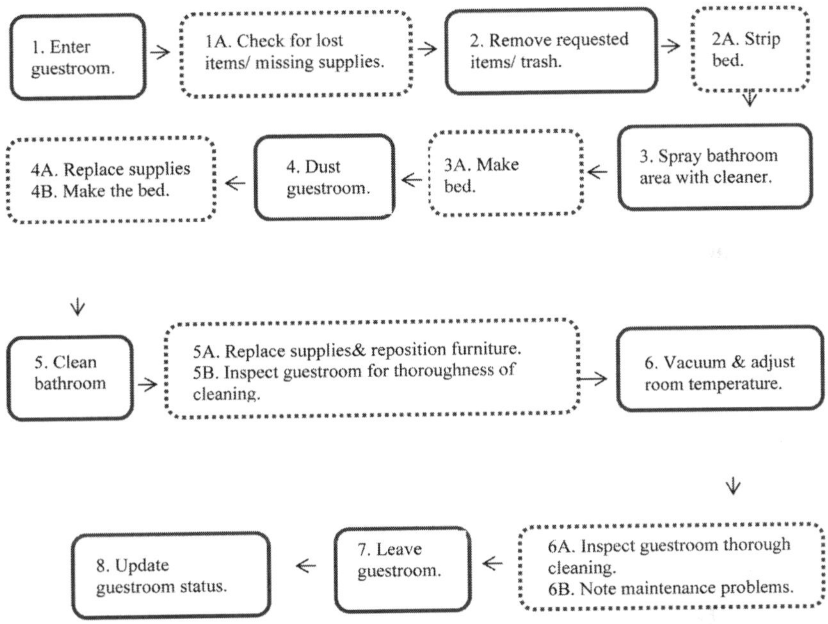

FIGURE 2 Cleaning Sequence for Hotel Guestrooms.

After analyzing the overall guestroom cleaning procedures, the researchers turned to the details of cleaning of bathroom. The reason for choosing this part of the guestroom was that, among the areas individually outlined in these two housekeeping manuals, the bathroom had the highest average of bacteria colony forming units per cm^2 (Stroia et al., 2012).

The similarities in the order of the bathroom cleaning procedures found in the manuals of the two major hotel chains were very few, although they both covered almost identical areas of the bathroom.

Cleaning the toilet as a second major step.

- Replacing the supplies and amenities before cleaning the door and the floor of the bathroom – special emphasis was placed in both manuals on folding the first sheet of the toilet paper to point.
- Cleaning the bathroom door, knobs and floor at the end.

In regards to similarities of the process of cleaning a bathroom, both manuals indicated that wiping surfaces till dry does help prevent mildew. Another interesting similarity was that both manuals referred to the toilet using the word "commode," a word not so commonly used in U.S. English.

The differences in the order of bathroom cleaning procedures found in the manuals of the two major hotel chains are as following:

- The timing of cleaning the tub/shower area – at the very beginning, or in the middle after cleaning the toilet;
- The timing of cleaning the toilet – after cleaning the tub/shower area or after cleaning the vanity, sink and mirror area;
- The timing of cleaning the vanity, sink and mirror area – before or after cleaning the toilet.

Also, only one of the housekeeping manuals mentioned the following:

- Using specific scrubbing pads (3M doodlebag or Scotchbrite #9030);
- Inspecting and removing any residuals from the bathroom ceiling;
- Cleaning the light switch;
- Not using linen as a cleaning rag for the mirror.
- Turning on the heat lamp to help dry the floor quickly.

In addition, one of the manuals intertwined safety procedures such as the use of protective equipment or the mat to prevent slipping, with the actual cleaning procedure. The other manual included a "Safety Tips" box at the very end of the step sequence, which included a more extensive list and a focus on each separate issue. Based on the content analysis, the following procedural model (Figure 3) for bathroom cleaning was developed:

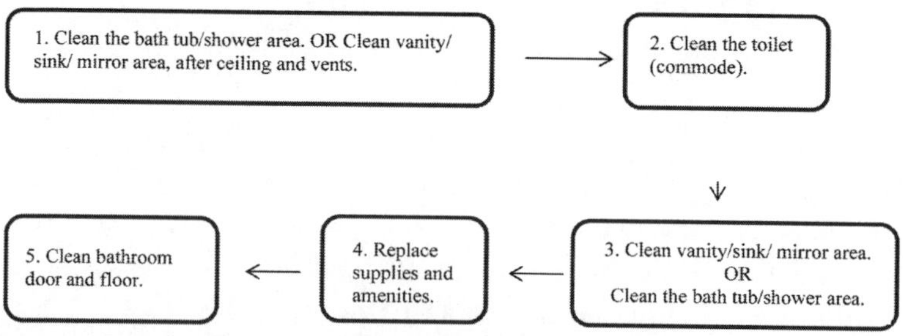

FIGURE 3 Cleaning Sequence for Hotel Bathrooms.

While the analysis of hotel guestroom cleaning procedures based on housekeeping textbooks and hotel manuals has identified similar procedures, it also pointed to the lack of an industry standard cleaning sequence. The order of the cleaning steps that would ensure the safest lodging environment for the guest, along with an efficient operation for the hotel, is still to be determined. In addition research has presented a content analysis that was based on housekeeping manuals from only two major hotels, thus the findings can be considered limited (Kline et al., 2012).

14.1.3 APPLYING FOOD SAFETY PRACTICES TO GUEST ROOM CLEANING PRACTICES

The hotel industry can benefit from practices used by the food service industry and hospital industry. One systematic approach that has been used for years by both hospitals and food service operations is Hazard Analysis Critical Control Points (HACCP). HACCP is a management system where food safety is addressed through the analysis and control of biological, chemical and physical hazards (US Food and Drug Administration, 2011).

The guest room cleaning HACCP model approach for cleaning a guest room bathroom was proposed by Yee (2004). The bathroom cleaning procedure is part of the entire cleaning process performed by room attendants. The follow chart in Figure 4 shows where the bathroom cleaning fits into a proposed cleaning procedural model for guest room cleaning (Yee, 2004). This decision was also based upon the high levels and increasing concerns for hygienic guestroom bathroom cleaning due to increasing customers concerns and negative impressions of dirty restrooms in a hotel or motel (Duncan, 2002).

Hazard analysis is the first part of the HACCP process where one would look at each step of the process and identify the hazards that are likely to be present and evaluate their significance to ensure that adequate measures for their control are in place. A hazard is considered to be significant if it is likely to cause harm to the consumer unless it is properly controlled. All significant hazards are managed through HACCP whereas nonsignificant hazards are controlled by other systems.

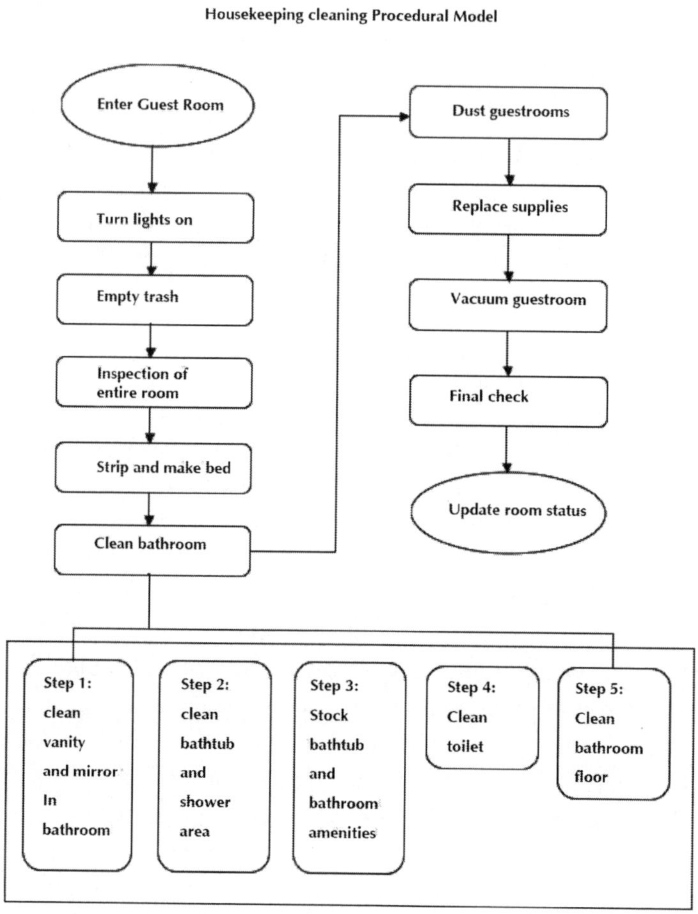

FIGURE 4 Housekeeping Cleaning Procedural Model.

The hazards may be found in the shower area and in the cleaning procedures for the toilet, the bathroom floor, and all the bathroom fixtures, such as sink handles, toilet handles, and light switches. The first hazard is found in cleaning the bathroom. The tub can receive a variety of microorganisms from skin, discharge from lesions and wounds, and occasionally some blood and semen. As stated by The American Podiatric Medical Association (2001), and Federation Internationale des Podologules (2004), *Tinea pedis* (athlete's foot) is a common skin infection caused by a fungus that may be contracted in a hotel shower or where bare feet may come in contact with the fungus. Due to these reasons, the bathtub is also a place where guests may be exposed to fungus and athletes foot (APMA, 2001). The second hazard, cleaning the toilet, is part of the process of cleaning the bathroom. Since the toilet is often exposed to body fluids, such as waste, saliva, urine, feces, semen, and discharges from any wounds or lesions, the toilet could become a hazard. The third hazard, found in cleaning the bathroom is the bathroom floor. The floor can be an area that can receive potential contaminants of microorganisms. Similar to bathtubs, athlete's foot could also be found on the floor and contracted where bare feet may come in contact with the fungus.

The fourth hazard which can be found in all the steps of cleaning the bathroom are the bathroom fixtures, which include sink handles, toilet handles, door handles, and light switches. Research shows that the most common way to get a cold is from someone who has a cold, sneezes on his hands, and then shakes yours. It is also possible to get a cold when a person blows his nose or coughs into a handkerchief and gets some of the germs on his hands, then touches a door knob, and hours later, you touch the door knob and put your fingers in your nose (Dykstra, 1990; Mirkin, 2005; Winthur et al., 2011).

After the analysis of the hazards in the bathroom cleaning procedures, one has to find and identify the critical control points. A critical control point is a step at which a control can be applied, and is essential to prevent or eliminate a food safety hazard or reduce it to an acceptable level. The critical control points identified for the bathroom cleaning procedures are shown in Figure 5.

Once the critical control points have been identified the criteria to distinguish between safe and potentially unsafe conditions for each critical control point needs to be decided. These are represented by parameters called critical limits or safety limits that must be met for each control measure at the critical control point. These factors are related to the type of hazards that the critical control point is designed to control and must be measurable parameters that can be determined and monitored through testing or observation. Criteria that is often used includes measurements of temperature, time, and/or moisture levels. Critical limits must be validated, meaning the criteria specified must control the identified hazard. Each of the critical control points corresponds to the hazards identified in the first step of HACCP. Cleaning the bathtub, toilet, bathroom floor, and bathroom fixtures are all critical control points found in the bathroom cleaning procedures. When setting up procedures for the critical control points, disinfectant should provide the highest level of cleanliness and sanitization so as to make the area as free as possible from disease causing microorganisms. The most commonly used disinfectant in the hotel and restaurant market, for example, is Ecolab's Oasis.

Critical Control Points (CCP) for Bathroom Cleaning Procedures

Shower Area CCP

Clean and Disinfect Bathtub

Make sure that proper disinfectant is used. While cleaning the bathtub, use a new cloth. Use disinfectants for the bathroom. Use single use, disposable gloves and wash hands after finished cleaning and disinfecting the bathtub. After disposing of gloves, hands should be washed with soap in 110° Fahrenheit water for 20 seconds.

Clean Toilet CCP

Clean and Disinfect Toilet

Make sure that proper disinfectant is used. While cleaning the toilet, use a new cloth. Use disinfectants for the bathroom. Use single use, disposable gloves and wash hands after finished cleaning and disinfecting the toilet. After disposing of gloves, hands should be washed with soap in 110° Fahrenheit water for 20 seconds.

Bathroom Floor CCP

Clean and Disinfect Bathroom Floor

Make sure that proper disinfectant is used. While cleaning the bathroom floor, use a new cloth. Use disinfectants for the bathroom. Use single use, disposable gloves and wash hands after finished cleaning and disinfecting the bathroom floor. After disposing of gloves, hands should be washed with soap in 110° Fahrenheit water for 20 seconds.

Fixture CCP
(Sink Handles, Toilet Handles, Door Handles, Light Switches)

Clean and Disinfect Bathroom Fixtures

Make sure that the sanitizing solution is measured properly with acceptable parts per million. While cleaning the bathroom fixtures, use a new cloth. Use single use, disposable gloves and wash hands after finished cleaning and sanitizing the bathroom fixtures. After disposing of gloves, hands should be washed with soap in 110° Fahrenheit water for 20 seconds.

FIGURE 5 Critical Control Points (CCP) for Bathroom Cleaning Procedures.

The next step is to establish a system to monitor control of the critical control points, which will determine if something is going wrong. Monitoring is conducted through a measurement or observation of the process to determine if it is operating within the critical limits at the critical control point. For the bathroom cleaning procedures, ideal testing or monitoring might include the use of chemical tests such as chlorine analysis, or pH and physical parameters such as temperature, time, moisture levels or black lights.

The next step in incorporating HACCP into bathroom cleaning procedures would be to establish corrective actions to be taken when monitoring indicates that a particular critical control point is not under control. When monitoring results show a deviation from the critical limits at a critical control point, corrective action needs to be taken. This would include an assessment of the procedures used, and where appropriate, the chemicals used in cleaning should be tested for proper strength.

Establishing procedures to confirm that the HACCP system is working effectively is the next step. The verification or validation step is used to obtain evidence that the elements of the HACCP procedure are effective. Verification activities could include auditing the HACCP system, reviewing and analysis of data, chemical product sampling, and testing and review of housekeeping training.

The last step in HACCP for cleaning the bathroom would be to establish appropriate documentation concerning all procedures and records appropriate to the HACCP principles and their application. The HACCP system for cleaning the bathroom should be documented and records maintained to demonstrate that it is both properly established and working correctly.

A possible HACCP model is suggested based on the literature review. It is proposed as a possible systematic approach for guest room cleaning. The HACCP procedures used today are mainly, intended for the foodservice industry to ensure food safety and prevent foodborne illness hazards before they happen. This chapter is meant to demonstrate the need for a safer more proactive approach to improve the cleaning procedures in the hotel guest-room and bathroom areas and similarly, would be expected to minimize the spread of disease hazards. This chapter suggests that it is possible to incorporate HACCP into the housekeeping cleaning procedures of the hotel industry, like the food service industry, so that it can provide greater assurance of customer safety and prevent hazards and illnesses before they happen.

A more recent study also explored the use of HACCP in the hotel bathroom cleaning procedures but had the added benefit of using information from the Kirsch et al.'s (2012) study that had scientifically identified and then recommended critical control limits based on hospital critical control points. In this study HACCP was applied to bathroom cleaning as follows:

(1) Hazard analysis is the first part of the HACCP process. It involves identifying the hazards likely to be present, evaluate their significance and ensure that adequate measures for their control are in place. The hazards found in cleaning the guest bathroom could be identified when the room attendant cleans the tub or the toilet, since they are often exposed to body fluids.

(2) Next, the critical control points (CCP) need to be identified. A CCP is an essential step at which control can be applied and is critical in preventing or eliminating a safety hazard or reducing it to an acceptable level (for the bathroom cleaning procedures, CCP should be determined by using a disinfectant for removing microorganisms and bacteria [Kline et al., 2012]).

(3) Critical limits must be established. There are no limits established for the hotel industry. Only one study has attempted to recommend a critical control limit for guest room surfaces. Kirsch et al. (2012) recommended a critical limit of 10 CFU/cm^2 that is double the critical limit for hospitals and food contact surfaces. Hospital critical limits of <5CFUs/cm^2 total aerobic plate counts and <1CFU/cm^2 coliform counts were established by a proposal for HACCP in hospitals (Dancer, 2004). These limits may be too stringent for hotels. To account for this, the Kirsch et al.'s (2012) study further evaluated against critical limits for total aerobic plate counts of <10CFUs/cm^2 and <50CFUs/cm^2 which represent levels 2x and 10x higher than the proposed limits for hospitals.

Table 1 shows the results from Kirsch et al.'s (2012) study where high touch areas of a hotel guest room were tested. This study has shown that it is possible to define critical control limits and test for them in a guest room environment.

(4) Then, a system to monitor control of the CCPs and operation within the critical limits are established. For the bathroom cleaning procedures, the use of ATP meters or environmental sampling for

microbial counts (Kirsch et al., 2012) and physical parameters such as temperature, time, moisture levels or black lights would be ideal.

(5) Following next, the HACCP model requires establishing corrective action to be taken when a deviation from CCPs is observed. An example would be where the appropriate bathroom-cleaning chemical tested for proper strength replaces the all-purpose cleaner.

(6) Establishing procedures for verification that the HACCP system is working effectively is the next step. Verification activities include auditing the HACCP system, review and analysis of data and chemical product sampling, and testing and review of housekeeping training.

(7) The last step in HACCP for cleaning the bathroom would be to establish appropriate documentation concerning all procedures and records appropriate to the HACCP principles and their application. The HACCP system for cleaning the bathroom should be documented and records maintained to demonstrate that it is both properly established and working correctly.

TABLE 1 Number of samples passed based on proposed critical limits for APC.

	#Items Pass/Total Items Sampled		
Critical Limit	**5 CFU/cm^2**	**10 CFU/cm^2**	**50 CFU/cm^2**
Type of Surface			
Room Door Handle	6/9	6/9	9/9
Main Light Switch	6/9	8/9	8/9
Entry Carpet	3/9	4/9	9/9
Headboard	9/9	9/9	9/9
Bedside Lamp Switch	5/9	5/9	7/9
Telephone Keypad	2/9	3/9	8/9
TV Remote Keypad	4/9	4/9	7/9
Bathroom Door Handle	8/9	8/9	9/9
Bathroom Floor	5/9	6/9	8/9
Bathroom Faucet	7/9	8/9	9/9
Bathroom Sink	4/9	6/9	8/9
Shower Floor	7/9	7/9	9/9
Toilet Paper Holder	4/9	7/9	9/9
Toilet Basin	5/9	6/9	7/9

TABLE 1 *(Continued)*

	#Items Pass/Total Items Sampled		
Critical Limit	**5 CFU/cm^2**	**10 CFU/cm^2**	**50 CFU/cm^2**
Mug	2/3	2/3	3/3
Glove From Maid Cart	2/9	4/9	9/9
Mop From Maid Cart	3/8	3/8	4/8
Sponge From Maid Cart	2/8	2/8	4/8
Curtain Rod	6/6	6/6	6/6

Kirsch et al., 2012

The HACCP procedures used today are mainly intended for the food service industry to ensure food safety and prevent hazards before they happen. Future research should explore the use of HACCP as a method to clean the entire guest room. The idea of using HACCP is an attempt to demonstrate that it is possible to provide a safer, more proactive way to improve the cleaning procedures in the hotel guestroom (Kline et al., 2012) and to begin to develop a systematic approach to cleaning a guest room. HACCP is just one possible systematic approach to guest room cleaning. Just as the food service industry's practice of food safety has evolved into a systematic approach to food sanitation so will the hotel industry evolve and develop systematic practices to guest room cleaning and sanitation.

KEYWORDS

- **cleaning solution**
- **commode**
- **food sanitation practices**
- **safety tips box**
- **severe acute respiratory syndrome virus**

REFERENCES

2011 U.S. Hotel Brands Survey results released. (2011). *Hotel Management.Net*. Retrieved from http://www.hotelmanagement.net/2011-us-hotel-brands-survey-results-released

Afiya, A. (2007). How clean is your hospitality business? *Caterer and Hotelkeeper*, 197(4474), 16–17. Retrieved from EBSCO*host*.

Almanza, B., Neal, J., Kline, S., Choi, J.K., Kirsh, K., Sirsat, S., and Stroia, O. (2011, September). *HACCP – It's not just for foods anymore*. Paper presented at the annual meeting of the Indiana Environmental Health Association, Angola, Indiana.

American Hotel and Lodging Associations (2012) Directory of Hotel and Lodging Companies. Available at www.strglobal.com. Accessed August 28, 2012.

American Podiatric Medical Association. (2001). Footfacts. Retrieved from http://www.apma.org/footfacts2001.pdfBell, D and World Health Organization.(2004) Public health interventions and SARS spread, 2003. *Emerging Infectious Diseases* 10(11) retrieved from http://wwwnc.cdc.gov/eid/article/10/11/04–0729_article.htm#suggestedcitation

Bell, D and World Health Organization.(2004) Public health interventions and SARS spread, (2003). *Emerging Infectious Diseases* 10(11) retrieved from http://wwwnc.cdc.gov/eid/article/10/11/04–0729_article.htm#suggestedcitation

Casado, M.A. (2011). *Housekeeping Management* (2nd ed.). Location: John Wiley and Sons, Inc.

Centers for Disease Control (CDC) (2012) *Legionella* (Legionnaires' disease and Pontiac Fever) Retrieved from http://www.cdc.gov/legionella/index.html

Centers for Disease Control (CDC) (2013) *Norovirus trends and outbreaks* Retrieved from http://www.cdc.gov/norovirus/trends-outbreaks.html.

Dancer, S., 2009. The role of environmental cleaning in the control of hospital-acquired infection. The Journal of Hospital Infection 73 (4), 378–385.

Dolnicar, S. (2002). Business travelers' hotel expectations and disappointments: a different perspective to hotel attribute importance investigation. *Asia Pacific Journal of Tourism Research*, 7(1), 29–35.

Duncan, L. (2002, November 6). Bathroom Is Barometer of Hotel Cleanliness According to New Survey. *Hotel Online.* Retrieved from http://www.hotel-online.com/News/PR2002_4th/Nov02_Bathrooms.html

Dykstra, J. (1990). Infection Control for Lodging and Food Service Establishments.
New York, NY: John Wiley and Sons, Inc.

Federation Internationale des Podologues (2004). Athlete's Foot. Retrieved from http://www.fipnet.org/index.php?option=com_content&view=article&id=116:athletes-foot&catid=42:foot-health-care&Itemid=69

Freije, M. (2006). Legionnaires' contamination preventable at reasonable cost. *Hotel and Motel Management*, 221(7), 8–17. Retrieved from EBSCO*host*.

James, M. S. (2003, May 19). Easterners Battle Bloodsucking Bedbugs. *ABC News*. Retrieved from http://abcnews.go.com/US/story?id=90005

Jones, T.J.A. (2007). *Professional Management of Housekeeping Operations* (5th ed.). Hoboken, NJ: John Wiley and Sons, Inc.

Jones, C. (2006) Hotel room health risks difficult to spot. Retrieved http://www.hotel-online.com/News/PR2006_3rd/Sep06_RoomSafety.html

Kirsch, K., Sirsat, S., Stroia, O. Almanza, B. Kline, S. and Neal, J. 2012. A Microbial Analysis of Environmental Surfaces in Hotel Rooms. American Society of Microbiology General Meeting, San Francisco, CA.

Kline, S., Neal, J. Almanza, B and Stroia, O (2012, October) A HACCP based model for hotel housekeeping. Euro-CHRIE Conference Lausanne Switzerland.

Lowe, C and Romney, M (2011) Bedbugs as vectors for drug-resistant bacteria. *Emerging Infectious Diseases* 17(6) Retrieved from http://wwwnc.cdc.gov/eid/article/17/6/10–1978_article.htm.

Magnini, V.P., Crotts, J.C., Zehrer A. (2011). Understanding Customer Delight: An Application of Travel Blog Analysis. Journal of Travel Research, 50(5), 535–545. DOI: 10.1177/0047287510379162.

Marco, Meg (2007, November 7) *Secret camera investigation: Every single hotel failed to wash your cups and glasses.* Retrieved from http://consumerist.com/2007/11/07/secret-camera-investigation-every-single-hotel-failed-to-wash-your-cups-and-glasses/

Mirkin, G. (2005). *Catch a cold.* Retrieved from http://www.drmirkin.com/morehealth/9941.html

Nitschke, A. A., Frye W.D. (2008). *Managing Housekeeping Operations* (3rd revised ed.). Orlando, FL: American Hotel and Lodging Educational Institute (EI).

Ritter, K. (24 August 2011) Guests slap resort with $337.5 million lawsuit over illness. *The Seattle Times* retrieved from http://seattletimes.com/html/nationworld/2015999894_apuslegionnairesdiseasearia.html

Shoemaker, D. (2007). Keeping Hotel Guests Healthy. *Executive Housekeeping Today,* 29(5), 5–10. Retrieved from EBSCO*host.*

Shoemaker, D. (2011). Singing the Bed Bug Blues. *Executive Housekeeping Today,* 33(5), 5–18. Retrieved from EBSCO*host.*

Smith, M (2012) Details given about Legionnaires' victims from Chicago outbreak Retrieved from http://articles.chicagotribune.com/2012–08–28/travel/ct-met-legionnaires-chicago-0829–20120829_1_legionnaires-disease-outbreak-legionella.

Stroia, O. V., Kirsch, K., Kline, S., Neal, J., Sirsat, S., Almanza, B. (2012, January). *Empirical Study on the Cleanliness of Hotels – Cleanest and Dirtiest Spots.* Poster presented at the 17th Annual Graduate Conference.

Torres, E. and Kline, S. F. (2006). From customer satisfaction to delight: A model for the hotel industry. *International Journal of Contemporary Hospitality Management. 18 (4),* 290–301.

The World Health Organization. (2004). *WHO Guidelines for the Global Surveillance of Severe Acute Respiratory Syndrome (SARS).* Retrieved from http://www.who.int/csr/resources/publications/WHO_CDS_CSR_ARO_2004_1.pdf

U.S. Food and Drug Administration. (2011). Hazard Analysis and Critical Control Points (HACCP). Retrieved from http://www.fda.gov/food/foodsafety/hazardanalysiscriticalcontrolpointshaccp/default.htm

Winther, B., McCue, K., Ashe, K., Rubino, J.R. and Hendley, J. O. (2007). Environmental Contamination With Rhinovirus and Transfer to Fingers of Healthy Individuals by Daily Life Activity. *Journal of Medical Virology, 79,* 1606–1610.

Winther, B., McCue, K., Ashe, K., Rubino, J.R. and Hendley, J. O. (2011). Rhinovirus contamination of surfaces in homes of adults with natural colds: transfer of virus to fingertips during normal daily activities. *Journal of Medical Virology, 83,* 906–909.

Yee, V. (2004) Proposed Model for using HACCP in Hotel Guest Room Cleaning Practices (Master's thesis). Purdue University.

CHAPTER 15

FOOD SAFETY IN SCHOOLS

KEVIN SAUER, PhD

Associate Professor, Kansas State University

JEANNIE SNEED, PhD, RD

Professor, Kansas State University

CONTENTS

Child Nutrition Programs are delivered in over 101,000 schools across the United States. These programs, administered by the United States Department of Agriculture (USDA) Food and Nutrition Service (FNS), include the National School Lunch Program, the School Breakfast Program, the Child and Adult Care Food Program, the Summer Food Service Program, the Fresh Fruit and Vegetable Program, and the Special Milk Program. Over 31 million school lunches and 12.1 million school breakfasts were served daily to children in 2011.

The goal of the Child Nutrition Programs is to serve nutritious and safe food that children will eat within an established budget. Over the past several years, there have been significant efforts in schools to enhance food safety efforts. Food safety in schools will be presented in this chapter through a discussion of the structure of Child Nutrition Programs, legislation for food safety in schools, surveillance and incidence of foodborne illness outbreaks in schools, research on food safety practices in schools, and resources related to school food safety.

15.1 STRUCTURE OF CHILD NUTRITION PROGRAMS

Child Nutrition Programs are federally funded and the Federal government has oversight for these programs. Consistent with the Federal model, state agencies administer programs in each state and programs are operated at the local level by a school food authority (typically a school district).

15.1.1 ROLE OF THE FEDERAL GOVERNMENT

The Federal government provides a set reimbursement for each of these programs to offset the costs of running the program. There are three categories of reimbursement for the meals programs: free meals; reduced-price meals; and full-paid meals. The Federal reimbursement and the money collected from the reduced-price and full-paid meals are used to cover the costs of the program. In addition, USDA provides USDA Foods (at a value of 22.75 cents per lunch meal served in FY 2012–2013) to support the program. Schools may also receive bonus USDA Foods when they are available from surplus agricultural production. It is estimated that about

15–20% of the food used in Child Nutrition Programs is USDA Foods, with the remaining products being purchased on the commercial market.

USDA Foods are purchased by either the USDA Agricultural Marketing Service (AMS) or Farm Service Administration (FSA). AMS purchases beef, pork, poultry, fish, fruits, and vegetables. FSA purchases products such as flour, peanut products, grain products, and dairy products. The source of foods used in Child Nutrition Programs is related to food safety, mainly due to the traceability of the product in the case of a food recall. In addition, USDA often has more stringent food safety requirements for foods purchased for use in Child Nutrition Programs than are established for the commercial market.

Food safety is a major concern to the purchasing agencies, and many food safety requirements are in place—some are general requirements and some are specific to a particular type of product. For example, all foods purchased by USDA must be produced in the U.S. and of domestic origin and all products must be traceable if there were to be a hold or recall on a product. Products of particular concern for safety (egg products, diced chicken, ground beef and bison, boneless beef, raw ground turkey, dried cherries and cranberries, trail mix, and dehydrated vegetable soup mix) must undergo microbial testing for indicator organisms. Products that test positive for *Salmonella* or *E. coli 0157:H7* are rejected. Another example of a specific requirement is that all fresh fruit and vegetable suppliers must pass an annual Good Handling Practices and Good Agricultural Practices audit. To access the entire document describing the food safety requirements for purchasing domestic commodities, go to the AMS website address: http://www.ams.usda.gov/AMSv1.0/getfile?dDocName=STELPRDC5080613.

15.1.2 ROLE OF STATE AGENCIES

The state agency for child nutrition may be located in a state department of education or a state department of agriculture, depending on the state. The state agency has the responsibility for administering the program in their state. This includes interpreting regulations, providing training, and evaluating whether or not the school food authority is following the program rules and regulations.

Specifically related to food safety, state agencies provide training and guidance related to food safety. Further, the state agency periodically reviews programs at the school district level to ensure that they are meeting any state or federal food safety requirements. The state agency also is tasked with reporting to USDA the number of health inspections each school district has received on an annual basis.

15.1.3 ROLE OF THE SCHOOL FOOD AUTHORITY (SFA)

The SFA is responsible for operating Child Nutrition Programs at local schools and school districts according to all federal and state rules and regulations. The school environment is multifaceted, which impacts traditional food safety efforts and standards. Often, on-site kitchens and cafeterias serve as the primary food outlet for breakfast and lunch meal service for K-12 children. Vulnerable groups, such as preschool children and pregnant teens and women, are served in many schools.

Within the school environment, many different food production systems are used:

1. Conventional. In these schools, food is stored, produced, and served at the same location.
2. Commissary. School districts that have a commissary system produce food at a central kitchen (commissary) and distribute the food to schools for service. Food may be distributed hot or cold (for which rethermalization would be needed) and either in bulk or preplated.
3. Assembly/Serve. Some schools do minimal preparation, basically they just assemble meals and serve them. This could be the case where a school receives food from a school district commissary. There are also schools that serve food produced at local restaurants or third-party vendors.
4. Combination systems. There are many examples of school districts that use combinations of the conventional, commissary, and assembly/serve systems. For example, in some schools a conventional system is used to cook and serve meals at the school and, in addition, they cook food and transport it to other schools or service locations.

In most situations, lunch meal service yields the highest meal participation compared to breakfast and snack offerings. Meal service periods for specific classes or groupings of students are relatively short in duration, and depending on the size of the school and type of menu structure, several hundred students and guests are served quickly through straight-line tray assembly, especially for elementary school children. Middle and high schools often provide service through a central food court or scramble food delivery system, allowing for greater selection of foods and more interaction among student customers. Some schools offer a la carte food options, referred to as competitive foods, sold to children separately from the foods that meet the standards for the National School Lunch Program. Clearly, as a focal point for large volumes of meals served, close personal contact and interaction, food safety and sanitation principles remain especially important.

In some situations, service and consumption of food may occur in the school classroom, on a school bus, or in a remote site such as a field trip or a food kiosk located away from the kitchen. In these cases, the school foodservice program may or may not be directly involved with the foods prepared and served. As such, food that is served outside of the traditional and controlled foodservice system requires additional and strict attention to food temperatures, handwashing, cross contamination, and food allergen control.

15.2 LEGISLATION FOR FOOD SAFETY IN SCHOOLS

Schools are required to follow the local or state Department of Health guidelines for inspections like any retail food establishment and they are inspected by the sanitarian assigned to the jurisdiction. In addition, schools have unique requirements because of federal legislation that imposes specific food safety guidelines that must be met. Child Nutrition Programs are reauthorized every five years, which provides a given time where programs are scrutinized and requirements updated. There are two Reauthorization Acts for the original Child Nutrition Act of 1966 that have been passed that impose food safety standards for schools: The Child Nutrition and WIC Reauthorization Act of 2004 (Public Law 108–265) and the Healthy, Hunger-Free Kids Act of 2010 (Public Law 111–296).

Section 111 of The Child Nutrition and WIC Reauthorization Act of 2004 made significant changes to food safety requirements for schools. The number of required food safety inspections was increased to two per year and the most recent inspection report was required to be publicly posted and provided to the public upon request. State and local governments are allowed to require more frequent inspections. States also were required to audit school districts on the number of inspections received and report those results to the Secretary of Agriculture (through the USDA FNS). Section 111 also added a requirement that school food authorities implement a "school food safety program for the preparation and service of meals that complies with any hazard analysis and critical control point system established by the Secretary." Section 125 of the Act added the following areas for which the National Food Service Management Institute was to provide training and technical assistance: hazard analysis and critical control point plan implementation, emergency readiness, responding to a food recall, and food biosecurity training.

The Healthy, Hunger-Free Kids Act was signed into law on December 13, 2010. This legislation made minor changes to food safety requirements in schools. Section 302 made the school food safety program applicable "to any facility or part of a facility in which food is stored, prepared, or served for the purposes of the school nutrition programs under this Act." This specifically means that if food is served by the school nutrition program in a kiosk, vending machine, or other service site, the food safety program based on HACCP principles must be followed at these service sites.

At a broader level, Section 308 was included to address the USDA Foods that are used in schools. USDA Foods are those that are purchased by USDA AMS or FSA for use in schools. This section states that within one year of enactment of the Act, the Administrator of the FNS, shall:

1. develop guidelines (in consultation with the Administrator of the AMS and the Administrator of the FSA) to determine the circumstances under which it is appropriate for the Secretary to institute an administrative hold on suspect foods purchased by the Secretary that are being used in school meal programs under this Act and the Child Nutrition Act of 1966 (42 U.S.C. 1771 et seq.);

2. work with States to explore ways for the States to increase the timeliness of notification of food recalls to schools and school food authorities;
3. improve the timeliness and completeness of direct communication between FNS and States about holds and recalls; and
4. establish a timeframe to improve the commodity hold and recall procedures of the Department of Agriculture to address the role of processors and determine the involvement of distributors with processed products that may contain recalled ingredients, to facilitate the provision of more timely and complete information to schools.

The Act also requires the USDA Food Safety and Inspection Service to revise their procedures to ensure that schools are included in effectiveness checks. It should be noted that these requirements are not applicable to the commercial food products purchased for use in schools, which accounts for 80–85% of all food used in Child Nutrition Programs.

15.3 SURVEILLANCE AND INCIDENCE OF FOODBORNE ILLNESS OUTBREAKS IN SCHOOLS

Foodborne illnesses attributed to food sources and foodservice operations are monitored through multiple expert sources, residing initially with The Centers for Disease Control and Prevention (CDC). State and local health departments conduct routine food safety inspections at school operations, providing additional localized surveillance. Additionally, state-affiliated child nutrition regulatory agencies often provide centralized food safety reviews, information, training, and current recommendations specific to federally funded school and child care foodservices. Upon the declaration of an outbreak by a local health department or other agency, a thorough investigation is initiated to determine the specific causes and further recommendations for the operations and victims impacted.

Although the CDC publishes a significant number of illnesses and hospitalizations due to foodborne diseases annually in the US, schools remain one of the safest foodservice environments. While the number of incidences of foodborne illness outbreaks in schools is less than home and restaurants, the number impacted during each outbreak is greater. Trend analyzes of current health inspection reports conducted by the Center of

Excellence in Food Safety Research for Child Nutrition Programs indicated comparative differences in behavioral, critical, and noncritical violations between school foodservice operations and restaurants. This and other data consistently show that schools incurred a fraction of the food safety and facility-oriented violations compared to their retail counterparts. Although violations and outbreaks are uncommon in schools, the majority of documented food safety violations are related to facility maintenance and sanitation.

Venuto et al. (2010) assessed data collected through the CDC's Electronic Foodborne Outbreak Reporting System (eFORS) in school settings to determine foodborne disease etiologies. The highest percentage of foodborne outbreaks in schools during 2000–2004 were of unknown or unconfirmed etiology. Outbreaks and illnesses during the same five-year period decreased in schools, however, the magnitude of outbreaks and illnesses attributed to norovirus increased. Among foodborne disease outbreaks of confirmed etiology due to common pathogens, chemical contamination, or metals, norovirus was attributed the highest percentage of outbreaks.

Most schools demonstrate a safe food environment, but also remain uniquely challenged in preventing the transmission of common infectious diseases among children. Environmental factors such as the close proximity, frequent personal contact between children, and hand-to-mouth contact all contribute to the spread of illnesses. As such, effective hand washing practices remain a key part of effective school health programs.

A number of communicable diseases afflict children in schools attributed to both bacterial and viral agents transmitted directly and indirectly. Widespread infections among children cause absenteeism and unfortunately, additional serious health complications. One environmental factor, hand contamination, is often associated with these pathogenic transfers and subsequent widespread infections.

More attention has focused on the transmission and consequences of norovirus infections among school children. Lee and Greig (2010) conducted a 10-year meta-analysis of gastrointestinal outbreaks in schools, reporting that 40% of school-related outbreaks were attributed to viruses. The majority (22%) of these same viral outbreaks were associated with norovirus, contributing to 2,791 illnesses, and 141 hospitalizations. Vigilant hand washing was reiterated as a predominate strategy to reducing illness outbreaks across the ten-year analysis of previous research (Lee and Greig, 2010).

In schools, norovirus is of particular concern not only because of the spread through foods and infected food handlers, but also because it is easily transmitted among children through poor personal hygiene; touching contaminated surfaces and one's mouth, sharing food or service ware with an infected individual, or consuming food and liquids already contaminated with the virus. Individuals also remain contagious soon after recovery. Therefore, effective hand washing practices, especially before, during, and after eating situations among children are paramount to prevent additional exposure and illness.

For foodservice staff and school personnel, traditional cleaning and sanitizing procedures are ineffective for suspected or known norovirus outbreaks. For example, nonvisible vomit particles may extend several feet from an exposed site, furthering the contagious nature and immediate spread of norovirus. In school cafeterias, exposed areas within 25 feet of vomit are considered contaminated. Children and others in the immediate vicinity should wash their hands, vomit should be covered and cleaned immediately with soap and water, and disinfectant applied with a 5,000 ppm bleach solution (National Education Association Health Information Network, in press). At this concentration level, personal protective equipment is needed to protect the individual doing the cleaning.

15.4 RESEARCH ON FOOD SAFETY PRACTICES IN SCHOOLS

There has been limited research specific to food safety in schools, but the body of research is growing. The next few sections will present school-based research related to employee food handling practices, implementation of food safety programs based on HACCP principles, cooling foods, and food allergies.

15.5 EMPLOYEE FOOD HANDLING PRACTICES

As in any foodservice operation, food-handling practices of employees have a major impact on food safety. There has been some research on food handling practices specific to school foodservice operations that is useful for developing training programs, improving employee supervision, and, in general, developing a strong food safety culture in schools. Research specific to handwashing and general food handling practices will be presented.

15.5.1 HANDWASHING

Handwashing is an important employee behavior needed to minimize the contamination of food and work surfaces, yet it is a behavior that is often not done when it is required or using appropriate techniques (Green et al., 2005; U.S. FDA, 2003). Strohbehn et al. (2008) examined handwashing behaviors of retail foodservice employees during food production, serving, and cleaning to identify the number of times that employees should have washed their hands, the number of times they actually did wash, and the methods used. The percentage of times employees were in compliance was determined for both the frequency and procedure for handwashing. School foodservice employees were in compliance 43% of the time for washing hands before engaging in food preparation and 63% of the time when entering the food preparation area. Hands were washed only 30% of the time before donning gloves and 8% of the time after handling potentially hazardous foods. During service, hands were washing only 11% of the time before donning gloves but 67% of the time after handling PHF. Based on these observational data, school employees should have washed their hands 11 times per hour during production, 12 times an hour during service, and 8 times an hour during cleaning to meet the FDA Food Code recommendations. It is apparent that employees need to wash their hands more often and use more thorough techniques. It is also apparent that work needs to be reconfigured to minimize the need for handwashing. For example, if a large number of sandwiches need to be assembled, assign one task to each employee and have a runner so that continuous production can be done with minimal handwashing needed when changing tasks.

15.5.2 OVERALL FOOD HANDLING PRACTICES

Henroid and Sneed (2004) conducted an evaluation of employee food handling practices in 40 schools in Iowa. The lowest compliance (less than

75%) was for the following practices: knowledge of proper cooking temperatures; cold foods being held at less than 41 degrees; cooking temperature being checked and recorded; use of calibrated thermometers; use of hair restraints; adequate handwashing; food stored in proper containers; daily checking of temperatures for refrigerators, freezers, and milk coolers; documentation of temperatures; and documentation of sanitation procedures. When these observations were made, food surfaces were swabbed to determine whether there was adequate cleaning and sanitizing (Henroid et al., 2004). Aerobic plate counts (APC), *Enterobacteriaceae,* and *Staphylococcus aureus* counts were done for samples of a work counter, piece of cooking equipment, serving trays, refrigerator or freezer handle, and handwashing sink handles—surfaces that could either come in direct contact with food or could cross contaminate. Four of the 40 schools met the standard for all surfaces for all tests. Refrigerator or freezer handles failed to meet the standard for APC in two-thirds of the operations. Most schools met the standard for *Enterobacteriaceae. S. aureus* was primarily found on handwashing sink handles or refrigerator/freezer handles, but only in 5 and 7 schools respectively. It was concluded that schools do a good job of sanitizing, but improvements could be made, especially for surfaces such as handles and knobs.

After collecting initial food handling behaviors, Sneed and Henroid (2007) conducted several interventions to improve food safety. Interventions included basic HACCP training in year 1, advanced HACCP training in year 2, educational materials (sample HACCP plan, sample SOPs, monthly training sessions), and onsite technical assistance as requested. Followup assessments were made in year 3. Participation in the study yielded improvements in food safety knowledge and attitudes of employees, and overall food handling practices.

A more recent study was done to examine cross contamination and temperature control in four retail foodservice environments, including schools (Strohbehn et al., 2011). School foodservice employees had significantly higher pre- and post-test scores compared to employees in assisted living, child care, and restaurants. Handwashing compliance was generally highest for employees in child care. This study also examined temperature control for a cold deli meat and found that schools had the best temperature control compared to other types of retail operations.

15.6 IMPLEMENTATION OF SCHOOL FOOD SAFETY PROGRAMS BASED ON HACCP PRINCIPLES

Although schools are required to follow food safety programs based on HACCP principles, little research has been conducted since 2004 mandate to evaluate how these food safety programs have been implemented nationally. In 2012, the Center of Excellence for Food Safety Research in Child Nutrition Programs (www.cnsafefood.k-state.edu) examined how schools had implemented food safety programs, including HACCP procedures, across 34 school districts in the seven USDA regions, including 11 small, nine medium, six large, and eight mega-sized school districts in seven states.

Results of this research found that food production and delivery systems in schools varied widely. In some schools, food was prepared and served on site, some received food from a school district central kitchen, while some received food from an outside caterer. This variation suggests that one generic food safety program is not adequate, but rather operations require unique food safety programs designed specifically for the operation. Some evidence showed state-wide food safety programs had not been adapted for specific schools, even though some of the program content was not applicable to the school. Food safety documentation in schools was evident, however, there were few examples of documentation of corrective actions.

Future research and initiatives will revolve around the customization of food safety programs for the unique food production and delivery systems in schools. In addition, very small schools and those with very simple menus and processes may be able to use a simple risk-based approach to food safety. In these situations, the plan would focus on food handling behaviors most associated with foodborne illness: handwashing; cleaning and sanitizing dishes, utensils, and work surfaces to reduce cross contamination; and temperature control.

15.7 COOLING FOODS

Improper or "slow cooling" of food has been implicated as the number one factor in outbreaks of *C. perfringins.* Cooling food was identified by some school foodservice directors as a challenge while they were developing their food safety programs based on HACCP principles. Some schools still do scratch cooking and cool food for service at a later time while others cool leftovers for use later. This is problematic in schools because of the short work day (schools typically serve breakfast and lunch, and all employees finish their work day by about 2 pm), which limits time to monitor the cooling process.

Krishnamurthy and Sneed (2011) conducted a survey of district school foodservice directors across the U.S. and found that more than half reported to cool foods such as taco meat filling, turkey (whole or roasts), and chili. More than a third cooled products such as spaghetti sauce, soup, macaroni and cheese, and roasts. Research has shown that it is difficult to chill these products within the FDA Food Code standards (Olds et al., 2006; Olds and Sneed, 2005; Roberts et al., 2013). Further, only 8% of directors reported to have a blast chiller and 30% did not have ice machines to support ice water bath cooling methods.

In terms of cooling practices, 82% reported to take temperatures of food during the cooling process and 70% recorded those temperatures (Krishnamurthy and Sneed, 2011). By examining when temperatures were taken, a different picture emerged. Only 17% took temperatures at 2 h after cooling begins, only 7.5% after 4 h, and only 11% take continuous temperatures. Based on these findings, many operators do not have sufficient data to support that adequate cooling has taken place.

Research shows that cooling food adequately is difficult and that there are several factors that impact the cooling process. One recent study (Roberts et al., 2013) examined the properties of chili and tomato sauce (marinara) cooled in a walk-in refrigerator, walk-in freezer, and ice water bath in a walk-in refrigerator with product depths of 2″ and 3." A 3-gallon volume was cooled with a chill stick. All products were cooled uncovered. The only cooling method that met both Food Code requirements (135 to 70 in two hours; 135 to 41 in six hours) for chili and tomato sauce was product at 2″ depth cooled in the walk-in freezer. Another study by Olds et al. (2013) examined the cooling properties of beef taco meat and steamed rice. Several cooling methods were compared, and the only method that was effective for beef taco meat was 2″ depth cooled in a walk-in freezer and the only effective method for rice was 2″ depth placed in an ice-water bath in a walk-in refrigerator. These studies demonstrate the difficulty in cooling foods to meet the recommendations in the FDA Food Code. Ideally, schools that cool foods should have a blast chiller, and if that is not available employees need to divide the food into pans of not more than 2″ depths, use freezers for cooling, and track temperatures throughout the cooling process.

15.8 FOOD ALLERGIES

Food allergies and subsequent reactions among children in schools are an important food safety concern and affect approximately 1 in 25 school-aged children (Sicherer, Mahr, and the Section on Allergy and Immunology, 2010.) Reports of food allergies have increased substantially in recent years, especially among children and the number of reactions occurring in schools. Sicherer, Mahar, and The Section on Allergy and Immunology (2010) reported that up to 18% of children with known food allergies have had a reaction while attending school.

Increasingly, all types of foodservice operations including schools are revising policies and procedures to address the increasing needs and expectations of consumers with food allergens. Accordingly, recent reports of litigation related to the responsibility of foodservice to accurately accommodate food allergies have brought national attention to the issue (U.S. Department of Justice, 2012).

An allergic reaction to a food substance can occur at any time during a person's life, and result in a range of reactions in children and young adults, from minor rashes or other short-term irritations to fatal, systemic reactions that occur soon after brief contact or ingestion of a specific food allergen. Individuals with food allergies must avoid food allergens to prevent allergic reactions. Some foods naturally contain allergens and others may become contaminated with these allergens through cross contact. Eight food items are associated with the majority of allergic reactions and include milk, eggs, peanuts, tree nuts, fish, shellfish, soy, and wheat. Many of these same food items are popular among children or commonly used as ingredients in recipes or purchased food products.

Food allergies and allergic reactions occurring in schools require attention and management of food and food allergens in the school environment as part of the overall food safety program. Because most schools receive financial assistance from the federal government, Section 504 of the Rehabilitation Act of 1973 protects students with life-threatening food allergies from being discriminated against due to specific food or eating needs. School foodservice providers develop specific plans for feeding individual children with food allergies through guidance provided by the USDA.

A school foodservice program may address several food allergies across the eight food types for one or many children concurrently. Therefore, the management of food allergens in the school environment requires close attention throughout the flow of food, including monitoring ingredient listings and product labels that might include confirmation of detailed product specifications with food manufacturers or suppliers. Managing food allergies is also multidisciplinary and may include frequent discussions and collaboration with the parents or guardians of children to review school menus, ingredients used, or alternatives. Foodservice staff, teachers, and school healthcare professionals also participate in the storage and accessibility of food allergy medications or allergic reaction response devices, such as an EpiPen, should a situation occur requiring immediate attention. School foodservice personnel, especially those in food production and service roles, may require specific training about food allergies. Training efforts may include food allergy recipe modifications, interpretation of fresh or pre-prepared product ingredient labels for aliases of ingredients, how to make appropriate product substitutions, and signs or symptoms of an allergic reactions among children.

15.9 RESOURCES FOR SCHOOL FOOD SAFETY

The USDA FNS provides many important food safety resources for schools. Their website (www.fns.usda.gov/food-safety) lists many of these resources, including those related to food safety for Child Nutrition professionals, food safety for child care, produce safety, food allergies, norovirus, food defense, and other food safety topics.

USDA also funds two important entities that support food safety efforts for schools—the National Food Service Management Institute (NFSMI) and the Center of Excellence for Food Safety Research in Child Nutrition Programs. The FNS Office of Food Safety, NFSMI, and the Center of

Excellence work together to plan and implement research and training so that training developed is based on sound research.

NFSMI (www.nfsmi.org) was established by Congress in 1989 and funded in 1991 to provide applied research, education and training, and technical assistance for Child Nutrition Programs. NFSMI is located at the University of Mississippi with the Applied Research Division located at the University of Southern Mississippi. NFSMI provides many educational resources through a variety of delivery methods. NFSMI also operates a help desk to provide answer to questions related to child nutrition program management. Funding for NFSMI comes primarily through USDA FNS.

In April 2011, the USDA FNS Office of Food Safety provided funding to establish the Center of Excellence for Food Safety Research in Child Nutrition Programs (www.cnsafefood.k-state.edu) at Kansas State University. The Center is located in the Department of Hospitality Management and Dietetics. The mission of the Center is to conduct food safety research that meets the needs of FNS's nutrition assistance programs using an interdisciplinary team approach and to disseminate results to a variety of targeted audiences including school foodservice directors, child nutrition program operators, scientists, policy makers, educators, and practitioners. Center research has focused on the efficacy of cooling methods used in schools, effectiveness of washing treatments on pathogen reduction in fresh produce, and the status of food safety programs based on HACCP principles. In addition, the Center has developed and pilot tested a 4-day educational program called, "Serving Up Science: The Path to Safe Food in Schools," which is intended to be delivered to school foodservice directors and state agency staff.

KEYWORDS

- **child nutrition programs**
- **electronic foodborne outbreak reporting system**
- **HACCP principles**
- **handwashing**

REFERENCES

Daniels N.A., Mackinnon L., Rowe S.M., et al. (2002). Foodborne disease outbreaks in United States schools. *Pediatric Infectious Disease Journal, 21*(7), 623–628.

GAO: School meal programs: Few outbreaks of foodborne illness reported, but opportunities exist to enhance outbreak data and food safety practices. U.S. General Accounting Office Web site. http://www.gao.gov/products/GAO-03–530. Published May 9, 2003. Accessed January 8, 2013.

Green, L., Selman, C., Banerjee, A., Marcus, R., Medus, C., Angulo, F., Radke, V., Buchanan, S., and EHS-NET Working Group. (2005). Foodservice workers' self-reported food preparation practices: An EHS-NET Study. *International Journal of Hygiene and Environmental Health, 208,* 27–35.

Henroid, D., Jr., Mendonca, A., and Sneed, J. (2004). Microbiological evaluation of food contact surfaces in Iowa schools. *Food Protection Trends, 24*, 682–685.

Henroid, D., and Sneed, J. (2004). Readiness to Implement Hazard Analysis Critical Control Point (HACCP) Systems in Iowa schools. *Journal of the American Dietetic Association, 104*, 180–186.

Institute for Food Safety and Health. *USDA-FNS/FDA/IFSH study on cooling of foods in school food service operations, final report.* Summit-Argo, IL: Illinois Institute of Technology; 2011:85–86.

Krishnamurthy, K., and Sneed, J. (2011). Cooling practices used in school foodservice. *Food Protection Trends, 31*, 828–833.

Lee, M.B., and Greig, J.D. (2010). Review of Gastrointestinal Outbreaks in Schools: Effective Infection Control Interventions. *Journal of School Health,* 80(12), 588–598.

National Education Association Health Information Network. *The stomach bug book.* Retrieved June 18, 2013. http://www.neahin.org/resources/Images/Stomach%20Bug%20 Book.pdf

Olds, D.A., Mendonca, A.F., Sneed, J., and Bisha, B. (2006). Influence of four retail foodservice cooling methods on the behavior of *Clostridium perfringens* ATCC 10388 in turkey roasts following heating to an internal temperature of 74°C. *Journal of Food Protection, 69*(1), 112–117.

Olds, D.A., Roberts, K.R., Sauer, K.L., Sneed, J., and Shanklin, C.W. (2013). Efficacy of cooling beef taco meat and steamed rice in United States School Foodservice Operations. *Food and Nutrition Sciences*, http://www.scirp.org/journal/fns

Olds, D.A., and Sneed, J. (2005). Cooling rates of chili using refrigerator, blast chiller, and chill stick cooling methods. *Journal of Child Nutrition and Management,* Available at: http://docs.schoolnutrition.org/newsroom/jcnm/05spring/olds/index.asp.

Pogostin LT, Ayers S, Gray T, Nguyen M, Lynch M, Williams I. School-associated foodborne outbreaks in the United States – 1998–2006. Poster presented at: 4th Annual OutbreakNet Conference; June 6, 2008; Denver, CO.

Roberts, K. R., Olds, D. A., Shanklin, C. W., Sauer, K., and Sneed, J. (2013). Cooling of foods in retail food establishments. *Food Protection Trends, 33,* 27–31.

Scallan E., Hoekstra R.M., Angulo F.J., Tauxe R.V., Widdowson M.A., Roy S.L., Jones J.L., Griffin P.M. (2011). Foodborne illness acquired in the United States — major pathogens. *Emerging Infectioous Diseases, 17*(1), 7–15.

Sicherer, S.H., Mahr, T., and The Section on Allergy and Immunology. (2010). Clinical Report – Management of Food Allergy in the School Setting. *Pediatrics, 126*(6), 1232–1239.

Sneed, J., and Henroid Jr., D. (2007). Impact of educational interventions on Hazard Analysis and Critical Control Point (HACCP) program implementation in Iowa schools. *The Journal of Child Nutrition and Management, 31*(1), http://docs.schoolnutrition.org/newsroom/jcnm/07s pring/sneed/index.asp.

Strohbehn, C.H., Paez, P., Sneed, J., and Meyer, J. (2011). Mitigating cross contamination in four retail foodservice sectors. *Food Protection Trends, 31*, 620–630.

Strohbehn, C., Sneed, J., Paez, P., and Meyer, J. (2008). Hand washing frequencies and procedures used in retail food services. *Journal of Food Protection, 71*(8), 1641–1650.

United States Department of Justice. (2012, December 20). Justice Department and Lesley University Sign Agreement to Ensure Meal Plan is Inclusive of Students with Celia Disease and Food Allergies. Available at: http://www.federalreserve.gov/boarddocs/hh/1998/july/fullreport.htm. Accessed June 18, 2013.

U.S. Food and Drug Administration. (2003). Hand hygiene in retail and foodservice establishments. U.S. Food and Drug Administration, Center for Food Safety and Applied Nutrition. Available at: http://www.cfsan.fda.gov/~comm/handhyg.html. Accessed June 13, 2013.

U.S. Department of Agriculture. (2001). Guidance for Accommodating Children with Special Dietiary Needs in the School Nutrition Programs. U.S. Department of Agriculture, Food and Nutrition Services. http://www.fns.usda.gov/cnd/guidance/special_dietary_needs.pdf. Accessed June 15, 2013.

Venuto, M., Halbrook, B., Hinners, M., Lange, A., and Mickelson, S. (2010). Analyzes of the eFORS (Electronic Foodborne Outbreak Reporting System) surveillance data (2000–2004) in school settings. *Journal of Environmental Health, 72*(7), 8–13.

CHAPTER 16

FOOD SAFETY AT FAIRS AND FESTIVALS

HUEY CHERN BOO, PhD

Associate Professor, Universiti Putra Malaysia

WEI LEONG CHAN, MS

Lecturer, YTL-International College of Hotel Management, Malaysia

CONTENTS

16.1 BACKGROUND

Fairs and festivals are among the fastest-growing types of tourism attractions. According to the International Association of Fairs and Expositions (2013), there are over 3,200 fairs in North America today. The number of community festivals has exceeded 20,000 (Janiskee, 1996) and is growing. Every year, hundreds of millions of visitors attend state fairs, making them one of the largest leisure spectator activities in the United States. In fact, more people attend fairs than baseball, football, and basketball games combined (Mihalik and Ferguson, 1994). For instance, the average number in attendance at the Feast of the Hunters' Moon, a 2-day annual event in Indiana, is approximately 40,000 (Christos, 2006). The world's largest food festival, Taste of Chicago, drew 1.2 million visitors in 2012 (Gatziolis and May, 2012). Likewise, the UK music festivals attracted more than 7.7 million attendees from all over the world in 2009 (Botelho-Nevers and Gautret, 2013).

Fairs and festivals are generally viewed as social or cultural celebrations. Based on the size of the attending population and the primary purpose for holding a festival, O'Sullivan and Jackson (2002) proposed three types of festivals: home-grown, tourist-tempter, and big-bang. Home-grown fairs and festivals are commonly held on a small scale. Fairs and festivals in this category primarily aim to provide cultural or entertainment benefits to the locals and visitors. Examples of home-grown fairs and festivals include the Montgomery Agricultural Fair in the United States and the Cheshire County Show in the United Kingdom. Tourist-tempter fairs and festivals are generally run in urban areas on a medium scale to boost the local economic development. The Modhera Dance Festival in India and the Water Festival (Songkran) in Thailand are examples of this type of fair and festival. Similar to the tourist-tempter fairs and festivals, big-bang fairs and festivals, such as Munich's Oktoberfest in Germany and the Hokkaido Snow Festival in Japan, are also held in urban areas, but they take place on a large scale to promote economic development via cultural or entertainment activities.

Regardless of their sizes, fairs and festivals attract a considerable number of people with various interests. Many countries increasingly view fairs and festivals as tourist attractions that have a significant economic impact on the surrounding region. In the late 1980s, Getzs (1989) highlighted the economic benefits of fairs and festivals through their role in attracting tourists, increasing their expenditures, and extending their length of stay. Mihalik and Ferguson (1994) also documented that the on-site expenditure by visitors at all of the American fairs in 1992 was estimated in the billions of US dollars. The median amount of dollars spent per group (with an average number of three people per group) at the fairground was $34, and the estimated total dollars spent was $6,349,366. In the UK, £150 million in extra spending was generated by overseas visitors from some 1,000 cultural events in 1995 compared to £40.6 million from 550 events in 1991 (O'Sullivan and Jackson, 2002). In Australia, the majority of fairs and festivals also reported earning significant profits (Gibson, Waitt, Walmsley, and Connell, 2010).

TABLE 1 Outbreaks of Food-borne Illness at Fairs and Festivals.

Reference	Event	Place	Year	Cases	Pathogen Involved	Vehicle	Factors
Barker and Runte (1972)	Dairy Princess Festival	Minnesota, US	1969	43	Tin[1]	Canned tomato juice	Corroded tin
Brown et al., (1987)	Pierce County Fair	Washington, US	1985	30	*Salmonella agona*	Bar-beque beef sand-wich	Improper temperature control Contamination from work surfaces and utensils Inadequate cleaning and sanitizing of pans and utensils
Hewitt et al. (1986)	County of Avon Festival	UK	1986	50	*Clos-tridium perfringens*	Boiled salmon	Improper cooling
Lee et al. (1991)	Michigan Women's Music Festival	Michigan, US	1988	3175	*Shigella sonnei*	Un-cooked tofu salad	Some workers prepared food while they were ill Unsanitary practices of food handlers Insufficient re-frigeration of large quantity

TABLE 1 *(Continued)*

Reference	Event	Place	Year	Cases	Pathogen Involved	Vehicle	Factors
Morgan et al. (1994)	Glastonbury Musical Festival	UK	1992	72	*Campylobacter*	Pasteurized milk	Milk was unpasteurized
Crampin et al. (1999)	Glastonbury Musical Festival	UK	1997	7	*Escherichia coli O157:H7*	Mud	Cattle grazed on fair site up to two days before the event and heavy rain. Mud was contaminated by cattle feces
Camps et al. (2005)	San Juan Festival	Spain	2002	1435	*Salmonella enterica*	Pastry with vanilla cream (coca)	(a) Improper temperature control (*b*) Contaminated source (*c*) Cross contamination of work surface (d) Over production in nondesignated facility
Christian et al. (2008)	Pumpkin Festival	South Carolina, US	2006	100	*Samonella* Thompson	Boiled peanuts	Improper temperature control Contaminated serving ladle
Food Poisoning Lawyer (2007a)	Isle of Eight Flags Shrimp Festival	Florida, US	2007	48	Norovirus	Lemonade	Contaminated water
Food Poisoning Lawyer (2007b)	Taste of Chicago	Illinois, US	2007	700	*Salmonella* Heidelberg	Hummus shirazi salad	(a) Unsafe source (*b*) Ill worker prepared food
Kitamoto et al. (2009)	Nagoya University Festival	Japan	2008	75	*Staphylococcus aureus*	Crepes	(a) Unsanitary practices of food handlers (*b*) Improper cooling (*c*) Long holding time

TABLE 1 *(Continued)*

Reference	Event	Place	Year	Cases	Pathogen Involved	Vehicle	Factors
Minnesota Department of Health, 2006	Smorgas-bord Event	Minnesota, US	2006	NA	*Esch-erichia coli O157:H7*	Potato salad and other ready-to-eat food	(a) Cross contamination from raw ground beef to potato salad (*b*) Cross contamination from work surfaces, uten-sils, or hands of volunteers during han-dling of raw ground beef

[1] The hazard is chemical-base.

16.2 OUTBREAKS OF FOODBORNE ILLNESS AT FAIRS AND FESTIVALS

Food is a common component of many fairs and festivals. With approximately one-quarter of fair and festival visitors (22%) indicating that eating was their second favorite activity after carnival rides (Mihalik and Ferguson, 1994) and an estimated two to three eating occasions per visit (Lee et al., 2010), substantial revenue can be generated from selling food. While fairs and festivals are supposed to benefit both the food vendors and consumers, the number of reported outbreaks of foodborne illness at fairs and festivals has not yet receded (Todd et al., 2007a). The Centers for Disease Control and Prevention (CDC) estimated 47 foodborne outbreaks and 1141 illnesses associated with fairs, festivals, and temporary mobile stands between 1998 and 2008 (CDC, 2013a). In just 2009–2010, 12 outbreaks and 220 illnesses were recorded (CDC, 2013b).

The incidents of foodborne illness at fairs and festivals presented in Table 1 clearly suggest that local regulatory agencies, organizers, vendors, and even consumers must direct greater attention to ensuring a satisfactory level of food safety and hygiene at these temporary foodservice settings. In view of the substantial number of fair- and festivalgoers, outbreaks of foodborne illness could adversely affect the profitability of the events and the image of the community sponsors. To that end, the following section will first examine consumers' perceptions of food safety and hygiene at fairs and festivals. It will be followed by discussions of the factors that contribute to outbreaks of foodborne illness and the effectiveness of training programs to improve the food safety and hygiene standard at fairs and festivals. The chapter will be concluded with future research avenues for fairs and festivals.

16.3 CONSUMERS' PERCEPTIONS OF FOOD SAFETY AND HYGIENE AT FAIRS AND FESTIVALS

The foodservice establishments at fairs and festivals are generally temporary foodservice facilities, such as food stands, food booths, and food carts, which operate at a fixed location for no more than 14 consecutive days at a time (Food and Drug Administration, 1997). Consumers have expressed greater concern about food safety and hygiene than healthfulness at these temporary settings (Boo et al., 2000). Food poisoning/spoilage (62.7%), dirt or dust contamination (30.7%), and insect contamination (25.3%) were the three major concerns. In addition, these consumers rated the food consumed at fairs and festivals as significantly less safe than that served at cafeterias and other types of restaurants. As a result, they also perceived a higher likelihood of contracting foodborne illness at temporary foodservice facilities compared with restaurants.

Contrary to the above findings, Worsfold (2003) revealed that none of those attending the Vale agricultural show in the UK_indicated any concerns about food safety and hygiene. The show attendees had failed to observe the barely adequate hand-washing facilities and the unacceptable surface finishes of some of the stalls. When prompted, the consumers expressed their beliefs that the traders/vendor were doing the best they could in the situation. These attendees claimed that they "relied on the displayed hygiene certificates and assumed that the traders/vendor would be trained in food safety and would be subjected to inspection" (p. 161). Considering that consumers may not possess the knowledge needed to judge the level of food safety (Fatimah et al., 2011), regulating agencies (i.e., health inspectors), organizers, and vendors play an extremely crucial role in upholding the food safety standards at fairs and festivals.

16.4 FACTORS THAT CONTRIBUTE TO OUTBREAKS OF FOODBORNE ILLNESS AT FAIRS AND FESTIVALS

Maintaining a high standard of food safety and hygiene at fairs and festivals is a great challenge to many organizers and vendors, given the nature of their structure and location. Bacteria may be introduced into the food through raw materials, unclean cooking utensils, and environmental contamination, or by the people handling the food during its preparation and sale. Boo et al. (2000) examined consumers' perceptions of the sources that contributed the most to foodborne illness (Table 2). Inadequate cooking (23.7%), improper cleaning of equipment and utensils (21.7%), and unhygienic practices (16.4%) were most often cited. This finding is consistent with the causes reported in previous research (Brown et al., 1987; Coté et al., 1995; Lee et al., 1991; Morgan et al., 1994) and by the CDC (Bean et al., 1996). Furthermore, approximately one-tenth (8.4%) of the respondents indicated that they thought prolonged holding of food was the most important cause, 17.1% thought that it was the second most important cause, and 12.4% rated it the third most important cause. A contaminated water supply, infected food handlers, and insects, dirt, or dust contaminations were perceived to be important by less than 10% of the respondents.

TABLE 2 The three most important concerns for customers with regards to food handling practices at fairs and festivals (N=299).

	Primary (%)	Secondary (%)	Tertiary (%)
Vendors didn't cook foods thoroughly	23.7	24.1	10.0
Vendor's equipment and utensils were not properly cleaned	21.7	14.7	10.0
Unhygienic practices of food handlers	16.4	11.0	19.4
Foods were spoiled before the vendor got them	13.7	3.0	3.7
Vendors held foods for too long a time	8.4	17.1	12.4
Cooked foods were contaminated by raw foods or dirty equipment and utensils	6.7	6.7	16.1
Vendor held foods at room temperature	5.7	9.7	11.7
Food handlers were sick or had sores	2.0	6.0	7.0

TABLE 2 *(Continued)*

	Primary (%)	Secondary (%)	Tertiary (%)
Foods were contaminated by insects, dirt or dust	1.2	6.4	8.4
Water supply was contaminated	0.3	1.3	1.3

Source: Boo et al. (2000).

Bryan (1981) categorized the sources into (a) potentially hazardous foods, (b) pathogenic foodborne organisms, (c) employee practices, (d) time-temperature combinations, and (e) environmental conditions. A review of prior research (Bean et al., 1990; Bryan, 1981, 1988, 1993; Fein et al., 1995; Knabel, 1995; Medeiros et al., 1996) suggests that sources of contamination could also be categorized into four factors: food, human, facilities and environment, and time-temperature control. These categories coincide with the top five risks for foodborne illness outbreak: food from unsafe sources, poor employee health and hygiene, dirty and/or contaminated utensils and equipment, improper hot/cold holding temperatures of potentially hazardous food, and improper cooking temperatures of food (Food and Drug Administration, 2000).

16.4.1 FOOD FACTOR

Historically, food obtained from dubious sources has been the least commonly reported factor in the outbreak of foodborne illness (Barker and Runte, 1972; Bean et al., 1990). More recently, a 1993 outbreak of enteric illness was traced to the contamination of apples with calf feces before or during harvest (Millard et al., 1994). The inoculum was believed to have been disseminated throughout a batch of cider because the contaminated apples were improperly washed before pressing. In the large gastroenteritis outbreak in Catalonia in June 2002, contaminated ingredients were identified as one of the many contributing sources. *Salmonella enteritidis* was isolated from samples of vanilla filling, egg products, and pasteurized egg yolk in an open container, and pine nuts (Camps et al., 2005). While vegetative cells may be destroyed during cooking, bacterial spores remain alive. Foods that are contaminated with spores continue to present a high risk. The spores will germinate when the conditions become conducive, especially during display.

As demonstrated by these two incidents, the safety of food sources has become increasingly crucial as raw and fresh foods gain greater popularity with consumers who do not usually understand the potential link between fresh produce and food poisoning (Sloan, 1995). Indeed, the foodborne illness outbreak during the Taste of Chicago in 2007 illustrates this danger. It was believed that the contaminated hummus shirazi, the only dish served at the Pars Cove booth, was responsible for the illness (Food Poisoning Lawyer, 2007b). The ingredients of this fresh herb-tomato-cucumber salad served over a bed of hummus were suspected as the source of contamination.

In addition to unsafe food sources, water has also been found to be a contributing factor. Contaminated water was responsible for the norovirus outbreak at the Isle of Eight Flags Shrimp Festival in Florida in 2007 (Food Poisoning Lawyer, 2007a). At least 48 people reported falling ill after consuming lemonade from a stand run by a group of cheerleaders.

While the above incidents were linked to microbiological hazard, hazards related to chemical-contaminated food have also been recorded. In 1969, many guests at a Dairy Princess Festival in Minnesota experienced acute abdominal bloating, vomiting, and diarrhea shortly after consuming tomato juice (Barker and Runte, 1972). The laboratory tests confirmed that the canned tomato juice was contaminated with a high level of tin from the cans' corroded tin coating. In sum, both microbiological and chemical hazards have contaminated food sources, leading to massive foodborne illness outbreaks.

16.4.2 HUMAN FACTOR

The problem with food safety is exacerbated when the individuals preparing food in these temporary settings are part-time, seasonal, or voluntary quantity cooks who rely on their knowledge of home food preparation. These volunteers often have a strong commitment, but they are less likely to participate in formal educational programs or have commercial foodservice experience when compared with cooks at restaurants and cafeterias. A study by Manning (1994) showed that foodservice workers in temporary foodservice operations did not appear to have a good understanding of safe food handling.

More recent studies have also supported the dearth of safe food-handling knowledge and practices among food handlers at fairs and festivals. For instance, Worsfold (2003) noted that fair and festival organizers commented that some vendors had insufficient hygiene awareness and made food safety/hygiene mistakes, while a few vendors lacked even an elementary understanding of the requirements of the relevant hygiene regulations. A deficiency of safe food-handling knowledge and practices and a lack of effective food safety education for food handlers in temporary foodservice settings have been cited as reasons why foodborne illness continues to be a problem (Manning and Snider, 1993; Williamson, Gravani and Lawless, 1992). Overall, several unsafe practices committed by food handlers include working with a poor health condition, poor personal hygiene, and unhygienic food handling.

16.4.2.1 EMPLOYEE HEALTH STATUS

Humans frequently carry enterotoxigenic staphylococcal strains in their noses and on their hands and skin and *Clostridium perfringens* in their intestine. Approximately 35% of human fingertips contain large amounts of bacteria (Tebbutt, 1991). Operators who are found to be infected with pathogens should be prohibited from handling foods. In 1988, a large foodborne outbreak of shigellosis at a festival was traced to a smaller outbreak among staff shortly before the festival began (Lee et al., 1991). The outbreak of typhoid fever at a Latin food festival was also attributed to potato salad that had been prepared by an infected but asymptomatic food handler (Coté et al., 1995). More recently, an ill food worker was implicated as a possible source for the contamination of food in the outbreak at the Taste of Chicago festival in 2007 (Food Poisoning Lawyer, 2007b).

16.4.2.2 PERSONAL HYGIENE

As consumers become increasingly aware of the potential health risk posed by sick employees and laws are passed to mandate that sick food handlers be prohibited from work, a more critical food safety aspect is foodservice workers' adherence to good personal hygiene and hygienic food-handling practices. Scant data have been collected on personal hygiene among fair and festival foodservice workers. The noted unacceptable personal hygiene practices include using nail polish, failing to use a hair restraint, failing to wash hands after taking a drink, and wiping the nose with the hand (Kitamoto et al., 2009; Lee et al., 2010). Not washing hands after visiting the lavatory was observed in one out of every three catering staff in one study (Food Standards Agency, 2002). Although foodborne illness caused by such misbehavior has not yet been reported at fairs and festivals, it certainly is a conceivable threat to food safety given that proper lavatory and hand-washing facilities are often lacking in such temporary settings.

16.4.2.3 FOOD HANDLING

Cross contamination associated with unhygienic food practices, such as handling food with bare hands, appears to be a common risk at fairs and festivals (Burt et al., 2003; Kitamoto et al., 2009; Lee et al., 2010; Lee et al., 1991; Manning and Snider, 1993; Willis et al., 2012; Worsfold, 2003). The outbreak of shigellosis at the outdoor musical festival in Michigan (Lee et al., 1991) was traced to the tofu being mixed by hand. The *Staphylococcus aureus* food poisoning at the Nagoya University Festival was partly attributed to handling food with bare hands (Kitamoto et al., 2009).

Although we know that the responsibility for handling food and money should be separated, a study conducted by Manning and Snider (1993) showed that 21% of workers and 36% of vendors violated this practice. More seriously, 44% of food workers were observed using their bare hands when handling foods despite their knowledge that touching cooked foods can lead to cross-contamination (Manning and Snider, 1993). In fact, the major hazards for cooked foods commence only after cooking. The foods are usually cooked to a high enough temperature, but the handling of foods after cooking introduces additional contamination. A study conducted by Bryan et al. (1992) reported that *Staphylococci* were found to contaminate cooked potatoes during peeling, cutting, shaping, and garnishing. Similarly, Kitamoto et al. (2009) asserted that the outbreak of *Staphylococcus aureus* food poisoning at the Nagoya University Festival in 2008 could possibly have been prevented if the cooks had used disinfectants or gloves when they wrapped the crepes.

Although some food handling after cooking is unavoidable, Tebbutt (1993) and Todd et al. (2007b) found that many potential food hazards were created by unnecessary food handling coupled with infrequent hand washing. Hand-washing hygiene failures among food handlers at large-scale open-air festivals were implicated in the transmission of gastrointestinal diseases (Botelho-Nevers and Gautret, 2013). In some cases, even though the workers indicated adequate knowledge of the importance of frequent and thorough hand washing, Burch and Sawyer (1991) discovered that only 15% of the employees frequently washed their hands, while Manning and Snider (1993) found that only 2% did. Today, the US Food Code clearly forbids bare hand contact to minimize employee contamination of food. Touching ready to eat food with employee's bare hands is prohibited. Alternatively, employees should use serving utensils, deli paper, and disposal gloves. Nevertheless, Burt et al. (2003) found that 67% of vendors served foods with their bare hands, 40% vended with visibly dirty hands or gloves, and none washed his or her hands or changed gloves during the 20-minute observation period. Overall, these findings suggest that lack of hand washing remains a critical issue to food safety at fairs and festivals at which hand washing facilities are significantly insufficient. Workers should also be taught to wash their hands as frequently as possible, especially before beginning work or handling cooked food, after handling raw foods of animal origin, using the toilet, coughing, sneezing, blowing, or touching sores or bandages.

16.4.3 FACILITY AND ENVIRONMENTAL FACTORS

Along with washing hands, cleaning surfaces between contact with raw and cooked food is also often neglected. This condition becomes more threatening when foods are prepared at premises that lack sanitary facilities, especially hand-washing sinks. The surfaces of equipment, such as cutting boards, slicers, grinders, knives, and storage containers, that have previously contacted raw foods are usually contaminated with pathogens. Without thorough cleaning, these pathogenic microorganisms can be transmitted to cooked foods. For instance, cross contamination of *Escherichia coli O157:H7* from raw ground beef to potato salad and other ready-to-eat food via surfaces and utensils was the cause of the outbreak of foodborne illness during the Smorgasbord Event in Minnesota in 2006 (Minnesota Department of Health, 2006). In the foodborne salmonella infection outbreak caused by egg-containing food at a festival in Catalonia in 2002, the vanilla filling was cross-contaminated when it was cooled on the work surface that was used to make the dough of *coca* (Camps et al., 2005). On the other hand, the salmonella foodborne outbreak at the Pumpkin Festival, South Carolina was associated with a contaminated serving ladle used for boiled peanuts (Christian et al., 2008).

Williamson et al. (1992) reported that only 54% of the respondents would wash knives and cutting boards with soap and water before chopping vegetables. The number of food workers who would use a disinfectant to clean the kitchenware or equipment might be expected to be even lower. Tebbutt (1991) has confirmed that 48% of cooked food surfaces were heavily contaminated with bacteria. Willis et al. (2012) also demonstrated that approximately half of the samples collected from large-scale events exhibited unsatisfactory microbial levels on chopping boards (57%) and work surfaces (48%). Although handling food with clean utensils rather than bare hands is known to minimize contamination, equipment has sometimes found to be improperly sanitized because of limited access to sanitizer and water. In fact, inadequate sanitary facilities were implicated as a cause of the shigellosis outbreak at the outdoor musical festival in Michigan (Lee et al., 1991).

Cross-contamination via dirty reusable wiping cloths has also been identified as a safety hazard (Little and Sagoo, 2009; Tebbutt, 1991, 1993). Thirty-three percent of cleaning cloths were found to contain more than 100 million bacteria, and 31% contained more than 10^3 *Escherichia coli* per cloth. The problem is further complicated by incorrect sanitizer use. Lee et al. (2010) documented that the most critical violation at fairs and festival was overuse of sanitizer (21.9% in 2006 and 21.4% in 2008). In the attempt to relate vendors' hygiene knowledge with satisfactory microbiological levels, Willis et al. (2012) revealed that more than half of the obtained sponge swab samples (52%) and cloths (73%) had unsatisfactory microbiological results when food workers at fairs and festivals were unaware of the correct dilution of disinfectant to use. Similarly, 50% of the swabs and 81% of the cloths that were tested had unacceptably high microbial counts when food handlers were unaware of the required disinfectant contact times.

In addition to inadequate and improper facilities, poor environmental conditions and animal/insect contamination are other possible contributing factors. Boo et al. (2000) showed that dirt or dust contamination and insect contamination were rated as the second and third greatest consumer concerns, respectively. Because fairs and festivals operate in the open air, the foods sold are inevitably exposed to environmental contaminants such as dust, dirt, and flies. There has been at least one outbreak linked to poor environmental conditions (Crampin et al., 1999). Prior to the start of the annual Glastonbury Music Festival in 1997, approximately 650 cows grazed on the site, which became a quagmire after heavy rain. The identified cases of *Escherichia coli O157:H7* infection reported no common food or water source and differing uses of sanitation facilities. However, all reported a high level of mud contamination, especially on hands and faces. Mud contaminated with cattle feces (which might get on hands, as well as faces if proper handwashing is not done) was suspected as the likely vehicle of infection. In addition to animal excrement, animal feed also poses potential risk, especially at agricultural fairs. Indeed, *Escherichia coli O157:H7* were detected in cattle manure piles at Minnesota county fairs (Cho et al., 2006) and contributed to outbreaks at petting zoos (Clark, 2012).

Pests (e.g., flies, rodents) and animals (e.g., dogs, birds) can also add to a serious sanitary deficiency. The relationship between houseflies and diarrheal disease has been well established (Cupp et al., 1992). Proper garbage, waste water, and refuse disposal systems are commonly lacking at fairs and festivals. In addition, the public may also feed the animals, causing food and waste to scatter on the ground. These materials deteriorate and provide food and harborage for insects and rodents, which spread pathogens to raw food and ready-to-eat food. Although information about foodborne illnesses directly associated with pests and animals is unavailable, the presence of flies and rodents signifies an unsatisfactory hygiene level.

16.4.4 TIME AND TEMPERATURE FACTORS

Time and temperature control during food production is paramount to safeguarding consumers' health. Boo et al. (2000) showed that consumers were primarily concerned about whether food was cooked thoroughly

(23.7%). The duration (8.4%) and temperature (5.7%) of food being held at fairs and festivals are two other aspects of concern. Regrettably, only a small number of vendors know the temperature danger zone (22% in 2006 and 14% in 2008), the reheating temperature requirement for food safety (22% in 2006 and 29% in 2008), and the cooking endpoint for pork (41% in 2006 and 43% in 2008) (Lee et al., 2010). Many foodborne outbreaks have been caused by time and temperature abuse during cooking, reheating, and holding (Bean et al., 1990; Brown et al., 1987; Morgan et al., 1994; Todd et al., 2007b).

16.4.4.1 COOKING

Inadequate cooking is frequently recognized in outbreaks of trichinosis, botulism, and sometimes salmonellosis. In 2011, the CDC estimated that there were over 1 million illnesses and approximately 400 deaths associated with *Salmonella*-contaminated food while more than 800,000 illnesses and 76 deaths with *Campylobacter* (CDC, 2013c). Majority of the cases are associated with eating insufficiently cooked or improperly treated food products. In an investigation conducted by Brown et al. (1987), inadequate cooking and reheating of chicken, barbecued beef and pork ribs was reported to be responsible for the outbreak at a large annual fair in Pierce County, Washington in 1985. At the end of June 1992, an outbreak of *Campylobacter enteritis* at a large musical festival was also documented (Morgan et al., 1994). The cause was linked to the consumption of unpasteurized milk. In 2002, the working group for the investigation of the salmonellosis outbreak in Torroella de Montgri found that the vanilla filling used to make *coca* was cooked in large containers in ovens with peripheral heat sources. Furthermore, there was no temperature control during cooking. Peripheral heating and the lack of temperature control during cooking were thought to contribute partly to the large *Salmonella* infection outbreak, which resulted in 1,435 infection cases and 117 hospitalizations (Camps et al., 2005).

Time and temperature control during cooking is critical. Vegetative forms of pathogenic bacteria may be present on incoming food. These microorganisms can survive if the foods are not thoroughly cooked. However, high-temperature cooking must be coupled with a sufficiently long heating interval. Vegetative pathogens can survive when either the cooking temperature or time is lower than the established protocols.

16.4.4.2 REHEATING

As discussed earlier, the contamination of cooked foods before reheating is relatively common and thus highlights the critical role of the reheating stage. Periodic reheating may prevent extensive bacterial growth. Nevertheless, any heat-resistant toxins, such as staphylococcal enterotoxins, that are already present would persist. Furthermore, as with cooking, reheating may only kill the vegetative forms of pathogenic bacteria. The spores germinate when conditions become suitable. A more critical hazard in reheating thus occurs when foods are not extensively heated. The risk of insufficient reheating would be greater because warming up cooked and chilled foods creates favorable conditions for the germination of spores and the multiplication of vegetative bacteria cells. Insufficient reheating has been reported as one of the factors that contributed to the salmonellosis outbreaks at a large annual fair in Pierce County, Washington in 1985 (Brown et al., 1987).

16.4.4.3 HOLDING

Holding food after cooking poses the greatest hazard of all. Research has shown that the major factor in foodborne illness outbreaks is the time and temperature abuse of food being held (Food and Drug Administration, 2000; Todd et al., 2007b). In fact, vendors at temporary foodservice establishments were found significantly lacking in their knowledge pertaining to cooling, holding, and the correct temperature for frozen foods (Neel, 2010).

Foods that are completely cooked and consumed immediately present little or no risk of foodborne illness. However, as the time between cooking and eating increases, foods stored at ambient temperature present a considerable risk. Many foods sold at temporary foodservice establishments, such as fairs and festivals, are prepared early in the morning or sometimes a day before and are displayed throughout the day until they are sold. This prolonged hot-holding is frequently identified in outbreaks caused by *Bacillus cereus*, *Clostridium perfringens*, *Salmonella*, *Staphylococcus* enterotoxin, and *Vibrio parahaemoliticus* (Bryan, 1988). Microorganisms multiply rapidly to dangerous levels when foods are held. A prior study conducted by Yang and Xu (1991) noted that the aerobic bacteria count of deep-fried foods rose more than 10 to 100 times during holding. Aerobic mesophilic colony counts could progressively increase from approximately 10^3/g after cooking to between 10^5–10^9/g during display (Bryan, 1993).

When cooked food is to be stored for later consumption, it should be cooled down within a short time. The product height or pan height is a critical factor to enhance the cooling process. Unfortunately, approximately 66% of the food workers at fairs and festivals are not knowledgeable about the appropriate pan height for cooling (Neel, 2010). In 1974, 50 guests who attended a festival in the County of Avon, UK were infected with *Clostridium perfringens* (Hewitt et al., 1986). Salmon mayonnaise served at the festival was identified as the vehicle. A large quantity of the fish had been boiled and left to cool overnight in its own juices. Although a refrigerator was used, apparently it did not function properly to cool the fish efficiently. Inefficient cooling of cooked food was also indicated as the major cause of the outbreak of *Staphylococcus aureus* at a university festival (Kitamoto et al., 2009). The crepes prepared by university students were wrapped before they had cooled down. Some were then stored in a home refrigerator, while others were kept at room temperature. An experiment reproducing the process showed that the crepes did not cool down to 25°C within a short time. It took at least three hours for the upper-layer crepes in the refrigerator to reach 25°C and more than 11 h for those kept at room temperature. Within six hours, the crepes began to produce enterotoxins. Meanwhile, the time lapse between cooking and selling the food was as long as 18.5 h.

Together with the time lapse between preparation and consumption, proper holding temperature plays an important role. Cooked foods are often kept warm in steam tables or hot-air cabinets. However, because of improperly designed or operated equipment at fairs and festivals, foods are sometimes found to fall within the danger zone. Ambient temperatures coupled with relative high humidity create a hotbed for the multiplication of pathogens. Unfortunately, 18% of the food vendors said they would not be concerned or were not sure about cooked meat being left at room temperature for more than four hours (Williamson et al., 1992).

As with hot food, improper holding temperatures for cold food also impose a health threat to fair attendees, though cold food holding temperatures are seldom a critical issue at temporary foodservice establishments (Bryan, 1988). Wyatt (1979) showed that 75% of the respondents did not know the acceptable temperature for holding frozen foods. The number remains low (62%) 30 years later (Neel, 2010). Burch and Sawyer (1991) also found that 15% or fewer of the food operators knew the correct temperature for holding chilled food. The safety risk associated with cold food may be heightened when foods are displayed at ambient temperature without proper mechanical refrigeration units (Worsfold, 2003).

16.5 IMPROVING FOOD SAFETY AND HYGIENE AT FAIRS AND FESTIVALS

It is apparent that both fair and festival organizers and food vendors should be motivated to ensure food safety. Food organizers are advised to provide sufficient space to work, suitable facilities for food handling with different risk levels, and adequate hand-washing units with appropriate water temperatures (Worsfold, 2003). Food handlers should be trained or educated to improve their knowledge of food safety and practices pertaining to good hygienic practices and the HACCP system (World Health Organization, 2000).

However, the findings on the effect of such knowledge were inconclusive. For instance, Lee et al. (2010) found that attendance at the educational workshop on food safety impacted the vendors' knowledge scores significantly in 2008. Conversely, the researchers also demonstrated that food safety knowledge was not significantly related to the number of violations, which substantiated earlier findings by Mathias, Riben, and Weins (1994). Similarly, Acikel et al. (2008) found that the knowledge that food safety trainees attained from the training programs or educational workshops was not reflected in all behavior, although other studies reported positive changes in the vendors' level of knowledge, skills, and attitude (Howells et al., 2008; Capunzo et al., 2005; Lillquist McCabe, and Church, 2005).

Binkley (2005) and Griffin and Neal (2000) also showed that food safety knowledge did not impact food safety performance significantly. In short, training may improve the food workers' safety knowledge, but not always their actual food safety practices (Angelillo et al., 2000; Stivers and Gates, 2000).

Several researchers have indicated that effective food safety training courses should not focus solely on the theoretical elements of food safety (DiPietro, 2006; Howells et al., 2008). Hands-on activities emphasizing personal hygiene and food handling are recommended (Sousa, 2008). More importantly, food safety training should be in accordance with the type of operation (Choi and Almanza, 2012) and be provided near the time of event so that the effect of training may be more evident (Lee et al., 2010). Researchers have highlighted distinct food safety problems at different types of foodservice establishments (Frash et al., 2003; Lee et al., 2010). For instance, limited production space and facilities and dirt and dust contamination are specific potential risks at fairs and festivals (Boo et al., 2000; Worsfold, 2003). Hence, it would be wise to develop training programs that address the specific food safety problems at fairs and festivals. In addition, the presence of a knowledgeable manager or person-in-charge may be more important than mandatory training for all food handlers (Lynch et al., 2003). This suggestion was echoed by Pilling et al. (2008). The researchers argued that the managers or persons-in-charge have the power to ensure adequate resources and reduce time pressures on handlers. From their interviews, Green and Selman (2005) also found that food handlers admitted that they were more likely to abide by the safety procedures when a reprimand from a manager was expected.

Clearly, effective employee training, especially for the people-in-charge, and certification of vendors can improve food safety standards and reduce consumers' health risk at fairs and festivals. The challenge lies in the training content and method of delivery.

16.6 CONCLUSION AND FUTURE RESEARCH IN FOOD SAFETY AT FAIRS AND FESTIVALS

Fairs and festivals are commonly associated with social and cultural activities. Millions of people are drawn to fairs and festivals domestically or abroad every year. Consequently, fairs and festivals have emerged as tourism products that contribute significantly to local and national economies.

Food tasting is one of the favorite activities of consumers at many fairs and festivals. Considering the high number of attendees and the frequency with which food is consumed, unsafe food sold during these events could have a profound negative impact. Unfortunately, the reports of foodborne outbreaks have not abated.

Food safety at fairs and festivals should be safeguarded by the organizers, health officers, and vendors. Food vendors play the most critical role. They should ensure safe food and water sources, uphold good hygienic practices, and adhere to proper time and temperature control. Additionally, health officers should enforce the food safety and hygiene standards by constantly inspecting the temporary foodservice premises. Finally, it is the responsibility of fair and festival organizers to provide proper structure and facilities to food vendors. Safe food-handling and certification should be mandatory.

It is important to note that the food safety knowledge and skills that vendors acquire from trainings and workshops may not always be translated into practice. The training programs' content and delivery method, the duration and time of training, and the responsibility of managers or people-in-charge warrant further research attention to determine their effectiveness at improving food handlers' safety and hygiene practices.

The review of foodborne outbreaks at fairs and festivals has generally underscored the microbiological hazards. The likelihood of chemical contamination or chemical risks in food, however, is often overlooked. For instance, while the cleaning and sanitizing of equipment surfaces to prevent cross-contamination is emphasized, incorrect sanitizer use appears to be a more common food safety violation at fairs and festivals (Lee et al., 2010). The illnesses caused by chemicals in food, however, may not occur immediately.

Besides incorrect chemical use, a number of other chemical-related food safety concerns have also arisen recently. For example, acrylamide was detected in foods that were subjected to high temperatures (Granda et al., 2004; Lingnert et al., 2002; Stadler et al., 2002; Williams, 2005). In addition, polycyclic aromatic hydrocarbons (PAH) have been found in grilled, smoked, fried, cured, and roasted foods (Sundararajan et al., 1999;

Perello et al., 2009; Chung et al., 2011). Foods typically served at fairs and festivals include barbecued pork, grilled beef, and French fries, and are thus an interest for future research as potential food safety threats. Future research may also wish to examine awareness among vendors and consumers, identify the chemical risks in foods sold at fairs and festivals, and develop relevant training programs to curb this emerging food safety issue.

KEYWORDS

- ***Campylobacter enteritis***
- ***Clostridium perfringens***
- ***Salmonella enteritidis***
- ***Staphylococcus aureus***

REFERENCES

Acikel, C.H., Ogur, R., Yaren, H., Gocgeldi, E., Ucar, M., and Kir, T. (2008). The hygiene training of food handlers at a teaching hospital. *Food Control, 19*(2), 186–190.

Angelillo, I.F., Viggiani, N.M.A., Rizzo, L., and Bianco, A. (2000). Food handlers and foodborne diseases: Knowledge, attitudes, and reported behavior in Italy. *Journal of Food Protection, 63*(3), 381–385.

Barker, W.H., and Runte, V. (1972). Tomato juice-associated gastroenteritis, Washington and Oregon, 1969. *American Journal of Epidemiology, 96*(2), 219–226.

Bean, N.H., Griffin, P.M., Goulding, J.S., and Ivey, C.B. (1990). Foodborne disease outbreaks, 5-year summary, 1983–1987. *CDC Surveillance Summaries, Morbidity and Mortality Weekly Report, 39*(ss1, March), 15–57.

Bean, N.H., and Goulding, J.S., Lao, C., and Angulo, F.J. (1996). Surveillance for foodborne-disease outbreaks – United States, 1988–1992. Morbidity and Mortality Weekly Report, *CDC Surveillance Summaries, 45*(5), 1–66.

Binkley, M.M. (2005). The impact of foodservice manager credentialing on food safety knowledge and health inspection scores. Unpublished doctoral dissertation, Purdue University, West Lafayette.

Boo, H.C., Ghiselli, R., and Almanza, B.A. (2000). Consumer perceptions and concerns about the healthfulness and safety of food served at fairs and festivals. *Event Management, 6*(2), 85–92.

Botelho-Nevers, E., and Gautret, P. (2013). Outbreaks associated to large open air festivals, including music festivals, 1980 to 2012. *Euro Surveillance, 18*(11), 1–9.

Brown, M.B., Veazie, M.A., and Harris, N. (1987). An outbreak of salmonellosis at a large fair in Washington State. *Journal of Environmental Health, 49*(4), 224–227.

Bryan, F.L. (1981). Hazard analysis of food service operations. *Food Technology, 35*(2), 78–87.

Bryan, F.L. (1988). Risks of practices, procedures and processes that lead to outbreaks of foodborne diseases. *Journal of Food Protection, 51*(8), 663–673.

Bryan, F.L. (1993). HACCP – Street vending in developing countries. *Food Australia, 45*(2), 80–84.

Bryan, F.L., Teufel, P., Raiz, S., Roohi, S., Qadar, F., and Malik, Z. (1992). Hazards and critical control points of street vended chat, a regionally popular food in Pakistan. *Journal of Food Protection, 55*(9), 708–713.

Burch, N.L., and Sawyer, C.A. (1991). Food handling in convenience stores. *Journal of Environmental Health, 54*(3), 23–27.

Burt, B.M., Volel, C., and Finkel, M. (2003). Safety of vendor-prepared foods: Evaluation of 10 processing mobile food vendors in Manhattan. *Public Health Reports, 118*(5), 470–476.

Camps, N., Dominguez, A., Company, M., Perez, M., Pardos, J., Llobet, T., Usera, M.A., Salleras, L., and The Working Group for the Investigation of the Outbreak of Salmonellosis in Torroella de Montgri. (2005). A foodborne outbreak of salmonella infection due to overproduction of egg-containing foods for a festival. *Epidemiology and Infection, 133*(5), 817–822.

Capunzo, M., Cavallo, P., Boccia, G., Brunetti, L., Buonomo, R., and Mazza, G. (2005). Food hygiene on merchant ships: The importance of food handlers' training. *Food Control, 20*(9), 807–810.

CDC. (2013a). Surveillance for Foodborne Disease Outbreaks – United States, 1998–2008.
Foodborne disease outbreaks – US. 2007. *Morbidity and Mortality Weekly Report, 62*(SSO_2), 1–34.

CDC. (2013b). Surveillance for Foodborne Disease Outbreaks — United States, 2009–2010. *Morbidity and Mortality Weekly Report, 62*(03), 41–47.

CDC (2013c). CDC Estimates of Foodborne Illness in the United States. Retrieve July 29 from http://www.cdc.gov/foodborneburden/2011-foodborne-estimates.html#annual.

Cho, S., Bender, J.B., Diez-Gonzalez, F., Fossler, C.P., Hedberg, C.W., Kaneene, J.B., Ruegg, P.L., Warnick, L.D., and Wells, S.J. (2006). Prevalence and characterization of *Escherichia coli O157* isolates from Minnesota dairy farms and county fairs. *Journal of Food Protection, 69*(2), 252–259.

Choi, J.K., and Almanza, B. (2012). An assessment of food safety risk at fairs and festivals: A comparison of health inspection violations between fairs and festivals and restaurants. *Event Management, 16*(4), 295–303.

Christian, K.A., Schlegel, J., Ard, L., Mays, E., Curry, P., and Davis, M. (2008, April 14–18). Outbreak of *Salmonella* serotype Thompson associated with boiled peanuts – South Carolina, 2006. Paper presented at the 57th Annual Epidemic Intelligence Service Conference, Atlanta, GA.

Christos, J. (2006, October 6). Amazing weather helps feast. *Journal and Courier.*

Chung, S.Y., Yettella, R.R., Kim, J.S., Kwon, K., Kim, M.C., and Min, D.B. (2011). Effects of grilling and roasting on the levels of polycyclic aromatic hydrocarbons in beef and pork. *Food Chemistry, 129*(4), 1420–1426.

Clark, B. (2012, October 21). Yet another E. coli outbreak – Is it time to ban petting zoos? *Food Poison Journal.* Retrieved June 13 from http://www.foodpoisonjournal.com/

foodborne-illness-outbreaks/yet-another-e-coli-outbreak-is-it-time-to-ban-petting-zoos/#.UYb_rbX-FqU

Coté, T.R., Convery, H., Robinson, D. Ries, A., Barrett, T., Frank, L., Furlong, W., Horan, J., and Dwyer, D. (1995). Typhoid fever in the park: Epidemiology of an outbreak at a cultural interface. *Journal of Community Health, 29*(6), 451–458.

Crampin, M., Willshaw, G., Hancock, R., Djuretic, T., Elstob, C., Rouse, A., Cheasty, T., Stuart, J. (1999). Outbreak of *Escherichia coli O157* infection associated with a musical festival. *European Journal of Clinical Microbiology and Infectious Disease, 18*(4), 286–288.

Cupp, E.W., Maré, C.J., Cupp, M.S., and Ramberg, F.B. (1992). Biological transmission of vesicular stomatitis virus (New Jersey) by Simulium Vittatum (Diptera: Simuliidae). *Journal of Medical Entomology, 29*(2), 137–140.

DiPietro, R.B. (2006). Return on investment in managerial training: Does the method matter? *Journal of Foodservice Business Research, 7*(4), 79–96.

Fatimah, U.Z.A.U., Boo, H.C., Sambasivan, M., and Salleh, R. (2011). Foodservice hygiene factors – The consumer perspective. International Journal of Hospitality Management, 30(1), 38–45.

Frash, R., Jr., Almanza, B., and Stahura, J. (2003). Assessment of food safety risk: A case study in Marion County, Indiana. *International Journal of Hospitality and Tourism Administration, 4*(4), 25–44.

Fein, S.B., Lin, C.T.J., and Levy, A.S. (1995). Foodborne illness: Perceptions, experience, and preventive behaviors in the United States. *Journal of Food Protection, 58*(12), 1405–1411.

Food and Drug Administration. (1997). Food Code. US Department of Health and Human Services: Washington, DC.

Food and Drug Administration. (2000). Report of the FDA retail food program database of foodborne illness risk factors. Retrieved July 29 from

http://www.fda.gov/Food/GuidanceRegulation/RetailFoodProtection/FoodborneIllness-RiskFactorReduction/ucm123544.htm

Food Poisoning Lawyer. (2007a, June 14). Lemonade source of norovirus outbreak. *Food Poison Journal*. Retrieved April 7 from http://www.foodpoisonjournal.com/foodborne-illness-outbreaks/lemonade-source-of-norovirus-outbreak

Food Poisoning Lawyer. (2007b, July 20). Salmonella cases linked to Taste of Chicago increase. *Food Poison Journal*. Retrieved April 7 from http://www.foodpoisonjournal.com/food-poisoning-watch/salmonella-cases-linked-to-taste-of-chicago-increase/#.UYi1OLX-FqU

Food Standards Agency. (2002). Largest ever survey of catering staff shows that one in three do not wash their hands after visiting the lavatory. Retrieved April 7 from http://tna.europarchive.org/20110116113217/http://www.food.gov.uk/news/pressreleases/2002/oct/handwash

Gatziolis, C., and May, M. (2012, July 16). A fresh taste draws 1.2 million attendees: Five-day festival attracts hungry crowd. Retrieved April 7 from http://www.cityofchicago.org/city/en/depts/dca/provdrs/chicago_festivals/news/2012/jul/a_fresh_taste_draws12m illionattendees.html

Getz, D. (1989). Special events: Defining the product. *Tourism Management, 10*(2), 125–137.

Gibson, C., Waitt, G., Walmsley, J., and Connell, J. (2010). Cultural festivals and economic development in nonmetropolitan Australia. *Journal of Planning Education and Research, 29*(3), 280–293.

Granda, C., Moreira, R.G., and Tichy, S.E. (2004). Reduction of acrylamide formation in potato chips by low-temperature vacuum frying. *Journal of Food Science, 69*(8), E405-E411.

Green, L., and Selman, C. (2005). Factors impacting food workers' and managers' safe food preparation practices: A qualitative study. *Food Protection Trends, 25*(12), 981–990.

Griffin, M.A., and Neal, A. (2000). Perceptions of safety at work: A framework for linking safety climate to safety performance, knowledge, and motivation. *Journal of Occupational Health Psychology, 5*(3), 347–358.

Hewitt, J.H., Begg, N., Hewish, J., Rawaf, S., Stringer, M., and Theodore-Gandi, B. (1986). Large outbreaks of *Clostridium perfringens* food poisoning associated with the consumption of boiled salmon. *Journal of Hygiene, 97*(1), 71–80.

Howells, A.D., Roberts, K.R., Shanklin, C.W., Pilling, V.K., Brannon, L.A., and Barrett, B.B. (2008). Restaurant employees' perceptions of barriers to three food safety practices. *Journal of the American Dietetic Association, 108*(8), 1345–1349.

International Association of Fairs and Expositions. (2013). The history of fairs. Retrieved April 7 from http://www.fairsandexpos.com/aboutiafe/history.

Janiskee, R.L. (1996). The temporal distribution of America's community festivals. *Festival Management and Event Tourism, 3*(3), 129–137.

Kitamoto, M., Kito, K., Niimi, Y., Shoda, S., Takamura, A., Hiramatsu, T., Akashi, T., Yokoi, Y., Hirano, H., Hosokawa, M., Yamamoto, A., Agata, N., and Hamajima, N. (2009). Food poisoning by *Staphylococcus aureus* at a University Festival. *Japan Journal of Infectious Disease, 62*, 242–243.

Knabel, S.J. (1995). Foodborne illness: Role of home food handling practices. *Food Technology, 49*(4), 119–131.

Lee, A.L., Ostroff, S.M., McGee, H.B., Johnson, D.R., Downes, F., Cameron, D.N., Bean, N.B., and Griffin, P.M. (1991). An outbreak of shigellosis at an outdoor music festival. *American Journal of Epidemiology, 133*(6), 608–615.

Lee, J.E., Almanza, B.A., and Nelson, D.C. (2010). Food safety at fairs and festivals: Vendor knowledge and violations at a regional festival. *Event Management, 14*(3), 215–223.

Lillquist, D.R., McCabe, M.L., and Church, K.H. (2005). A comparison of traditional handwashing training with active handwashing training in the food handler industry. *Journal of Environmental Health, 67*(6), 13–16.

Lingnert, H., Grivas, S., J□gerstad, M., Skog, K., T□rnqvist, M., and Åman, P. (2002). Acrylamide in food: Mechanisms of formation and influencing factors during heating of foods. *Scandinavian Journal of Nutrition, 46*(4), 159–172.

Little, C., and Sagoo, S. (2009). Evaluation of the hygiene of ready-to-eat food preparation areas and practices in mobile food vendors in the UK. *International Journal of the Environmental Health Research, 19*(6), 431–443.

Lynch, R., Elledge, C., Griffith C., Boatright, D. (2003). A comparison of food safety knowledge among restaurant managers, by source of training and experience in Oklahoma County, Oklahoma. *Journal of Environmental Health, 66*(2), 9–14.

Manning, C.K. (1994). Food safety knowledge and attitudes of workers from institutional and temporary foodservice operations. *Journal of the American Dietetic Association, 94*(8), 895–897.

Manning, C.K., and Snider, O.S. (1993). Temporary public eating places: Food safety knowledge, attitudes and practices. *Journal of Environmental Health, 56*(1), 24–28.

Mathias, R., Riben, P., and Weins, M. (1994). The evaluation of the effectiveness of routine restaurant inspections and education of food handlers: Restaurant inspection survey. *Canadian Journal of Public Health, 85*(2), 61–66.

Medeiros, L.C., George, R.T., Brusns, K., Chandler, C., Crusey, S., Fittro, J., Hill, M., Jess, M., Miller, C., Reid, J., and Welker, E. (1996). The safe food handling for occasional quantity cooks curriculum. *Journal of Nutrition Education, 28*(1), 39–43.

Mihalik, B.J., and Fergusaon, M.F. (1994). Visitor profile analysis: A pilot market research study of an American state fair. *Journal of Travel and Tourism Marketing, 3*(4), 85–103.

Millard, P.S., Gensheimer, K.F., Addiss, D.G., Sosin, D.M., Beckett, G.A., Jankoski, A.H., and Hudson, A. (1994). An outbreak of cryptosporidiosis from pressed apple cider. *Journal of the American Medical Association, 272*(20), 1592–1596.

Minnesota Department of Health. (2006). An outbreak of *Escherichia coli O157:H7* infections. Retrieved April 15 from http://www.marlerblog.com/files/2013/02/LONGVILLE-final-writeup.pdf

Morgan, D., Gunneberg, C., Gunnell, D., Healing, T.D., Lamerton, S., Soltanpoor, N., Lewis, D.A., and White, D.G. (1994). An outbreak of *Campylobacter* infection associated with the consumption of unpasteurized milk at a large festival in England. *European Journal of Epidemiology, 10*(5), 581–585.

Neel, P. (2010). Temporary public eating places: Food safety knowledge and practices. Unpublished master's thesis, University of Missouri, Columbia.

O'Sullivan, D., and Jackson, M.J. (2002). Festival tourism: A contributor to sustainable local economic development? *Journal of Sustainable Tourism, 10*(4), 325–342.

Perello, G., Marti-Cid, R., Castell, V., Llobet, J.M., and Domingo, J.L. (2009). Concentrations of polybrominated diphenylethers, hexachlorobenzene and polycyclic aromatic hydrocarbons in various foodstuffs before and after cooking. *Food and Chemical Toxicology, 47*(4), 709–715.

Pilling, V., Brannon, L., Shanklin, C., Roberts, K., Barret, B., and Howells, A. (2008). Food safety training requirements and food handler's knowledge and behaviors. *Food Protection Trends, 28*(3), 192–200

Sloan, A.E. (1995). Feeling safe about food safety. *Food Technology, 49*(6), 21.

Sousa, C.P. (2008). The impact of food manufacturing practices on food borne diseases. *Brazilian Archives of Biology and Technology, 51*(4), 815–823.

Stadler, R.H., Blank, I., Varga, N., Robert, F., Hau, J., Guy, P.A., Robert, M.C., and Riediker, S. (2002). Acrylamide from Maillard reaction products. *Nature, 419*, 449–450.

Stivers, T.L., and Gates, K.W. (2000). Survey of grocery store seafood employees. *Dairy Food Environment Sanitation, 20*(10), 746–752.

Sundararajan, N., Nadife, M., Basel, R., and Green, S. (1999). Comparison of sensory properties of hamburgers cooked by conventional and carcinogen reducing 'safe grill' equipment. *Meat Science, 51*(4), 794–797.

Tebbutt, G.M. (1991). Assessment of hygiene risks in premises selling take-away foods. *Environmental Health, 99*(4), 97–100.

Tebbutt, G.M. (1993). Risk-assessment analysis in premises selling raw and cooked meats. *International Journal of Environmental Health Research, 3*(4), 217–224.

Todd, E.C.D., Greig, J.D., Bartleson, C.A., and Michaels, B.S. (2007a). Outbreaks where food workers have been implicated in the spread of foodborne disease. Part 2. Description of outbreaks by size, severity and settings. *Journal of Food Protection, 70*(8), 1975–1993.

Todd, E.C.D., Greig, J.D., Bartleson, C.A., and Michaels, B.S. (2007b). Outbreaks where food workers have been implicated in the spread of foodborne disease. Part 3. Factors contributing to outbreaks and description of outbreak categories. *Journal of Food Protection, 70*(9), 2199–2217.

Williamson, D.M., Gravani, R.B., and Lawless, H.T. (1992). Correlating food safety knowledge with home food preparation practices. *Food Technology, 46*(5), 94, 96, 98, 100.

Williams, J.S.E. (2005). Influence of variety and processing conditions on acrylamide levels in fried potato crisps. *Food Chemistry, 90*(4), 875–881.

Willis, C., Elviss, N., Aird, H., Fenelon, D., and McLauchlin, J. (2012). Evaluation of hygiene practices in catering premises at large-scale events in the UK: Identifying risks for the Olympics 2012. *Public Health, 126*(8), 646–656.

World Health Organization. (2000). Foodborne diseases: A focus on health education. Geneva, Switzerland.

Worsfold, D. (2003). Food safety at shows and fairs. *Nutrition and Food Science, 33*(4), 159–164.

Wyatt, C.J. (1979). Concerns, experiences, attitudes and practices of food market managers regarding sanitation and safe food handling procedures. *Journal of Food Protection, 42*(7), 555–560.

Yang, M.L., and Xu, S.Y. (1991). Investigation of bacterial contamination of street-vended foods. *Dairy, Food and Environmental Sanitation, 11*(12), 725–727.

CHAPTER 17

FOODSERVICE AND SENIORS IN MEALS ON WHEELS PROGRAMS AND CONGREGATE MEAL SITES: A SERVICE WITH CHALLENGES

LIONEL THOMAS, PhD

Assistant Professor, North Carolina Central University

CONTENTS

As a part of the Older Americans Act (1965), State and Area Agencies on Aging provide nutrition and supportive services to eligible members of the older population (Department of Health and Human Services, 2006). The purpose of the Administration on Aging has been to allocate funds to Area Agencies on Aging to provide nutrition assistance to eligible individuals.

According to the amended Older Americans Act (2000), qualified individuals must be:

1. 60 years of age or older to qualify for services in general;
2. a caregiver of someone 60 years of age or older or an older individual caring for a child 18 years of age or younger to qualify for Family Caregiver Supports;
3. 55 years of age or older and have an adjusted income at or below 125% of the Federal Poverty Level to qualify for Older Worker Employment Services; and
4. in the greatest social or economic need, with a special focus on low-income minority individuals and individuals living in rural areas.

Older individuals specifically targeted by the USDA Administration on Aging (2004) and the U.S. Department of Health and Human Services (2003) are described as follows:

1. persons 60 years or older in greatest social and economic need (with particular attention to low-income minority adults and older adults residing in rural areas)
2. older adults with severe disabilities, and/or limited English-speaking ability
3. and/or Alzheimer's disease or related disorders with neurological and organic brain dysfunction (and the caretakers of such individuals)
4. those participants who meet eligibility criteria and are at greatest nutritional risk

Individuals under the age of 60 are allowed to participate at congregate meal sites if they pay full cost, have a spouse that qualifies for the program, or if they are a volunteer at the site.

According to the National Restaurant Association Education Foundation (NRAEF) (2013), children, older individuals, pregnant individuals, and individuals with chronic conditions or otherwise compromised immune systems are especially susceptible to foodborne illness. Prevention of foodborne illness takes on an increased emphasis because the majority of the clients served by congregate meal sites and the Meals on Wheels programs are elderly individuals who also have a diminished ability to combat illness (U.S. Food and Drug Administration, 2011). The ability to acquire and prepare meals that are nutritious and appealing, eat independently, dine in an environment that promotes proper caloric intake, and receive dietary assistance contribute to an adequate diet for elderly Americans (Payette and Shatenstein, 2005). Equally as important as providing a nutritious meal to elderly individuals is ensuring their meals are safe to eat. Failure to hold food at the proper temperature is one of the five most common factors responsible for foodborne illness as outlined by the US Food and Drug Administration's Food Code (U.S. Food and Drug Administration, 2011).

Two of the more well-known components of the Administration on Aging Nutrition Services as outlined by Title III of the Older American Act are Meals on Wheels (MOW) programs and the congregate meal sites programs. The Meals on Wheels program is designed to help combat hunger and poor diets for the homebound, disabled, and frail as well as individuals who are at risk socially, physically, nutritionally, and economically (Johnson and Fischer, 2004; Meals on Wheels Inc. of Tarrant County, 2004; Wellman and Kamp, 2004). According to 2010 figures, the federal nutrition services programs for the elderly annually provide approximately 242 million meals (including 96,426,593 meals at congregate sites and 145,454,44 home delivered meals) to a total of 2.6 million participants (including 1,733,176 congregate meal participants and 868, 076 home delivered meal participants) (Administration on Aging, 2013). Funding for these programs has remained relatively consistent according to 2008 to 2010 data. In 2008 these combined programs accounted for 35.48% ($493,610,508) of the total expenditures of State Units on Aging ($1,391,384,078) with congregate meal sites accounting for $265,468,813 and MOW accounting for $228,141,695. In 2010 these programs accounted for 36.88% ($522,447,211) of the total expenditures of State Units on Aging ($1,416,673,830) with congregate meal sites accounting for $278,662,972 and MOW $243,784,239.

The importance of both of these programs is clearly important for the large number of clients that they serve and from the government's perspective of dollars spent. They are also important from a food safety perspective. For the Meals on Wheels programs, temperature during delivery is a critical issue. According to the FDA (2011), food must maintain an internal temperature of at least 135°F while in transport to an offsite location. If time is used instead of temperature as the means of controlling the safety of the food item, it must be consumed within four hours of leaving the facility (U.S. Food and Drug Administration, 2011) Some sites use more conservative guidelines and set the maximum time for delivery of a meal to a client of two hours (Brovont, 2005). In reality, some sites report that it takes an average of an hour and a half from the time a meal leaves the facility for a driver to reach their last client in good driving conditions (Brovont, 2005). Given that the actual time of consumption once meals leave the facility is unknown, this raises concern as to the safety of the food being consumed,

The temperature of the food during delivery to clients is generally not an issue for congregate meal sites. Congregate meal sites provide eligible participants with hot or otherwise appropriate, nutritionally balanced meals served in a setting which promotes social interaction among the individuals (Bureau of Aging and In-Home Services, 2003; USDA Administration on Aging, 2004). Such facilities include community/senior centers, senior citizens' apartment complexes, retirement facilities, high schools, Salvation Army Centers, churches, and fire stations (Niesz, 2006). Participation in congregate meal programs has been specifically shown to enhance (1) daily nutrient intake, (2) nutrition status, (3) amount of social interactions, and (4) functionality of older adults (Wellman and Kamp, 2004). Meals provided at congregate meal sites are designed to account for one-third of the daily caloric intake or recommended dietary allowance for 70-year-old males. The delivery temperature of meals may be an issue however for meals prepared at a satellite facility and delivered to the congregate meal site where they must maintain their temperature integrity until they are consumed by the site participants.

17.1 MEALS ON WHEELS CONCERNS: A CASE STUDY

MOW operations bear the responsibility to protect their high risk clients from harm by maintaining proper food safety procedures throughout the flow of food (Bertagnoli, 1996). This means that they must ensure that food stays out of the temperature range conducive to bacterial growth (temperature danger zone): 41°F (6°C) to 135°F (57°C). This service takes on a particular challenge in that this food must be protected during preparation at a satellite facility and delivery to clients in rural areas or to a congregate meal site. The FDA mandates that hot food be kept above 135°F (57°C) and cold food be kept below 41°F (6°C) throughout the service process. If hot food items are below 135°F (57°C) for a period of four hours, the food item is to be discarded due to the increased potential for the rapid growth of bacteria (US Food and Drug Administration Center for Food Safety and Applied Nutrition, 2005). Maintaining food temperature during transportation from facility to client is a critical component of the Meals on Wheels flow of food due to the elderly population's susceptibility to foodborne illness. Some states such as New York, have implemented a two hour maximum delivery time due to the increased susceptibility of the target population to foodborne illness (New York State Department of Health, 2010).

One example of a facility that provides both home delivered and congregate meal service is Mid-Land Meals in Lafayette, Indiana. Mid-Land Meals was formed in 1974 with a simple mission to provide congregate meals for poverty level senior citizens in Indiana. It formed relationships with community and religious site sponsors to carry out this mission. Initially it produced about 30 meals per day, for a total of approximately 6,000 meals in the first year. Since the date of its inception, Mid-Land Meals has met the original goals of the corporation, and steadily expanded its services, especially in the area of home delivered meals. Over its years of activity it has produced and served more than 10 million meals. It is now a full partner agency with the Area Four Agency on Aging, providing assistance for client eligibility, nutrition reviews for menus, client record keeping and monitoring. With assistance from grants in 2002, Mid-Land Meals built a new expanded central kitchen and warehouse designed to prepare up to 2,000 meals per day. This kitchen was designed to prepare and ship hot bulk meals for each day of service. Due to increasing community need and company efforts, Mid-Land Meals now serves nearly 500,000 meals annually, more than 80 times its original level of service.

This facility has used many means of transporting meals to clients and congregate meal site facilities including the use of heated trucks, passenger vans, and individual vehicles (Brovont, 2005). Given this wide range of transport vehicles, some of which are not conducive for maintaining food temperature over an extended period of time, the actual containers used to transport the meals take on added significance. If the transport container fails to function as intended, there is the possibility that much of the time, energy, and expense used in the production of the food product will be wasted, and the health of the recipient could be placed in jeopardy (Robertson, 1993).

Food transport containers come in various shapes, sizes, colors, and employ various types of insulation and padding to help maintain food temperature. Also affecting the ability of these containers to maintain food temperature are the types of sealing techniques employed such as buckles or latches, zippers, and Velcro. Insulated nylon bags, insulated hard plastic containers, corrugated paper boxes, plastic bags, and standard thermal coolers are some of the more typical MOW transport containers (Brovont, 2005). Finally, the ability to clean the containers is also critical in maintaining the safety of foods transported in them. Some of the rigid plastic containers are more easily cleaned, but corrugated paper boxes cannot be cleaned and this may become important if they are dirty and reused for meal transport.

Because of the variability in the containers used to hold and transport seniors' meals, Thomas, et al. (2009) conducted a study of the more widely used transport containers at the request of Midland Meals, Inc. This study measured the ability of the containers to maintain heat throughout the transport process. The findings of this study raised concerns about the abilities of certain containers to maintain appropriate temperatures while meals are being delivered to the elderly clients and helped in their selection of the appropriate transport containers to use.

In general, the type of transport container was found to be very important for maintaining temperature and food integrity during transport. Unfortunately, the high cost associated with some of the better performing containers was found to make them too expensive for many "budget-strapped" Meals on Wheels providers (Thomas et al., 2009). Although all MOW operators desire to have the best transport units available; choice is dependent primarily upon cost – and then ability to maintain temperature, functionality, be cleaned, and durability.

The results of this study also showed that the number of meals in a given container had a significant impact on the ability of the container to maintain heat. As meals were removed from the containers, not only was cooler air introduced, but heat was removed. This was evidenced by the fact that none of the containers were able to keep the food outside the temperature danger zone when they were two-thirds (or less) full. This has serious implications since the temperature maintenance capabilities of the containers are significantly decreased as meals are being delivered given the length of some of the routes (Thomas, et al., 2009).

Perhaps even more surprising was the determination of when meals first entered the temperature danger zone of 41°F (6°C) to 135°F (57°C). Results showed that hot foods dropped below 135°F (57°C) as they were being portioned into the individual food containers that would be delivered to clients (Thomas, et al., 2009). Meals that were placed into hot holding units prior to being picked up by delivery drivers did regain temperatures above 135°F (57°C) before they were transported. Unfortunately, the results of this study suggested that meals that are not placed into hot holding units after portioning might start their delivery routes at the wrong temperature so that no matter what transport container was used, there might be a danger of being at the wrong temperature for the entire delivery time period.

In addition, it was found that not all clients were eating their meals immediately upon delivery. If a client meal were to be delivered prior to lunch, such as 11 AM, and the client did not eat the meal until 1 PM, the temperature concern would be extended, particularly if the clients did not store the meal at the right temperature. In fact, clients commonly stated that they stored the meal on the counter prior to eating it.

17.1.1 BETTER PRACTICES

To ensure that food temperature and integrity is properly maintained during delivery, there are a number of things that MOW operations can do even if they cannot afford the more expensive heated units. Inexpensive transport containers such as cardboard boxes and plastic bags do not provide adequate barriers to heat loss nor do they provide adequate protection from spilling or contamination. If MOW programs can afford electrical transport containers that actively maintain the hot temperatures, this may provide a safe alternative, particularly with long delivery routes. Unfortunately, electrical transport containers are expensive. Other alternatives may include nylon, microfiber, or plastic-types of durable insulated containers. Heat may not be added with these units, but these units will help to maintain the heat that accompanies the meals while protecting the contents from being damaged.

Another selection criteria should be the size of the container, which should be matched to the number of meals on the delivery route. Research results suggest that an appropriately sized container (in other words, that is completely filled with meals) will maintain the temperature of the meals inside for the longest time possible.

After selecting the best container, operations should consider the best ways to handle food to help ensure that the food stays out of the temperature danger zone during delivery. First, they should plan routes so that they are as short as possible. By shortening the routes, there is less time for the food to lose heat and drop in temperature. Another important practice is to ensure that the food is as hot as possible when it is placed into the containers. In the temperature ranges seen during delivery, the rate of heat loss and temperature drop will be relatively constant. By starting with a higher temperature, the food will take longer to enter the temperature danger zone.

In addition to a higher starting temperature, there are a number of other things MOW programs can do to help maintain the meal temperature. First, the containers can be preheated before food is added to them. Preheating the containers will reduce the initial heat lost by the food as the container temperature equalizes to that of the food. This can be done by placing plastic containers of hot water in the containers prior to adding the food. Operators also have the option of adding heat to containers in order to help maintain temperature. By adding items with high heat capacity such as nontoxic gel packs or metal or ceramic plates, the temperature drop of the food will be slowed and it will stay out of the danger zone longer.

For the example of Midland Meals, the focus in terms of temperature control of home delivered meals has changed over the last 10 years.

Production has been shifted to provide more frozen meals. The result of this change is that production has become more efficient, the desirability of the meals has been reported to have increased, and nutritional content of the food is also improved. The change has also required adjustments in staffing and equipment. Included in the frozen production of meals is the idea of blast chilling and freezing of the meals, which can then be shipped frozen via a Federal Express program straight to the client's door step. An average of just under 500 meals per day are currently being packaged for this program. Most importantly, this delivery method results in fewer concerns in regards to food safety because frozen meals remain in a safe temperature range for a far longer time period as compared to meals that are delivered hot. In addition, this is projected to save in both personnel costs and delivery expenses.

17.1.1.1 CHALLENGES

There are many overall challenges when it involves the provision and consumption of the meals provided by the Elderly Nutrition Services to an individual's home. Given the lack of financial resources to purchase containers capable of sustaining the heat of the meals, the idea of reducing the length of the routes would seem logical. However, the majority of the individuals delivering for Meals on Wheels are volunteers. Thus, there is often a shortage of drivers and many of them take on multiple routes.

In addition, the average portion size for the hot food items provided by Elderly Nutrition Services ranges from four to six ounces for the protein food item and four ounces for both the starch and vegetable food items. Increasing the temperature of a meal with these small quantities of food prior to transportation might have a negative impact on the quality of the food. This might also make the food less appealing and some protein food items difficult to consume, particularly in light of the possible complications some members of this population have with chewing. This in turn could lead to the individual not consuming the meal, which is the primary purpose of the program.

Another major concern, as discussed previously, is in regards to how clients handle the meals after they are delivered. On average, clients have been found to consume their meals 1.22 h after the food is delivered (Almanza et al., 2007). Although approximately 63 percent of MOW clients were found to consume their meals upon delivery, and the majority of individuals that did not consume their food immediately did place their hot and cold food in refrigerators, some clients left their meals out in the temperature danger zone until consumption. Careful attention should be paid to the entire period of time from onsite preparation to offsite consumption. Clients should be encouraged to either consume or refrigerate meals appropriately to keep them safe.

In summary, the success of home-delivered meal programs among older Americans is dependent on multifactorial collaborations. At the same time, clients need more information about how to handle meals in their homes and a better understanding of the importance of proper handling for prevention of foodborne illness (Almanza, et al., 2007). Home delivered meal providers might also wish to consider alternative temperature control methods, such as frozen meals rather than hot meals. Challenges will continue to occur in the provision of safe and nutritious meals, but continued efforts from foodservice providers in safe handling of home-delivered meals are important in helping to protect older Americans.

17.2 CONGREGATE MEAL SITES

If hot meals are transported from the facility to the meal sites, many of the same obstacles that the MOW division experiences are also encountered. The one benefit to transporting meals to the congregate meal sites (as compared to the home delivered meals) is that the meals are transported in bulk and do not have the issue of the transport container being opened and closed allowing for heat to leave the container. The additional concerns associated with meals served at congregate meal sites are that the meals are delivered prior to the meal service time; however, some of the facilities used to host the meal gatherings may not have the proper hot and cold holding equipment. While the actual delivery time and holding time until clients arrive for consumption is unknown, the fact that the meals (in most cases) remain in the delivery unit until consumption is cause for concern. The individuals receiving the meals at the facilities are generally volunteers who have received limited training. This means that understanding how to keep the meals as hot or cold as possible until consumption may not be a primary concern for the service provider making the actual temperature of the food upon consumption unknown – placing the safety of the food item in jeopardy.

As previously mentioned, Midland Meals of Lafayette, Indiana provides meals for 30 congregate meal sites across the State of Indiana and directly manages 12 sites (MidLand Meals, 2013). Through their program, they provide almost half a million meals annually to clients (including the home delivered meals). Their coverage area includes Benton, Carroll, Clinton, Fountain, Montgomery, Tippecanoe, Warren and White Counties. The meals produced at Midland Meals are distributed to community/senior centers, senior citizens' apartment complexes, retirement facilities, high schools, Salvation Army Centers, churches, and fire stations across Area IV. In line with the new focus on frozen meals, a new goal for Midland Meals is to obtain equipment at each congregate meal site so that all meals can be shipped in a frozen form to the meal sites, then stored, re-thermalized, and served. Delivery of the congregate meals in a frozen form is valuable in that it is likely to create fewer concerns in regards to food safety. As discussed for home delivered meals, frozen meals remain in a safe temperature range for a far longer time period as compared to meals that are delivered hot. In addition, the delivery of frozen congregate meals decreases personnel costs and delivery expenses.

In summary, more and more challenges are occurring today in the provision of congregate meals (as well as home delivered meals). Many financial challenges occur, as well as operational challenges in efficiently delivering and safely providing meals with limited financial, equipment, and personnel resources. Congregate meal programs are essential in providing safe and nutritious meals to older Americans and continued efforts from foodservice providers in safe handling of congregate meals is important in protecting older Americans.

KEYWORDS

- **budget-strapped**
- **congregate meals**
- **frozen meals**
- **Meals on Wheels**

REFERENCES

Administration on Aging. (2013, 4/24/3013). Nutrition Services (OAA Title IIIC) Retrieved 7/15, 2013, from http://www.aoa.gov/AOARoot/AoA_Programs/HCLTC/Nutrition_Services/index.aspx#home

Almanza, B. A., Namkung, Y., Ismail, J. A., and Nelson, D. C. (2007). Clients' Safe Food-Handling Knowledge and Risk Behavior in a Home-Delivered Meal Program. *Journal of the American Dietetic Association, 107*(5), 816–821. doi: http://dx.doi.org/10.1016/j.jada.2007.02.043

Bertagnoli, L. (1996). Taking it to the Streets. *Restaurant and Institutions, 106*(6), 112–113.

Brovont, E. (2005, July 15, 2005). [Interview at Midland Meals, Inc of West Lafayette].

Bureau of Aging and In-Home Services. (2003). Congregate Meal Sites and Home Delivered Meals Directory Retrieved October 21, 2006, from http://www.state.in.us/fssa/elderly/aging/pdf/nutrition%20directory%202003%20m aster.pdf

Department of Health and Human Services. (2006). State and Area Agencies on Aging Retrieved October 31, 2006, from http://www.aoa.gov/eldfam/How_To_Find/Agencies/Agencies.asp

Johnson, M. A., and Fischer, J. G. (2004). Eating and Appetite: Common problems and practical remedies. *Generations San Francisco, 28*(3), 11–17.

Meals on Wheels Inc. of Tarrant County. (2004). Eligibility Retrieved November 27, 2005, from http://www.mealsonwheels.org/eligibility/default.asp

MidLand Meals. (2013). MidLand Meals Retrieved 7/15, 2013, from http://www.midlandmeals.com/Default.aspx

New York State Department of Health. (2010). 14–1.40 Food protection, potentially hazardous food, temperature and refrigeration requirements Retrieved 7/15, 2013, from http://www.health.ny.gov/regulations/nycrr/title_10/part_14/subpart_14–1.htm#s80

Niesz, H. (2006). Elderly nutrition programs and nutrition services incentive program *OLR Research Report* (pp. 4). Washington, DC: US Department of Agriculture.

Payette, H., and Shatenstein, B. (2005). Determinants of healthy eating in community-dwelling elderly people. *Canadian Journal of Public Health, 96*(July/August), S 27–32.

Thomas, L., Nelson, D., Almanza, B., and Binkley, M. (2009). The use of thermal capacity in measuring the effectiveness of meals on wheels transport containers. *Journal of Florida International University Hospitality Review, 18*(4), 401–413.

U.S. Food and Drug Administration. (2011, 08/19/2013). Supplement to FDA Food Code 2009: Annex 4 – Management of Food Safety Practices – Achieving Active Managerial Control of Foodborne Illness Risk Factors Retrieved 9/4, 2013, from http://www.fda.gov/food/guidanceregulation/retailfoodprotection/foodcode/ucm272584.htm

US Food and Drug Administration Center for Food Safety and Applied Nutrition. (2005). Food Code Retrieved August 15, 2006, from http://www.cfsan.fda.gov/~dms/fc05-toc.html

USDA Administration on Aging. (2004). Nutrition services incentive program: Total meals served Retrieved November 28, 2006, from http://www.fns.usda.gov/pd/nsipmeals.htm

Wellman, N., and Kamp, B. (2004). Federal Food and Nutrition Assistance Programs for Older People. *Generations, 28*(3), 78–85.

PART 4

GLOBAL ISSUES

CHAPTER 18

DEALING WITH DISASTERS

SANDRA SYDNOR, PhD

Assistant Professor, Purdue University

CONTENTS

18.1 INTRODUCTION

A little story...

> *The majority of us make implicit assumptions that the rate of change in the happenings of our daily lives is fairly static. We get up, brush our teeth, think about breakfast, make coffee, get dressed, and dash out of the house or apartment to work or school with little variation. Our assumption is that we will, with very little deviation, repeat this scenario over and over until we intentionally plan to change it. And most of the time, we're right. But sometimes we're not. So even thinking about preparing for the big-impact-but-unlikely to happen events is itself unlikely, or as Seth Godin quipped, "Resilience, the ability to shift and respond to change, comes way down the list of the things we often consider" (Seth Godin's blog, April 26, 2013). And resilience, the ability to thrive, is precisely what we need in uncertain times such as these.*

Would you…or your team know what to do about your restaurant and its food inventory under the threat of an impending natural disaster?

Natural disasters reportedly inspire havoc not only of the built environment but of other channels as well, foodservice and food systems included (Sydnor-Bousso et al., 2011). In truth, contrary to popular media images and disaster movies that promulgate mass hysteria, people don't panic when disaster strikes. Rather, an eerie calm sets in, numbing the unprepared into a docile naivety regarding the danger confronting them (Ripley, 2005), making preparation more essential than ever. Nonetheless, floods, droughts, earthquakes, tornadoes…any disaster has the capacity of creating congested staging and/or living conditions, utility interruptions, and the proliferation of disease. Rates of foodborne illnesses are on the rise and overall, current data show a lack of progress in reducing foodborne infections (Fitzgerald, 2013).

How serious is the spread of infectious disease? Well, from a food and agriculture perspective there are approximately 2.17 million farms, covering more than 1 billion acres of land, with at least 1 million firms engaged in the supply chain (USDA, 2013). That encompasses possible contagion points cutting a wide and comprehensive swath across the United States. As an example of how pathogens can travel (Figure 1) from a source, any source, and the graphic below demonstrates a pathway for the spread of the common virus, influenza:

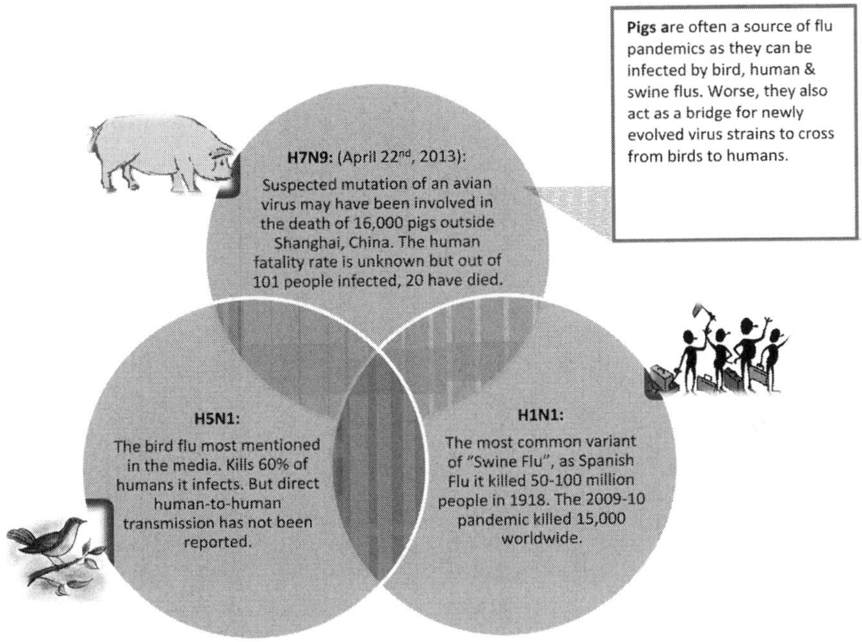

FIGURE 1 Who can catch the Flu.

Excerpted from: http://www.fastcodesign.com/1672417/infographic-which-flu-strain-is-most-likely to-bring-the-apocalypse?partner=newsletter

18.1.1 THE PROBLEM

When food supplies become compromised it is not only our physical health at stake; the entire food industry system becomes economically vulnerable as well as the individual firms that participate in food's production including family and commercial farms, processing and packaging, retailing, and distribution. The Hazard Analysis and Critical Control Points (HACCP) manual, published by the Food and Drug Administration (FDA) for foodservice and retail establishments suggests the following regarding the economic impact of foodborne illnesses:

> *"Foodborne illness in the United States is a major cause of personal distress, preventable death, and avoidable economic burden. Mead* et al. *(1999) estimated that foodborne diseases cause 76 million illnesses, 325,000 hospitalizations, and 5,000 deaths in the United States each year… The annual cost of foodborne illness in terms of pain and suffering, reduced productivity, and medical costs is estimated to be $10–83 billion."*

The economic fallout from foodborne illnesses on the food industry and its businesses ranges from damaged brand images and lost revenue to negative abnormal returns, particularly for small businesses (Seo, 2013).

This chapter is all about the most appropriate set of preparations and responses *your* business or operation can undertake to lessen the effects of natural disasters' impact on your food supply. The types of foodservice operations addressed here involve retail and institutional foodservice operations. They include:

- Back-country guided trips for groups
- Bakeries
- Bed and breakfast operations
- Cafeterias
- Camps – recreational, children's, etc.
- Casinos, bares, and taverns
- Child and adult day care
- Church kitchens
- Commissaries
- Community fund raisers
- Convenience stores
- Fairs
- Food banks
- Grocery stores with specialized departments
- Deli
    - In-store prepared foods
    - Produce
    - Meat and seafood
    - Health care facilities
- Interstate conveyances
- Markets
- Meal services for home-bound persons

- Mobile food carts
- Penal institutions
- Restaurants
  - Chains
  - International specialties
  - Quick Service Restaurants (QSR)/Fast food
  - Full service
  - Independent operations
  - Road-side stands
- Schools
- Snack bars
- Temporary outdoor events
- Vending machines

This list, generated from the HACCP manual produced by the FDA, includes a wide range of industries and employee resources, from highly trained executive chefs to entry-level front line employees. The employees encompass a broad range of education levels and communication skills. Accommodating such a wide repertoire is challenging; in-house training and maintaining a high level of trained staff may be difficult due to high industry turnover, low profit margins, scale economies, and language fluency. Additionally, many of these operations may be small, start-up businesses operating without the benefit of a large corporate infrastructure. One of the challenges associated with any recommendations for these operations is that collectively, they present an endless number of production techniques, products, menu items, ingredients used, and suppliers, and all of these characteristics may frequently change.

Given the sheer number and types of foodservice establishments, it isn't surprising then that each situation is different and each circumstance and operation will demand a nuanced scheme; there is no *one-size-fits-all* kind of recommended response to natural disasters' disruptions of a food supply. Nonetheless, there *is* a set of common actions foodservice operators can undertake to ensure the operation is doing its best to preserve high quality food supplies and protect customers from foodborne illnesses. The key is to *prevent, prepare for,* and *respond to* (Figure 2) natural disasters in ways that minimize damage and accelerate recovery. While these steps are being presented linearly, each step impacts the step before *and* after and the system exhibits high a high level of interdependence:

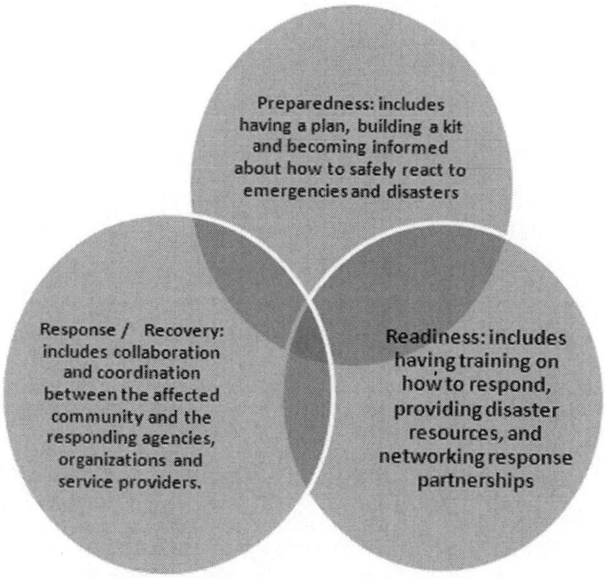

FIGURE 2 Prevent, Prepare, and Respond: The 3 P's and interconnectedness of the Disaster Cycle.

The first step in *preparing* for a disaster is to refine the slate of all possible disaster occurrences and then assess the likelihood of a disaster occurring in your specific area. Develop a plan, communicate the plan, and *rehearse* the plan. Make certain you have provisions capable of meeting at least 3 days worth of needs, including 1 quart of drinking water per person, per day. Be cognizant that *you are your own first responder* and supplies/suppliers may not be able immediately available; a 7-day supply is recommended (Beach, 2013).

When *in* an emergent crisis, response recommendations ranging from institutional school food service and large-scale health systems to Mom and Pop restaurants have a similar, 3-step process:

1. Assess the magnitude of the disruption,
2. Modify your emergency readiness plan accordingly,
3. Communicate what's happening to predesignated personnel. Remember, telephones, cell phones, e-mail, and computers may not operate in the event of an emergency. Consider alternate commu-

nication measures such as message carriers, drivers, or radio communication such as a walkie-talkie system.

(National Food Service Management Institute, 2003)

Post-disaster activities include assessing needs, reducing suffering, limiting the spread and consequences of the disaster, taking pictures, documenting everything, and paving the way for restoration.

In this chapter we will cover the nature of natural disasters, preventive strategies, preparedness activities, and suggested responses to the possibility of a natural disaster's impact on foodservice systems. What we do not cover in this chapter is intentional, manmade threats to food supplies such as agri-terrorism. Manmade disasters however are very real and increasingly relevant; for a well-versed treatment of food systems and *manmade* disaster vulnerability, preparation, surveillance, response, and recovery see the USDA's Food Defense and Emergency Response publications, as well as Kennedy, 2012.

18.2 A SHORT HISTORY ABOUT NATURAL DISASTERS

Natural hazards are naturally occurring physical phenomena and include drought; earthquake; flooding; fog; hail; heat; hurricane; landslide; lightning; severe storm; thunderstorm; tornado; tsunami; volcano; wildfire; wind; and winter weather (Hazards and Vulnerability Research Institute, 2013). Not all hazards result in natural disasters. Natural disasters result from the interaction of natural hazards and populations vulnerable to the consequences of extreme weather events, such as those presented by the natural hazards presented here (Cutter, 2003). Man-made disasters are specific events where an anthropogenic hazard has come to fruition. Anthropogenic disasters always involve intentional malfeasance, system failure, negligence, or error. Examples of man-made disasters include sociological hazards such as acts of terrorism, war, arson), technological hazardous (power outages, material spills such as oil), nuclear accidents and transportation accidents that occur in and on air, space, road, and by rail. While we may think of typical disaster impacts relative to lives impacted or asset and dollar damages suffered, because of its importance to food supply systems, food crops are also regularly accounted for in disaster damage calculations. Figure 3 illustrates natural disasters' place-based impact on food-growing regions of the United States during the year 2011.

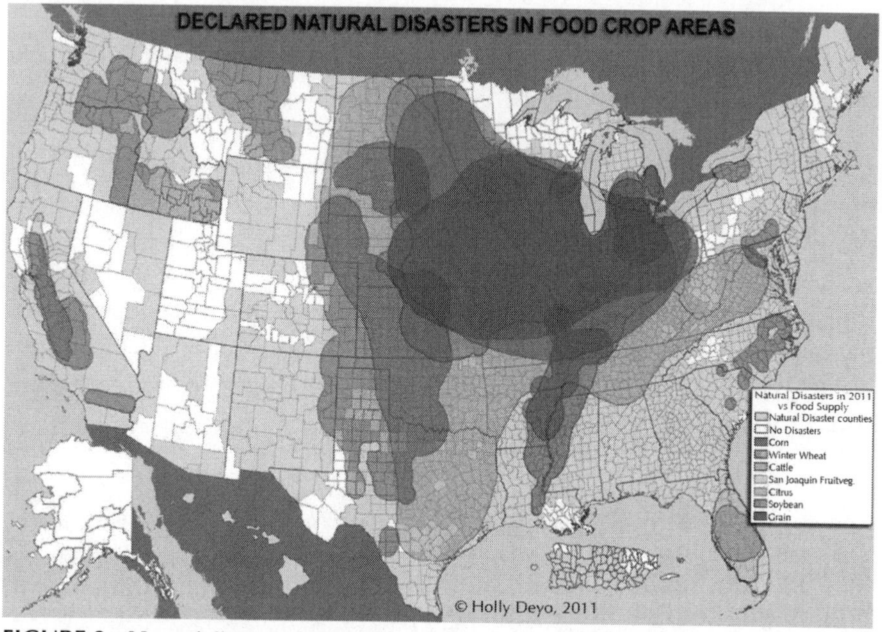

FIGURE 3 Natural disasters in food-growing regions of the U.S. (2011).

(Excerpted from http://www.standeyo.com/NEWS/11_Food_Water/110516.nat.diz-crop.impact.html).

18.2.1 THE FOUR STAGES OF A NATURAL DISASTER

Much of the government and business literature devoted to natural disasters suggests four distinct stages of a disaster, with each stage requiring a certain set of responses. Figure 5, below, shows the interconnectedness of disaster stages. The *mitigation* stage emphasizes efforts to reduce disaster effects over the longer term via organizational and community learning such as enforcing building codes and zoning laws, conducting vulnerability analyzes, and drawing upon lessons learned from prior disasters. The *preparedness* stage focuses on planning how to respond to future disasters such as developing preparedness plans, staging emergency training and exercises. The *response* stage of emergency management includes attempts to minimize the hazards created by a disaster such as providing emergency relief and organizing search and rescue teams. The final stage, *recovery*, highlights a return to some level of functioning, often representing a 'new' normal such as establishing temporary housing, accessing needed medical care, and applying for and the receiving of financial aid.

These four disaster stages (**M**itigation, **P**reparedness, **R**esponse, **R**ecovery) correspond quite well to the 3 USDA tenants for emergency preparedness: *Prepare for, Respond to,* and *Recover from*.

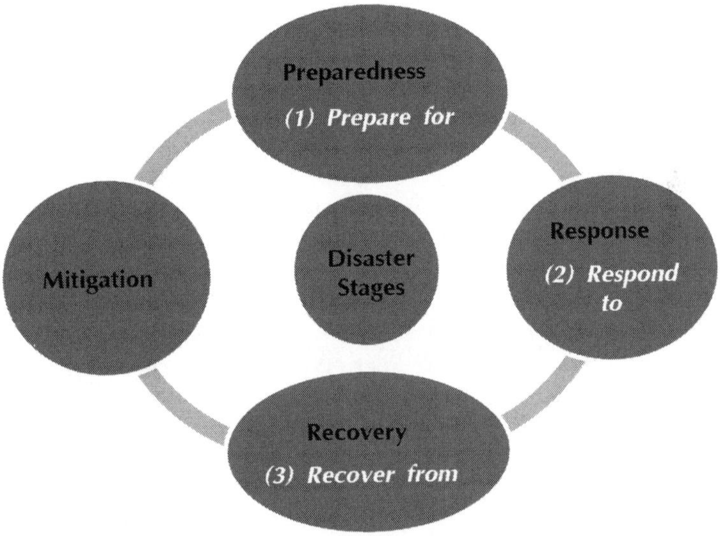

FIGURE 4 The Four Stages of a Disaster and Three Phases of Food Protection.

18.2.2 A BRIEF HISTORY ABOUT FOOD PROTECTION AND DEFENSE

The multidisciplinary National Center for Food Protection and Defense (NCFPD) was officially launched as a Homeland Security Center of Excellence in July 2004 and addresses the vulnerability of the nation's food system to attack through intentional contamination with biological or chemical agents. Food defense is the protection of food products from intentional adulteration by biological, chemical, physical, or radiological agents.

1. PREVENTING AND PREPARING TO LESSEN THE EFFECTS OF DISASTERS ON FOOD-STOCKS: THE FIRST STAGE IN RISK REDUCTION

"The experience and ultimate recovery of those affected by disaster depends less on the performance of a specific agency or organization than it does on the combined efforts of the community and the response and recovery system as a whole." *In other words, you and your neighbors are your own first responder.*
(Association Healthcare Foodservice Conference, New Orleans, 2013)

Although natural hazards cannot be prevented, it is entirely possible to lessen the impact of a hazard and or retard the possibility of a natural disaster. Just mentioning 'disaster' invokes images of chaos, lack of control, and helplessness. But being prepared for a disaster by developing an emergency plan lessens the impact of negative consequences disasters. What is the difference between prevention and preparedness? The Transitional Government of Ethiopia (1993), heralding from a rich tradition of planning for and managing disasters, define prevention undertakings as those activities taken to prevent natural hazards from having harmful effects on people, landscaping and the built environment, and similar economic assets. Prevention strategies are developed based on risks associated with a potential hazard and assessing the capacity (of a group, business, community, etc.) of managing the likely consequences of the hazard.

When natural disasters occur, food safety is catapulted to an important public health issue due to possible contamination. Natural disasters can make proper sanitation methods nearly impossible and lack of safe drinking and potable water in food establishments have led to mass outbreaks of foodborne diseases. The goal then, of prevention strategies is to reduce the risk of foodborne disease by ensuring reasonable protection from contaminated food and improving the sanitary condition of food establishments. A concurrent goal is, naturally, to protect as much as reasonably possible, food inventories.

TACTICAL PREPARATIONS

When the threat of a natural disaster is nearby, individuals and businesses alike will undergo similar food preservation procedures (consider how personal preparedness translates to the workplace), albeit in different magnitudes. The U.S. Food and Drug Administration, 2007 (FDA) (Table 1) suggest the following protocol:

TABLE 1 FDA Food and Water Safety during Hurricanes, Power Outages, and Floods.

a) Make sure you have appliance thermometers in your refrigerator *and* freezer.
b) Check to ensure that the freezer temperature is at or below 0°F, and the refrigerator is at or below 40°F.
c) In case of a power outage, the appliance thermometers will indicate the temperatures in the refrigerator and freezer to help you determine if the food is safe.
d) Freeze containers of water for ice to help keep food cold in the freezer, refrigerator, or coolers in case the power goes out. If your normal water supply is contaminated or unavailable, the melting ice will also supply drinking water.
e) Freeze refrigerated items such as leftovers, milk, and fresh meat and poultry that you may not need immediately. This helps keep them at a safe temperature longer.
f) Group food together in the freezer. This helps the food stay cold longer.
g) Have coolers on hand to keep refrigerated food cold if the power will be out for more than 4 h.
h) Purchase or make ice cubes in advance and store in the freezer for use in the refrigerator or in a cooler. Freeze gel packs ahead of time for use in coolers.
i) Check out local sources to know where dry ice and block ice can be purchased, just in case.
j) Store food on shelves that will be safely out of the way of contaminated water in case of flooding.
k) Make sure to have a supply of bottled water stored where it will be as safe as possible from flooding.

The World Health Organization's (WHO) (Table 2) specification of preventive food safety measures to undertake in preparation of exposure to a natural disaster suggests safe warehouse management and recommends:

TABLE 2 WHO, 2005, Food Safety in Natural Disasters.

• Storage structures should have good roofs and ventilation. Products should be kept at away from walls and off the floor. Pallets, boards, heavy branches, bricks, or plastic bags or sheets should be placed underneath them. Bags should be piled two-by-two cross-wise to permit ventilation.
• Spilled food should be swept up and disposed of promptly to discourage rats.
• Fuel, pesticides, bleach and other chemical stocks should never be stored together with food.

TABLE 2 *(Continued)*

• If spray operations for pest control are needed, they should be carried out by qualified technical staff, under the close supervision of the national authority (Department of Health / Department of Agriculture). Wearing of protective gear to reduce exposure of the operators to toxic chemicals is essential.

Additionally, the National Restaurant Association (NRA) publishes recommendations and tips (Table 3) for its membership on how to prepare for natural disasters.

TABLE 3 Hurricane Preparedness Tips.

Safety of employees and guests is first and foremost. Keep up to date with the latest developments through state and local government sources to determine if and when to close your business.
Account for all hazards. Plan in advance to manage any emergency situation
Assess the situation. Use common sense and available resources to take care of yourself, your co-workers and your business's recovery
Think first about the basics of survival. Make sure you have access to fresh water, food, clean air and warmth
Talk to your people. Involve all co-workers in all levels of emergency planning
Carefully evaluate internal and external operations. Determine which staff members, materials, procedures and equipment are absolutely necessary for the business to keep operating
Internal communications. Detail how you will be in contact with employees, customers and others during and after a disaster. Make sure you have a staff phone tree, including cell phone numbers, as well as management contact information. Also have benefit contact information and vendor contact information available.
External communications. Make sure you have signs available indicating your restaurant is closed, cancel all reservations and parties and post your closing information on the home page of your website.
Take steps to ensure your safety. Have the upper hand in responding to medical emergencies
Shelter in place or evacuate. Make sure you plan for both possibilities
Conduct a room-by-room walk-through. Determine what needs to be secured and then take steps to secure those physical assets
Secure the area. Board up the windows, place sandbags by the doors/entrances to reduce flooding, remove ice from the ice bins or ice machines and unplug all appliances and electronics

TABLE 3 *(Continued)*

Prepare for utility and essential services disruptions. Know what to do for extended outages during and after a disaster. This could include electricity, potable water, gas and phone service.
Reduce food supplies. If you find yourself in need of evacuation and you have food on hand, consider donating it to the local fire department, shelters and emergency facilities.
Cyber security. Make sure your data and information technology systems are protected.

Finally, HACCP's seven principles (Table 4) for retail and foodservice establishments provide guidance, if not strict adherence, for being prepared for disaster and disaster-like emergencies. While HACCP recognizes the impracticality of all seven principles being implemented in an across-the-board fashion for the industry, the guidelines offer a sensible approach to identifying and managing risk factors in retail and foodservice operations.

TABLE 4 The Seven Hazard Analysis and Critical Control Points (HACCP) Principles.

Perform a Hazard Analysis. The first principle is about understanding the operation and determining what food safety hazards are likely to occur. The manager needs to understand how the people, equipment, methods, and foods all affect each other. The processes and procedures used to prepare the food are also considered. This usually involves defining the operational steps (receiving, storage, preparation, cooking, etc.) that occur as food enters and moves through the operation. Additionally, this step involves determining the control measures that can be used to eliminate, prevent, or reduce food safety hazards. Control measures include such activities as implementation of employee health policies to restrict or exclude ill employees and proper hand-washing.
Decide on the Critical Control Points (CCPs). Once the control measures in principle #1 are determined, it is necessary to identify which of the control measures are absolutely essential to ensuring safe food. An operational step where control can be applied and is essential for ensuring that a food safety hazard is eliminated, prevented or reduced to an acceptable level is a critical control point (CCP). When determining whether a certain step is a CCP, if there is a later step that will prevent, reduce, or eliminate a hazard to an acceptable level, then the former step is not a CCP. It is important to know that not all steps are CCPs. Generally, there are only a few CCPs in each food preparation process because CCPs involve only those steps that are absolutely essential to food safety.
Determine the Critical Limits. Each CCP must have boundaries that define safety. Critical limits are the parameters that must be achieved to control a food safety hazard. For example, when cooking pork chops, the Food Code sets the critical limit at 145°F for 15 sec. When critical limits are not met, the food may not be safe. Critical limits are measurable and observable.

TABLE 4 *(Continued)*

Establish Procedures to Monitor CCPs. Once CCPs and critical limits have been determined, someone needs to keep track of the CCPs as the food flows through the operation. Monitoring involves making direct observations or measurements to see that the CCPs are kept under control by adhering to the established critical limits.
Establish Corrective Actions. While monitoring CCPs, occasionally the process or procedure will fail to meet the established critical limits. This step establishes a plan for what happens when a critical limit has not been met at a CCP. The operator decides what the actions will be, communicates those actions to the employees, and trains them in making the right decisions. This preventive approach is the heart of HACCP. Problems will arise, but you need to find them and correct them before they cause illness or injury.
Establish Verification Procedures. This principle is about making sure that the system is scientifically sound to effectively control the hazards. In addition, this step ensures that the system is operating according to what is specified in the plan. Designated individuals like the manager periodically make observations of employees' monitoring activities, calibrate equipment and temperature measuring devices, review records/actions, and discuss procedures with the employees. All of these activities are for the purpose of ensuring that the HACCP plan is addressing the food safety concerns and, if not, checking to see if it needs to be modified or improved.
Establish a Record Keeping System. There are certain written records or kinds of documentation that are needed in order to verify that the system is working. These records will normally involve the HACCP plan itself and any monitoring, corrective action, or calibration records produced in the operation of a the HACCP system. Verification records may also be included. Records maintained in a HACCP system serve to document that an ongoing, effective system is in place. Record keeping should be as simple as possible in order to make it more likely that employees will have the time to keep the records.

Retrieved from http://www.fda.gov/Food/GuidanceRegulation/HACCP/ucm077964.htm#use

Preparedness is a process, a way of thinking and planning about how you will make it and, optimally thrive, through the process. Preparing for natural disasters is one of the single most effective activities hospitality organizations and tourism companies can do to mitigate the negative consequences of a natural disaster. Being prepared and having a plan: increases the chance that your organization and its assets, human and otherwise, will stay safe; increases your capacity to be self-sufficient in the case of no immediate outside assistance; and will help your company bounce-back faster, or in other words, increases the capacity for resilience.

Preparedness strategies involve building up of capacities prior to a disaster; preparedness activities lower the negative impact of a disaster.

Installing a backup generator won't stop a disaster from happening but it will extend the shelf life of your inventory and significantly decrease the possibility of foodborne bacteria infecting your employees and customers. Buying the backup generator *prevents* the restaurant from a power interruption and simultaneously *prepares* and assists the restaurant for continuation of services during and after natural disasters.

However, preparedness efforts are meaningless if the organization's human resources are unable to respond. After having a plan, the single most important aspect of disaster preparation: practice, practice, and practice.

2. RESPONDING TO THE EFFECTS OF DISASTERS ON FOOD-STOCKS: THE SECOND STAGE IN RISK REDUCTION

> Post-disaster images captured by media often invoke a sense of natural disaster events being beyond our control and chaotic. More likely, our behaviors during natural disasters are a far cry from chaotic and uncontrollable (Ripley, 2005). Rather, when confronted with a threat, our first response is disbelief, incredulous disbelief, that the event is even happening at all.
>
> *"Most of us become incredibly docile. We are kinder to one another than normal. We panic only under certain rare conditions. Usually, we form groups and move slowly, as if sleepwalking in a nightmare. People caught up in disasters tend to fall into three categories. About 10% to 15% remain calm and act quickly and efficiently. Another 15% or less completely freak out–weeping, screaming or otherwise hindering the evacuation. That kind of hysteria is usually isolated and quickly snuffed out by the crowd. The vast majority of people do very little. They are 'stunned and bewildered'.*
>
> (Excerpted from "How to Get Out Alive", by Amanda Ripley, 2005)

The response stage in food disaster management is concerned with enacting your already established disaster management plan. Inspecting and salvaging food become the primary activities in the immediate aftermath of your operation's exposure to the risks from a natural disaster.

TACTICAL RESPONSES

After human safety and preservation, maintaining utilities and power becomes a priority.

Often of course, power lines will become compromised and fail; without a backup generator, which also sometimes fail, precautions must be taken to preserve food for as long as is possible. Basic tips for keeping food safe include:

- Keeping the refrigerator and freezer doors closed as much as possible to maintain the cold temperature. The refrigerator will keep food cold for about 4 h if it is unopened.
- A full freezer will keep the temperature for approximately 48 h (24 h if it is half full) if the door remains closed.
- Buy dry or block ice to keep the refrigerator as cold as possible if the power is going to be out for a prolonged period of time. Fifty pounds of dry ice should hold an 18 cubic foot, fully stocked freezer cold for two days.
- To eat refrigerated or frozen meat, poultry, fish or eggs while it is still at safe temperatures, it's important that each item is thoroughly cooked to its proper temperature to assure that any foodborne bacteria that may be present are destroyed. However, if at any point the food was above 40 °F for 2 h or more — discard it.
- Wash fruits and vegetables with water from a safe source before eating.
- For infants, try to use prepared, canned baby formula that requires no added water. When using concentrated or powdered formulas, prepare with bottled water if the local water source is potentially contaminated.

It may be difficult to determine conditions for safely consuming food. The rule of thumb advice for damaged food products following a disaster, is "When in doubt, throw it out." The Centers for Disease Control (CDC) recommends the disposal of food that *may* not be safe to consume. Further, the guidelines provide for:

- Throwing away food that may have come in contact with flood or storm water.
- Throwing away food that has an unusual odor, color, or texture. When in doubt, throw it out.

- Throwing away perishable foods (including meat, poultry, fish, eggs and leftovers) in your refrigerator when the power has been off for 4 h or more.
- Thawed food that contains ice crystals can be refrozen or cooked. Freezers, if left unopened and full, will keep food safe for 48 h (24 h if half full).
- Throwing away canned foods that are bulging, opened, or damaged.
- Food containers with screw-caps, snap-lids, crimped caps (soda pop bottles), twist caps, flip tops, snap-open, and home canned foods should be discarded if they have come into contact with floodwater because they cannot be disinfected.
- If cans have come in contact with floodwater or storm water, remove the labels, wash the cans, and dip them in a solution of 1 cup (8 oz/250 ml) of bleach in 5 gallons of water. Re-label the cans with a marker. Include the expiration date.
- Not using contaminated water to wash dishes, brush your teeth, wash and prepare food, wash your hands, make ice, or make baby formula.

Retrieved from http://emergency.cdc.gov/disasters/foodwater/facts.asp

The World Health Organization's (WHO) provide additional guidelines and behaviors to assure safe food handling during food distribution and preparation in natural disasters.

- Water for drinking and food preparation should be treated as contaminated unless specifically confirmed as safe. Therefore, all water should be boiled or otherwise made safe before it is consumed or used as an ingredient in food.
- Identify what agricultural production has been adversely affected and what areas exist where food can still be harvested or where food has been safety stored after harvesting.
- Assess what agricultural produce may be contaminated with microorganisms (from raw sewage or decaying organisms) and potentially hazardous chemicals. Note that while it is sometimes possible to eliminate potentially hazardous microorganisms by thoroughly cooking or disinfecting the produce, such actions may not completely remove chemical hazards.
- If crop fields have been contaminated, an assessment should be carried out to establish measures to reduce the risk of transmitting pathogens and hazardous chemicals.

During emergency response operations, large-scale distribution of imported or locally purchased food items as well as mass preparation of cooked foods frequently occurs. In this context special attention must be brought to the following:

- There must be adequate facilities for waste disposal.
- Water and soap must be provided for personal cleanliness, and detergent for cleaning utensils and surfaces, which should also be sanitized with boiling water or a sanitizing agent, e.g., bleach solution.
- Foods should be stored in containers that will prevent contamination by rodents, insects, or other animals.
- Hot and/or cold holding of food may have to be improvised.

Retrieved from http://www.ct.gov/dph/lib/dph/environmental_health/food_protection/pdf/fda_food_and_water_safety_duringfloods_and_hurricanes_for_residents.pdf)

Sometimes, due to prolonged and or immediately unsafe circumstances, the foodservice operation will need to close. This is often the best decision management can make; if this happens the National Restaurant Association provides the following checklist (Table 5):

TABLE 5 Checklist for emergency closing of a restaurant.

Reduce and/or reschedule purchases and deliveries to reduce waste.
Sandbag the entrance to the restaurant.
Board up store fronts and windows.
Stock up on dry ice and take ice out of ice machines to fill plastic containers. Place the containers in the walk-in refrigerator. Move food products in the walk-in closer to bottom shelves and floor level. If the power goes out, the ice-filled containers will serve as a big ice chest.
Consider donating surplus food to local shelters and emergency facilities if you are closing your business, as well as when you reopen. Restaurants are often on the front lines in disaster recovery efforts by providing food and water for first responders or displaced residents. Check with disaster relief organizations such as the American Red Cross, local fire departments and your state restaurant associations.
Shut down computers and equipment.
Turn off gas lines – this is very important due to the risk of damage.
Secure rooftop air conditioning units with ropes and/or straps.
Obtain a generator and fuel for reentry into the restaurant for cleanup.

NRA, 2011. Retrieved from http://www.restaurant.org/News-Research/News/Checklist-for-emergency-closing-of-a-restaurant

3. RECOVERY: THE THIRD STAGE IN RISK DEDUCTION

Recovery is generally depicted as the process of rebuilding and restoring after the effects of a natural disaster. The recovery phase of natural disasters can be protracted and represent a return to a "new normal" (Norris et al., 2008). Transitioning from the response stage to the recovery stage can be fuzzy and determining when an operation has stabilized or is 'all clear' may not be easily sensed. The operation may have been heavily damaged as a result of direct effects of the disaster, such as physical damages to buildings, equipment, and inventories. Indirect effects may also be experienced; they tend to be more widespread and difficult to measure such as neighboring closed businesses (e.g., banks), emotional damage, the disruption of employees' personal lives, and vendor's and supplier's own interruptions.

As Figure 5 (the four stages of a disaster) suggests, each phase of a disaster impacts and is impacted by the other stages, suggesting that the process is never really quite finished. Unlike the response phase, decisions and issues called for in the immediate afterwards of a disaster dictate what the recovery phase will entail. This is the time to develop mitigation measures that might be difficult to make relevant when the results of a disaster are not top of mind for everyone. In the transition from response to recovery, specific tasks should be considered, including:

- Determining when a situation has stabilized
- Declaring an 'all clear' and establishing a disaster recovery center if needed (depending on size of organization and the scale of the disaster)
- Notifying responders and redeploying resources or staff
- Initiating a plan to resume normal operations
- Scheduling a response phase de-brief meeting
- Lifting emergency measures, as appropriate

There will never be a time with the felt immediacy of a disaster like the postdisaster phase, so this is also the time to write down everything deemed important to recover and respond to the next possible disaster. This includes:

- Documenting the chronology of events and the decisions made about those events;
- Assessing and photographing any damage;

- Identifying labor, equipment, and materials used in the disaster response phase;
- How your operation resumed operations of safe food production;
- And how you updated and revised the operations disaster plan.

If a Business Continuity Plan (BCP) isn't in place and hasn't been developed, you should begin one now. The 4 key steps in a BCP include establishing the planning team; identifying and analyzing capabilities and hazards; developing the plan; and implementing the plan. An expanded and detailed BCP planning process and suite of planning tools developed by FEMA can be retrieved at on the FEMA website (www.ready.gov).

Excerpted from A Coordinated Response to Food Emergencies: Practice and Execution, 2010.

The level of preparedness of communities, their businesses, families, and individuals has a direct bearing on the degree of damage and the pace of recovery. Most natural disasters of any consequence involve one or several utility disruptions. Once and when the operation has power restored, one of the first activities to undertake is determining the safety of your food.

Guidelines from the FDA suggest:

- If an appliance thermometer was kept in the freezer, check the temperature when the power comes back on. If the freezer thermometer reads 40°F or below, the food is safe and may be refrozen.
- If a thermometer has not been kept in the freezer, check each package of food to determine its safety. You can't rely on appearance or odor. If the food still contains ice crystals or is 40°F or below, it is safe to refreeze or cook.
- Refrigerated food should be safe as long as the power was out for no more than 4 h and the refrigerator door was kept shut. Discard any perishable food (such as meat, poultry, fish, eggs or leftovers) that has been above 40°F for two hours or more.

Keep in mind that perishable food such as meat, poultry, seafood, milk, and eggs that are not kept adequately refrigerated or frozen may cause illness if consumed, even when they are thoroughly cooked (Food Facts, USDA, 2007).

18.3 CONCLUSION

An effective emergency readiness plan is essential to businesses; natural disasters could affect any business at any time and when business is disrupted, income streams can be disrupted as well. Lost revenues and extra expenses equal reduced profits. An effective emergency management plan helps reduce the negative consequences possible after experiencing the effects of a natural disaster. The steps of a disaster / effective emergency readiness plan include:

a. First determine whether a disaster plan already exists! If so, review the plan to ensure it is current and reflects current realities. Make the sure the plan is widely circulated and understood. Practice, practice, practice.
b. Identify positions, assign roles, and develop a directory in the event of a natural disaster or emergency.
c. Conduct a hazard analysis of likely disaster scenarios and emergencies that may negatively impact your foodservice operation and determine responses to the parts of the operation that may be compromised.
d. Develop a list of tasks that are needed to prepare for activities that will need to be completed during a disaster and during disaster recovery.
e. Teach the emergency readiness plan to assigned foodservice staff; review the entire plan with all foodservice personnel.
f. Rehearse and practice emergency readiness drills.
g. Continually update the plan to match current realities.
h. (Edelstein, 2008)

Natural disasters are rarely formulaic and workplaces should all have site-specific emergency plans depending on the level of tolerable risk. Natural disasters may completely overwhelm local systems and emergency responders, making you and your employees your own first responder. Increasingly, organizations offering products and service in the risk management industry are piloting early warning system software and services that deliver site-specific hazard warnings and standard operating procedural instructions to key operational human resources, tailored to their clients' directives. Be prepared and have a plan in hand.

KEYWORDS

- **agri-terrorism**
- **HACCP manual**
- **manmade disasters**
- **natural hazards**

REFERENCES

Beach, Major R. J., II CCFP, Orleans Parish Sheriff's Office (2013). Remarks from Association of Healthcare Foodservice Conference, New Orleans, LA.

Cutter, S.L., B.J. Boruff, and W.L. Shirley, 2003. Social Vulnerability to Environmental Hazards, *Social Science Quarterly, 84*(1): 242–261.

Edelstein, S. (2008). *Managing Food and Nutrition Services for the culinary, hospitality, and nutrition professions*. Sudbury, MA: Jones and Bartlett Publishers.

EHA Consulting Group (Feb 26, 2010). *Disaster Planning and Response for Food Processors.* Retrieved from http://www.ehagroup.com/crisis-management/disaster-planning-response-food-processors/

Federal Emergency Management Agency (FEMA) (2012). Business Continuity Planning Suite. Retrieved from http://www.ready.gov/business-continuity-planning-suite.

Fitzgerald, K. (2013, April 18). "Rates Of Foodborne Illness Cases On The Rise." *Medical News Today*. Retrieved from http://www.medicalnewstoday.com/articles/259334.php.

Hazards and Vulnerability Research Institute (2013). *The Spatial Hazard Events and Losses Database for the United States,* Version 10.1 [Online Database]. Columbia, SC: University of South Carolina. Retrieved from http://www.sheldus.org.

Hennessey, M., Kennedy, S., and Busta, F. (2010). Demeter's resilience: An international food defense exercise. Journal of Food Protection, 73(7), 1353–6. Retrieved from http://search.proquest.com/docview/608927138?accountid=13360

Kennedy, S. (2012). Emerging Global Food System Risks and Potential Solutions. Improving Import Food Safety, 1–20.

National Food Service Management Institute. (2003). *Emergency readiness plan: guide and forms for the school foodservice operation.* University, MS: National Food Service Management Institute. Publication Number: ET 43–03.

Norris, F. H., Stevens, S. P., Pfefferbaum, B., Wyche, K. F., and Pfefferbaum, R. L. (2008). Community resilience as a metaphor, theory, set of capacities, and strategy for disaster readiness. *American journal of community psychology, 41*(1–2), 127–150.

Ripley, A. (April 25, 2005). How to get out alive. *Time.* Retrieved from http://www.time.com/time/magazine/article/0,9171,1053663,00.html#ixzz2UgzU9dOC.

Seo, S., Jang, S., Miao, L., Almanza, B., and Behnke, C. (2013). The impact of food safety events on the value of food-related firms: An event study approach. *International Journal of Hospitality Management, 33* (June), 153–165.

Sydnor-Bousso, S., Stafford, K., Tews, M., and Adler, H. (2011). Toward a Resilience Model for the Hospitality and Tourism Industry, *Journal of Human Resources in Hospitality and Tourism, 10* (2), 195 — 217.

Transitional Government of Ethiopia (TGE) (1993). National Policy on Disaster Prevention and Management (Unpublished Document), Addis Ababa – Ethiopia.

U.S. Food and Drug Administration (2007). Food Facts: Food and Water Safety during Hurricanes, Power Outages, and Floods. Retrieved from http://www.ct.gov/dph/lib/dph/environmental_health/food_protection/pdf/fda_food_and_water_safety_during-floods_and_hurricanes_for_residents.pdf

U.S. Department of Homeland Security, (2010). A Coordinated Response to Food Emergencies: Practice and Execution. Baton Rouge: National Center for Biomedical Research and Training.

U.S. Food and Drug Administration (2013). Hazard Analysis and Critical Control Points (HACCP) Manual. Retrieved from http://www.fda.gov/Food/GuidanceRegulation/HACCP

U.S. Department Agriculture (2013). Home / USDA Emergency Preparedness and Response. Retrieved from http://www.usda.gov/wps/portal/usda/usdahome

U.S. Department Agriculture (2013). Agricultural Productivity in the US. Retrieved from http://www.usda.gov/wps/portal/usda/usdahome

World Health Organization. (2005). Food Safety in Natural Disasters, *International Food Safety Authorities Network (INFOSAN), Note 5/2005,* September, 13. Retrieved from http://www.who.int/foodsafety/fs_management/No_05_NaturalDisasters_Sept05_en.pdf.

Wilson, M. (2013). Who can catch the flu (Infographic). Accessed on April 25, 2013 at: http://www.fastcodesign.com/1672417/infographic-which-flu-strain-is-most-likely to-bring-the-apocalypse?partner=newsletter.

APPENDIX

Related Resources:

GOVERNMENTAL AGENCIES/WEBSITES CONCERNED WITH FOOD SAFETY

1. Federal Emergency Management Agency (FEMA). Covers a host of natural disaster mitigation, preparation, response, and recovery guides for individuals, families, businesses, and communities.
   a. http://www.fema.gov
   b. http://www.ready.gov/business (resources specifically tailored for businesses)
2. Hazard Analysis and Critical Control Points (HACCP): HACCP is a management system in which food safety is addressed through

the analysis and control of biological, chemical, and physical hazards from raw material production, procurement and handling, to manufacturing, distribution and consumption of the finished product.

a. http://www.fda.gov/Food/GuidanceRegulation/HACCP/default.htm
b. FDA Retail Food Protection
c. http://www.fda.gov/Food/GuidanceRegulation/RetailFoodProtection/default.htm

Governmental Agencies/Websites concerned with Food Defense

3. Food and Drug Administration (FDA) of the Department of Health and Human Services. The FDA primarily (but not entirely) focuses on the processing of food products; that is, postharvest. Overview and History of FDA and the Center for Food Safety and Applied Nutrition
4. Food Safety and Inspection Service (FSIS) of the U.S. Department of Agriculture
   "The Food Safety and Inspection Service (FSIS) is the public health agency in the U.S. Department of Agriculture responsible for ensuring that the nation's commercial supply of meat, poultry, and egg products is safe, wholesome, and correctly labeled and packaged."
5. The Office of Data Integration and Food Protection (ODIFP) oversees all food defense activities of the USDA developing and implementing procedures to prepare for, respond to, and recover from intentional and unintentional contamination and significant food emergencies and natural disasters affecting meat, poultry and processed egg products. ODIFP is also responsible for coordinating all of the Agency's data collection, analysis, and integration activities across program areas. ODIFP closely collaborates with other offices within FSIS to ensure adherence to emergency management policies, food defense directives, and the consistency and quality of data analyzes.
6. National Center for Food Protection and Defense (NCFPD)
7. Food Defense

FDA works with other government agencies and private sector organizations to help reduce the risk of tampering or other malicious, criminal, or terrorist actions on the food and cosmetic supply.
Food Defense Assessment Staff (FDAS)
Functional Statement – The Food Defense Assessment Staff provides the scientific and technical basis to support preparedness, response, and recovery initiatives. Further, FDAS provides advice and consultation to other government agencies and industry on all threats affecting the food supply. Specific key functions include conducting vulnerability assessments, providing guidance to industry on food defense plans and counter-measures, and interfacing with the intelligence community on gathering threat information. Through outreach and education, FDAS works cooperatively with industry to promote the adoption and application of food defense programs, best practices and guidelines, and helps small and very small plants develop their own prevention, response, and recovery strategies.
Private Resources concerned With Business Continuity

8. http://www.disastersafety.org/wp-content/uploads/OFB-EZ_Toolkit_IBHS.pdf
   a. The Insurance Institute for Business and Home Safety (IBHS)
   IBHS has developed a new streamlined business continuity program for small businesses that may not have the time or resources to create an extensive plan to recover from business interruptions. Open for Business (OFB) is the Institute's comprehensive business continuity planning program, and the new OFB-EZ tool is a streamlined kit for small businesses with fewer than 10 employees. This is an excellent resource for any small business!

CHAPTER 19

FOOD DEFENSE CONCEPTS

SARAH SLETTE, MS
Enteric Epidemiologist, Indiana State Department of Health

MARY STIKER
Food Defense Coordinator, Indiana State Department of Health

A. SCOTT GILLIAM, MBA, CP-FS
Director, Food Protection Program, Indiana State Department of Health

CONTENTS

19.1 OVERVIEW

As discussed in other chapters, food safety is the prevention of any type of accidental food contamination. Food defense is taking it a step farther and involves the prevention of intentional food contamination. Over the years the food industry and regulators have been reminded of these kinds of occurrences, such as when the cyanide tainted Tylenol capsules in 1982 were intentionally tampered with by a deranged individual and the pesticide laced hamburger was contaminated by a disgruntled employee in 2003, it is important to remember the risk is always there.

For many years regulators relied on the Federal Anti-Tampering Act of 1983 to address the intentional contamination of consumer products (food, drug, medical device or cosmetics) that can cause harm to the public. Even if the intent is not to cause bodily injury, but to gain attention or notoriety, the law still has strict guidelines for punishment for any violations. Many persons have been prosecuted under this act, but society needed more protection than what this act alone could do.

After the attacks on September 11, 2001, a new federal law was adopted to address possible threats to the United States. The Patriot Act was the first of a series of law enactments to thwart any kind of terrorism threat. In 2002, Congress passed the Public Health Security and Bioterrorism Preparedness and Response Act (Bioterrorism Act) that mandated the federal government, with state and local governments, take action to address possible terrorist threats, both foreign and domestic, in a variety of key areas, including chemical, biological, radiological and nuclear (CBRN) threats. Our focus in this chapter will be on this law and the more recently adopted Food and Drug Administration (FDA), Food Safety Modernization Act (FSMA) of 2011 that for the first time set requirements for implementation of food defense measures by the food industry.

Preparedness for these kinds of threats is very complicated and involves many different parts of the government, economic sectors, academia and the public. The food industry is widely varied and has a complex distribution system that can be difficult to trace back and trace forward. Many foods are imported to the United States from foreign countries in a complicated web of producers, brokers, distributers, manufacturers and retailers. Foods produced in another country have been linked to food borne disease outbreaks, such as the large multistate *Salmonella* Saintpaul outbreak eventually traced to fresh jalapeño and Serrano peppers grown, harvested and packed in Mexico and imported for further distribution. http://www.cdc.gov/salmonella/saintpaul/jalapeno/ In this case, the contamination was accidental, or possibly negligent, but it was not intentional, yet many persons fell ill to this disease.

What could happen if something like this was done with the purpose of causing harm in order to make a political statement or for revenge? The results could be disastrous! Sadly, in addition to the previously mentioned tampering cases there have been several others of similar severity and seriousness. One of the most famous was the Rajneeshee cult group in Oregon whose plan was to culture and grow *Salmonella* for the purposes of contaminating restaurant salad bars in their area in an effort to sicken consumers and impact a local election. Over 700 persons fell ill to this despicable act, but because they failed to properly gauge the incubation times it did not affect the election outcomes. These kinds of threats exist and it is critically important for all workers in the food industry to have awareness of the risks and take responsibility to prevent and report potential adverse events to management for proper action.

The Bioterrorism Act and FSMA established new requirements for the food industry. Initially, a new registration requirement was enacted requiring all firms that manufacture, process, pack, transport, distribute, receive, hold, or import food complete a basic registration with the FDA so that a comprehensive database of firms will be available and current at all times. Under FSMA these firms were required to reregister providing additional new information, which also includes importers of food. In addition, importers are required to provide prior notice of shipments into the United States so that FDA has the ability to decide whether or not to inspect these shipments before they enter into commerce. The new law also requires these firms to maintain records of the movement of food and to make them available to FDA upon request. These requirements do not include retailers or farms.

New guidance and education measures were created for the retail food industry to assist them with being prepared to prevent and effectively address a security breach either by someone from the outside or within the organization. One of the first was a guidance document published in 2007 called Retail Food Stores and Food Service Establishments: Food Security Preventive Measures Guidance with a checklist to be used to evaluate the readiness of the establishment to address food security as it was then called. The state of Indiana conducted its own statewide survey using a version of this checklist and found substantial compliance; however, there were several areas that needed improvement.

These new laws also provided the FDA new administrative detention and recall authority that may be used if an adulterated product is in interstate commerce. This is somewhat similar to what most states have in the form of "embargo" power to withhold product from commerce if it is suspected of being adulterated or misbranded. The FDA also created the Reportable Food Registry (RFR) that requires firms that handle FDA regulated product to file a report in the RFR if they find an issue with the product that may be a risk for human injury or death and for which they have lost control because it is already in the distribution chain.

The United States Department of Agriculture's (USDA) Food Safety Inspection Service (FSIS) has also taken steps toward implementation of food defense in the many facilities regulated under their Meat Inspection and Poultry Inspection Acts, in addition to egg processors and areas covered under the Animal, Plant Health Inspection Service (APHIS). All agencies involved with protecting the food supply encourage the food industry to conduct vulnerability assessments of their operations to identify any areas for improvement. There are various tools available, such as the Food and Agriculture Sector Criticality Assessment Tool (FASCAT) and CARVER + Shock, that will be examined in more detail later in this chapter.

Anytime an adverse event of any kind is determined to be a terrorist act, it immediately falls under the jurisdiction of the Federal Bureau of Investigation (FBI). They then control the criminal investigation going forward.

Lastly, it is important to remember that food defense is critical in any area where food is being provided to the public, especially in large mass gatherings. Examples of these kinds of measures will also be examined later in the chapter in addition to understanding the concept of the National Incident Management System (NIMS) that is used to manage many different kinds of situations and should be considered for use by the food industry.

19.2 BACKGROUND

The Council to Improve Foodborne Outbreak Response (CIFOR) developed Guidelines for Foodborne Disease Outbreak Response that outlines the detection, investigation, control, and prevention of foodborne illness. These guidelines were developed to aid government agencies in preventing and managing foodborne disease. They focus on local and state agencies, including public health, environmental health, agriculture and other agencies responsible for food safety, because they investigate almost all of the foodborne disease outbreaks in the United States. The concepts are the same for all outbreak investigations, regardless of the intent of contamination.

Any events of intentional contamination will be detected at the local and state levels and initially investigated at this level. There are many public health surveillance systems in place to detect an unusual disease occurrence. There are a variety of sources of information for public health surveillance sources including human disease, environmental data, population characteristics, animal data, and health care services data. Public health surveillance is the foundation for communicable disease epidemiology, which is a critical component for a food protection program. Surveillance data can reveal the burden of a particular disease in the community or the presence and scale of a possible outbreak. This data may offer clues into the source of and contributing factors to disease outbreaks. Over time, surveillance data can identify disease and behavioral trends and learn more about the diseases being tracked and ways to prevent the diseases. These systems are much broader and expand beyond foodborne disease surveillance. Systems used include disease specific notifiable condition surveillance, foodborne disease complaint systems, and reports of outbreaks.

The Council for State and Territorial Epidemiologists (CSTE) works to advance public health policy and epidemiologic capacity. The CSTE is an organization of member states and territories representing public health epidemiologists, which establishes the nationally notifiable conditions, including enteric diseases likely to be foodborne, for which health departments provide information to the Centers for Disease Control and Prevention (CDC). Once a health-care provider diagnoses, or a laboratory detects, a reportable condition, they are to notify public health officials of the illness. Once public health officials are notified, the instance is compared with other similar reports. Most states and territories send enteric disease surveillance information to the CDC by the National Notifiable Disease Surveillance System (NNDS). State public health laboratories also contribute to surveillance by the Public Health Laboratory Information System (PHLIS) by reporting laboratory-confirmed isolates, such as *Salmonella* and shiga toxin-producing *E. coli* (STEC). The notifiable disease surveillance systems are a critical component to food defense by allowing background surveillance of enteric conditions that are typically associated with foodborne illness. When there is an unusual disease occurrence, public health officials will investigate and will try to determine the source or contributing factors of the illness. If suspicious or criminal intent is suspect in the investigation, law enforcement authorities may be brought into the investigation.

Another form of surveillance is obtained through direct foodborne disease complaints from the public, an important system every local or state public health agency has in place to detect foodborne illness in the community. Each agency has different means of processing complaints and requirements for follow-up on the implicated food establishments or food products. These systems are important for detecting a foodborne illness outbreak as it occurs instead of the delay that occurs with physician or laboratory reporting. They also detect contaminates that may not be nationally notifiable, but have resulted in a foodborne disease outbreak. This is the most common means of detecting a tampering incident.

PulseNet is a national network of local, state or territorial, and federal laboratories coordinated by CDC that allows comparison of subtypes of pathogens isolated from humans, animals, and foods across multiple jurisdictions. PulseNet has a database of molecular fingerprints of bacteria determined by the laboratory method, pulsed-field gel electrophoresis (PFGE) and is one of the most important aspects to foodborne investigations, especially multistate investigations. This technique allows investigators to focus their outbreak investigations on a specific subgroup of persons infected with the same pathogenic strain to more quickly identify the source of an outbreak. PFGE may also be used to characterize bacterial strains in food or the environment to determine if that same strain has caused human illness. PFGE allows public health officials to detect an outbreak of the same strain of bacteria among a group of people that may not otherwise be compared to each other. This is extremely important in intentional outbreaks because it is able to quickly identify a biological link between cases. PulseNet has standardized the PFGE methods used by laboratories to distinguish strains of STEC, *Salmonella*, *Shigella*, *Listeria*, and *Campylobacter*. It also maintains a national database of the strains to be able to compare what is circulating nationally, which allows investigators to detect and investigate relatively small outbreaks, but make a large public health impact.

Another significant surveillance system in foodborne disease, whether by intentional or unintentional means, is syndromic surveillance using the public health emergency surveillance system (PHESS). Real-time or near real-time emergency department (ED) data is uploaded into PHESS and helps in early detection of disease outbreaks, environmental health hazards and chemical exposures, and acts of bioterrorism. There are limitations to the data received so investigations by the public health agencies are warranted once an alert has been received. For example, an alert from the emergency department may be having a chief medical complaint labeled as gastrointestinal illness that is above the expected number for that day, but the actual illness is later determined to be related to a chronic condition, such as Crohn's Disease. Establishing relationships with the hospital infection preventionists and infectious disease physicians is crucial to deciphering PHESS data. PHESS is crucial in detecting an intentional contamination event, since alerts are established for unusual ED visits and/or a large influx of visits. The public health agency has the ability to be notified as soon as an ill person checks into their local ED and can act appropriately, immediately upon notification to gather information to prevent any additional cases.

There are many different systems to detect unusual occurrences of disease. However, how the public health agency investigates the reports of disease is critical to determining if the implicated source of illness was intentionally contaminated. The public health agencies have established roles at the local, state, and federal levels for routine outbreak investigation, as well as a food defense investigation with an intentional contamination component. Once intentional contamination is suspected, it becomes a criminal investigation and the proper law enforcement entity needs to be involved. Every outbreak investigation should have proper chain of custody and documentation that could contribute to a criminal investigation even before the law enforcement authorities are involved. Not only are the government agencies to have their established roles, the food industry plays a key component in the investigation, as well. The involvement of food industry members early in an outbreak investigation is the quickest way to accurately and effectively stop an outbreak with the lowest economic burden to the industry, as well as, to catch the culprit of the act.

Once an incident has been detected and investigated the public health agency needs to take steps to control the incident. This may not be a linear process, the decisions made to control an outbreak need to be implemented rapidly even if all of the answers are not available yet. The CIFOR guidelines call this information-based decision-making where control measures must be implemented concurrently with investigations. Waiting for laboratory results, diagnoses, or other pieces of the investigation is not necessary to initiate control measures. Practices that led to the outbreak, whether a contaminated food product, ill food employee, or malicious attack, are likely to continue unless an intervention stops them.

Once control measures are in place Incident Command Structure (ICS) should be used. The National Incident Management System (NIMS) allows for response to any type of incident, natural, manmade, or accidental, or from acts of terrorism. All Federal departments and agencies are required to adopt NIMS. NIMS is a national approach to incident management. NIMS covers all jurisdictional levels, across all levels of disciplines. NIMS provides the "whole" system approach versus looking at one simple component. NIMS emphasizes functional exercises, training, and practical applications. NIMS allows for flexibility of working with small or large scale incidents and organizational structures that are standardized to increase interoperability.

The NIMS has six (6) major components. The six (6) components include Command and Management, Preparedness, Resource Management, Communications and Information Management, Supporting Technologies, and Ongoing Management and Maintenance. With these components the ICS has several basic concepts. The basic concepts include that incidents may be managed locally, use measureable objectives, are scalable, and are interactive in the management components. The ICS emphasizes management by objective, manageable span of control, integrated communications, chain of command, unity of command, controlled and managed deployment of resources, and modular organizational. Under the Incident Command Organizational Chart, there will be an Operations Section Chief, Planning Section Chief, Logistics Section Chief, Finance and Administration Section Chief, and the Public Information Officer, Safety Officer, and Liaison Officer.

The Incident Management System relies closely on the Incident Action Plan (IAP). The IAP addresses incident goals and objectives. It should determine specific goals and the approximation of the length of the operational period. The IAP should designate individual roles, specific tasks, and strategies. As with most documents, the IAP should be updated and reviewed periodically so consistency is maintained.

In food defense incidents, the use of NIMS provides an important link with the National Response Plan (NRP). The NRP provides for proactive application of federal resources. NIMS provides control and command structure for the management of an incident. Response to such incidents allows for the partners involved to respond and mitigate the situation or incident quickly. Incident readiness is important in food safety and food defense. One of the aspects of readiness is partnering with other available resources such as additional state, local and federal agencies.

During the investigation messages need to be communicated to the public as information becomes available. One of the biggest challenges is communicating to the public during a fluid investigation, especially one with food defense implications. In any food defense incident, if a public message is warranted, the public will have a fear of the unknown and having a collaborative public message involving the government agencies and industry will be the most effective means of having a clear and accurate message.

Just as there is a fluid investigation, there are fluid control measures, as well. Communication between all parties involved, most importantly, among the incident command, is critical to success. Each jurisdiction and stakeholder involved have differing priorities, so challenges arise when trying to determine what control measures to implement and when it's time to change the focus. Communication is key to success of any control measure. Many factors may rapidly change the focus of control measures such as laboratory results, additional cases, environmental assessments, and other medical diagnoses. These changes need to be communicated as rapidly as possible to implement changes that incident command has determined are best for the investigation.

The last component to an outbreak response is prevention; however, this may arguably be the first component. Prevention is the only component of a food defense system across all levels of government and the food industry where there may be a direct impact in stopping a food defense incident. The government and food industry may take steps to improve detection, investigations, and control measures, but ultimately, the intent of the individual will not falter by these steps. This is where the preparedness measures that a food distributor or manufacturer already has in place will help play a major role in deterring an incident or preventing an individual with ill intent from entering their system.

The National Infrastructure Protection Plan (NIPP) provides a framework that integrates a range of efforts designed to enhance the safety of the nation's critical infrastructure. The NIPP's goal is to prevent, deter, neutralize, or mitigate the effects of a terrorist attack or national disaster, and to strengthen national preparedness, response, and recovery in the event of an emergency. The NIPP has identified critical infrastructure sectors that have its own Sector-Specific Plan (SSP) that is unique to that sector. The following sectors have a SSP:

- Financial Services
- Chemical
- Commercial Facilities
- Communications
- Critical Manufacturing
- Dams
- Defense Industrial Base
- Education Facilities
- Emergency Services
- Energy
- Food and Agriculture
- Government Buildings

- Healthcare and Public Health
- Information Technology
- National Monuments and Icons
- Nuclear Reactors, Materials, and Waste
- Postal and Shipping
- Transportation Systems
- Waste and Wastewater Systems

As outlined in the NIPP, government and the private sector are jointly responsible for protection of the critical components that make up the designated Critical Infrastructures and Key Resources (CIKR). As part of the private sector responsibilities, the FDA's FSMA requires the food industry to conduct hazard analyzes of their supply, processing and production chains, to document these hazards, as well as implement control measures to protect their critical systems. First, the critical components need to be identified and in the food and agriculture infrastructure, due to its unique, globally distributed, and highly integrated nature as a systems of systems. As a result, the systems and subsystems that make up that unique infrastructure need to be identified before any prevention measures may be implemented. This is a way to maximize resources and measures across the food distribution system for a particular commodity.

The Food and Agriculture Sector Criticality and Assessment Tool (FASCAT) is designed to create a CIKR Information Sharing Environment (ISE) to assist a food firm or state government, in partnership with both the private sector and other regional states, as appropriate, to share critical sector information. This allows states to contribute an effective response to United States Department of Homeland Security (DHS) National Data Calls for information on critical infrastructure components for food and agriculture. FASCAT will assist states in prioritizing vulnerability assessments and even possible protective measures. There is also a means to export assessment results into the DHS Automated Critical Asset Management System (ACAMS), which is a tool for governments to build critical infrastructure protection programs.

The food and agriculture private sector will benefit by meeting the hazard identification, assessment and documentation requirements of the FDA's FSMA. In addition, they will be better able to determine the information to support vulnerability assessments, develop and implement preventive controls, and develop mitigation and response planning and other risk reporting and documentation requirements.

The FDA developed a risk assessment tool called CARVER + Shock, which is intended to assist food processors in protecting their products from deliberate contamination. The CARVER concept was originally developed by the U.S. military to identify areas that may be vulnerable to an attacker. Subsequently, the FDA and the United State Department of Agriculture (USDA) have adapted it for the food and agriculture sector. CARVER stands for the six (6) attributes that are used to evaluate targets for an attack:

- Criticality: What impact would an attack have on public health and the economy?
- Accessibility: How easily can a terrorist access a target?
- Recuperability: How well could a system recover from an attack?
- Vulnerability: How easily could an attack be accomplished?
- Effect: What would be the direct loss from an attack, as measured by loss in production?
- Recognizability: How easily could a terrorist identify a target?

The "shock" aspect is the seventh attribute that evaluates the psychological impacts of an attack. The CARVER + Shock tool is an online software system that the food industry can use to evaluate their processes.

In 2004, the Indiana State Department of Health (ISDH), Food Protection Program conducted a food defense survey of Indiana food manufacturers, processors, and distributors and based on response data, ISDH has identified key areas of interest. The ISDH Food Protection Program encourages managers, owners, operators, or other responsible staff to enhance and improve the food defense of their operation as appropriate and practical. While the survey was specific to Indiana, these areas may assist other states or organizations in development of food defense programs. The key areas of interest identified and recommendations are:

1. Food Defense Plans: The three basic steps to develop a plan are: (1) conduct a vulnerability assessment; (2) minimize weaknesses identified in the first step through security upgrades, modification of operational procedures, and/or policy changes; and (3) prepare and implement the food defense plan. The plan should cover all ap-

plicable areas: general inside, processing, storage, general outside, shipping, receiving, and transportation securities.

2. Food Defense Training: Modify employee training regimens to incorporate food defense. Train employees to courteously, but firmly challenge unaccompanied visitors on the property.
3. Employee Hiring Practices: Examine the background of applicants as appropriate to the position for which they are being considered. The U.S. Citizenship and Immigration (USCIS) E-Verify tool is an internet-based system that allows businesses to determine the eligibility of their employees to work in the United States or the USCIS Systematic Alien Verification for Entitlements (SAVE) program that's an inter-governmental information service initiative, which verifies the immigration status of benefit applicants.
4. Access Controls: Limit staff access to only those areas that are necessary for their job functions. Ensure that access controls are functional and are not intentionally disabled for convenience (i.e., self-closing door propped open, etc.). Limit and monitor visitor and other outsider access. Utilize at least one means of rapid visual identification (i.e., identification badges, color-coded hairnets, caps, uniforms, hard hats, smocks, etc.). Vendors and contractors who routinely visit service accounts, such as uniform laundry, bottled water, propane for forklifts, snack food, etc. should be restricted to the areas requiring their services. Restrict access to bulk food storage areas.
5. Management Activities: Local law enforcement authorities should be contacted immediately in the event or threat of deliberate contamination. Supervisors and managers should be alert to suspicious or unusual employee behavior, as well as, an atypical health condition. Develop a Standard Operating Procedure for addressing suspicious behavior. Routine, random food defense inspections should be conducted.
6. Physical Facility Security Measures: Be aware of who can access the facility and by what means. Cameras and adequate lighting are important deterrents to trespassers (even if the cameras do not work), in addition to motion lights. The appearance of security is an important motivating factor to not trespass and tamper with the

facility's food products. Doors with alarms are highly recommended.

7. Food Transportation Security: Food transportation security has been an increasing concern and the facility is responsible for the safety of their food product once received. Requiring shipment to be locked before accepting and unloading is a simple means of ensuring product safety.
8. Evacuation and Response Plans: An established protocol for evacuation and response is highly recommended. Hazardous substances should be identified on-site with material safety data sheets. Employee reassembly points and off-site contact lists to keep informed if business halts should be created and employees properly trained.

A food defense team should be composed of a variety of employees from various disciplines to implement food defense prevention strategies. A vulnerability assessment may be conducted by this team, but might also include the use of resources at a government agency or a third party contractor with food defense expertise.

The FDA has several initiatives intended to raise awareness of state and local government agencies and industry representatives regarding food defense issues and preparedness. The concepts are generic enough to apply anywhere along the farm-to-table supply chain. ALERT identifies five key points that industry and businesses can use to decrease the risk of intentional food contamination. The key points are as follows:

- **Assure** – How do know your suppliers and ingredients are safe?
- **Look** – How do you look after your security measures?
- **Employees** – what do you know about your employees and other people who visit your facility?
- **Report** – Can you provide reports about the security of your product?
- **Threat** – What do you do and who do you notify if you have a THREAT or issue at your facility, including suspicious behavior?

The FDA has also developed the Employees FIRST initiative that food industry managers can include in their ongoing food defense training programs. The program is intended to educate front-line food industry workers from farm-to-table about the risk of intentional food contamination and the actions they can take to identify and reduce these risks. These concepts are as follows:

- **F**ollow your company food defense plans and procedures
- **I**nspect your work area and surrounding areas.
- **R**ecognize anything out of the ordinary
- **S**ecure all ingredients, supplies and finished products.
- **T**ell management if you notice anything unusual of suspicious

There are many resources available to prevent a food defense incident and everyone from the government agency to the private sector is responsible to prevent such events.

ACAMS: http://www.dhs.gov/automated-critical-asset-management-system-acams

FASCAT: http://www.foodshield.org/fascat/docs/v3g uidance.pdf

NIPP: http://www.dhs.gov/national-infrastructureprotection-plan

NIPP SSP: http://www.dhs.gov/xlibrary/assets/nipp-ssp-food-ag-2010.pdf

CARVER Shock http://www.fda.gov/ForConsumers/ConsumerUpdates/ucm094560.htm

Although in United States we have increased our awareness of Food Safety and Food Defense, we still have not been one hundred percent successful. From specific threats that have occurred, we have been able to educate ourselves and learn from those incidents. Developing strong partnerships with state, local and federal agencies is a must in meeting the objective of minimizing contamination of the food supply. The Federal Bureau of Investigations (FBI), United States Department of Agriculture (USDA), and the Department of Homeland Security (DHS) developed a Strategic Partnership Program Agro terrorism (SPPA) initiative. This initiative is to set up to for the food and agriculture sector to provide collaboration with both private industry and the local and state agencies. Setting up partnerships with the intention of supporting assessments allows for coordinated infrastructure protection. This is stated in the National Infrastructure Protection Plan (NIPP) and the Homeland Security Presidential Directive 9 (HDSP-9).

Partnerships with local, federal, and state agencies allow specific federal agencies, such as the FDA and the CDC to collect necessary data to identify lapses and gaps missing in the food agriculture vulnerabilities.

This allows for better identification of key resources, development of early warning capabilities, and enhancement of response for protection of the nation's food supply. In addition, the USDA, FDA, and DHS has created a council made of private industries and another of governmental agencies to assist with this effort. The Food and Agriculture Sector Coordinating Council (FASCC) has members from both private companies and associations that represent various entities of the food supply chain. The Government Coordinating Council (GCC) has members from state, federal, local and tribal entities. Both of these Councils work closely with the aim of working as one voice to protect the nation's food supply. Additional partnerships involving local, state and federal entities in regards to food defense are the USDA Food Safety and Inspection Service (FSIS), the Council to Improve Foodborne Outbreak Response (CIFOR), the Association of Food and Drug Officials (AFDO) and The Environmental Protection Agency (EPA) Food Emergency Response Network (FERN).

One key partnership between the U.S. Food and Drug Administration (FDA) and the state agencies is the Rapid Response Teams (RRT) Project. This project is an FDA initiative with state programs to build food safety infrastructure. The RRT employs the "all-hazard" approach and uses the Rapid Response Team (RRT) FDA Best Practices Manual 2011. The FDA Manual develops emergency core and target capabilities. The FDA RRT Manual bases much of the information on the National Response Framework (NRF).

Emphasizing the importance of community partnerships for food defense activities only strengthens the collaboration between state, federal and local entities. Communication flow is enhanced from this as well. The trust gained from consumers and industry demonstrates another benefit. Knowledge that can be shared with these specific partnerships is a must for everyone working toward improving the overall aspect of food defense. One example of combining resources and partnerships that allows for sharing of ideas and brainstorming issues is the use of Food Safety and Defense Task Forces.

As of 2012, there are 26 Food Safety and Defense Task Forces. These represent 25 states and the District of Columbia. The Food Safety and Defense Task Force is funded by the FDA under a grant made available to state and tribal entities that are valuable stakeholders in food defense and food safety. Members of the task force include academia, agriculture, food distributors, as well as many private wholesale or retail establishments.

The Food Safety and Defense Task Force works as a group to hold educational seminars on various issues in the food defense area as well as food/feed safety. The Food Safety and Defense Task Force allows for communication and sharing of information among various local, state, and federal partners. Technical expertise with other members may include academia, consumers, first responders, law enforcement, and those in the health care field. The Food Safety and Defense Task Force is not limited to any specific association or industry. Instead, this task force acknowledges and educates individuals from many areas on food defense issues so food threats may be addressed with the best possible outcome in regards to food safety and food defense.

Food partnerships and cooperation have resulted in a greater awareness of future vulnerabilities. In order to address these vulnerabilities even more, it is imperative all stakeholders in the food supply chain create better systems to respond to such outbreaks or contamination issues. This means all stakeholders need to be educated on not only food supply issues, but also current threats that are happening around the world. The knowledge of such threats allows all individuals involved to be prepared and react properly.

Short term goals should include looking at or developing core emergency response capabilities. Basic development of the "All-Hazard" approach would provide a set of concepts to institute this. In keeping with the all-hazards approach, individuals and stakeholders need to follow the National Response Framework (NRF) and the National Preparedness Guidelines. These documents allow one to pursue risk-based capacity development using their outlined principles. The use of documents such as the NRF and the National Preparedness Guidelines also build stronger relationships both interagency and agency wide because they help clearly define roles and relationships.

Additional short-term and long-term goals to address food defense readiness are to create Communication Standard Operating Procedures (SOPs) that all partners may share, use and update. Responsibility and roles need to be addressed and evaluated as to which agency or stakeholders will be involved. As with any SOP, proper maintenance of this document needs to happen so the response and readiness is up-to-date.

Training is one of the most essential aspects to providing a strong food defense foundation. Each stakeholder or agency must identify those individuals in the response role and fit their educational and training needs to that specific role. The framework for food defense response is built on the knowledge, education and training of those individuals that will be involved with a food safety or food defense response. The training should include at a minimum ICS, sampling technique, communication, foodborne illness investigation, and trace back and evidence preservation. Along with these types of training, hands-on training is essential for all individuals involved in the food defense response. Participation in table top exercises or preplanned exercises allows individuals to practice what they have learned and evaluate what training is still needed. The hands-on training allows individuals to actually learn and practice skills obtained. Along with the importance of training, one must remember continuing education is a critical component. Continuing education will expand knowledge and allow individual responders to keep up-to-date on new developments. All stakeholders need to take the opportunity to improve the overall understanding of food defense response. Placing emphasis on training is a vital part of defending the food supply in United States.

19.3 EXAMPLES

Super Bowl 46

In 2012, the NFL Super Bowl XXXXVI was held in Indianapolis, Indiana. The planning for this event started years in advance with the naming of the host city and the establishment of the host committee. Once this happens, 18 additional subcommittees are established to address a variety of different areas of concern. One such committee was the "Public Health Committee" where officials from the state and local agencies participated to provide input and concerns about many different aspects of public health, including food defense. The overall agencies in control of these processes are the host city police department and the FBI. These overseeing agencies understand the importance of food safety and defense, but they must also view the broader picture to avoid and minimize damage from "any" kind of negative event, such as a riot, bad weather, power outages, disease outbreak, etc.

The ISDH and the Marion County Health Department (MCHD) were on this committee and were asked by the FBI to develop and submit a "concept of operations" plan so they could gain an understanding of the needs of food safety officials. This was completed and then the two agencies worked together over the next year and a half with planning, training and exercising to be as prepared as possible for the event. It should be understood that this event is almost eight days long with many thousands of people attending multiple venues and activities starting several days in advance of game day.

As part of the planning, ISDH and MCHD staff conducted food safety training for the industry at multiple hotel venues for hundreds of food workers about a month before the event. Staff also conducted site specific vulnerability assessments with the management staff of the five main hotels and the Lucas Oil Stadium regarding food defense issues. Examples including restricting staff access to only those areas where they had a reason to be and discussing the concepts of employees taking responsibility to report occurrences that seemed out of the ordinary. Management staff was very appreciative of learning about food defense and measures they would need to implement to minimize risks from any intentional food contamination.

In addition, the ISDH hosted two federal trainings for their staff and the surrounding counties where additional event activities may occur. The FDA provided a newly developed course just for this event regarding food safety and defense at large events. This was based on their experiences with conducting these activities for the high profile events such as the Democratic and Republican Conventions and the Presidential Inaugurations. The second training was on handling a food emergency during large mass gatherings. Finally, a third event was hosted by the MCHD that was a table top exercise involving all of the participants, such as inspectors, laboratories, epidemiologists and public affairs individuals. This exercise involved an outbreak of STEC associated with ground beef during the Super Bowl and the measures instituted to identify the source, implement controls and mitigate further injury.

During the actual Super Bowl XXXXVI event there was a recall of chopped egg product due to potential salmonella contamination that was distributed to some of the venues. Because of the planning and training going into the event, the situation was quickly brought under control and no illnesses occurred during the event.

Preparations by all involved contributed to the successful event without any foodborne disease outbreaks.

Indiana State Fair

For several years, the Food Defense Section of the ISDH has conducted food defense audits on the grounds of the Indiana State Fair, both during the fair and before. At the same time food safety inspections are also being conducted. Many times this means visiting the grounds very early in the morning when no one is around to see how the food vendors left their operations the night before. Keep in mind that most of these are temporary food establishments with some of them being tent set-ups. It has been observed that many coolers and storage areas were not secured or locked. Many of these operators have been doing this at fairs for many years and are accustomed to a casual manner in regards to protection for food defense. As a result, it has been stressed to them that they should reconsider how they treat their assets to avoid theft and pilferages and at the same time protect the public from potential threats from intentional contamination. The fair is also a high profile event with about 1 million people visiting each year, so there is a risk for something negative to occur. The ISDH continues to work with these operations and the state fair toward continued improvements.

KEYWORDS

- **Bioterrorism Act**
- **chemical, biological, radiological and nuclear threats**
- **Rajneeshee cult group**
- **PulseNet**

CHAPTER 20

GLOBAL ISSUES

JIN-KYUNG CHOI, PhD
Assistant Professor, Woosong University, Korea

JEFF FISHER, MS
Purdue University

CONTENTS

About 17% of the population suffers from foodborne illness in the United States (Scallan et al., 2011) and about 70% of foodborne illnesses are linked to food prepared away from home (Hensen et al., 2006; Jones et al., 2004). Food service inspection is a tool used by health inspectors to communicate the safety of the facility to management. Publicizing the inspection results is a way to communicate with consumers, by providing information about the food safety practices of the establishment. Consumers want to know if their food is being prepared safely. Because they are not present in the kitchen during food preparation, they must rely on the disclosure of food safety information.

Food poisoning is a preventable illness and inspections that identify potential risks are a key preventative measure. Transparency linked to the quality of the products and processes is often lacking. As a result, consumers are not fully informed (Bavorova and Hirschauer, 2012). Previously, consumers with a lack of nutrition knowledge, argued obesity and other harmful diseases were the result of not disclosing the nutritional information of food. In the same sense, sanitation or food safety literacy is needed for consumers to protect themselves from the possibility of contracting a foodborne illness by consuming food unsafely prepared. Kimura (2011) suggested that food literacy is an approach that is needed for consumers' food education.

Many countries have initiated and developed regulations to protect their nation's health and quality of life, however, carrying out and applying such regulations is not an easy task. Additionally, most regulations are in need of revising to keep up with the fast evolutionary environment of society. Published in 2010, the European Union's communication "Smart Regulation" was aimed at reducing the regulatory burden on small to medium size enterprises (EU Commission 2010, Bavorova and Hirschauer, 2012). Foodborne disease outbreaks were apparent at the community level in 2004 (Anonymous, 2005) and became mandatory to report in 2005 (Much, 2009). The three primary locations of exposure in Europe were restaurants/hotels (19.8%), households (46.4%) and unknown (12%) (Anonymous, 2007). At the same time governments were trying to establish regulations to prevent foodborne disease outbreaks, consumers' concern for food safety information had increased. In Belgium, expenditures for food away from home increased from 14% in 1980 to 23% in 2000

(Vandevijvere, et al., 2009). A survey by the German polling institute Emnid indicated that 87% of the respondents supported the idea of the public announcement of food service inspection results (Maurin, 2009). This chapter examines the forms used by health inspectors and how restaurant inspection results in various countries are communicated to the public. Because developed countries have more established food safety programs than do developing countries, member countries of the Organization for Economic Cooperation and Development (OECD) have been considered. In particular, some of the countries with OECD memberships were examined and their restaurant inspection systems were reported.

20.1 AUSTRALIA

Australia established a surveillance system and has conducted risk assessments (Greig and Ravel, 2009). In 2010, this system reported 30,035 notifications of diseases or conditions that were transmitted by contaminated food (The OzFoodNet Working Group, 2012). In New South Wales, food outlets that do not "apply" appropriate hygiene standards cause about 1 million cases of food poisoning each year. Contaminated food puts about 18,000 Australians in hospitals and causes 120 deaths annually. Nevertheless, the government has rejected amendments to the Food Act that would require inspection notices to be posted on the Internet and on a prominent place at the restaurant's entrance. There are some restaurants that have not been inspected for years.

The city of Greater Dandenong, a local government area in Victoria, Australia located in the south-eastern suburbs of Melbourne, adopted a scoring system that allows the Council of Environmental Health to quickly identify food establishments that are not practicing proper food safety. The possible scores range from unacceptable (1) to excellent (5) and are awarded by category such as cleaning, record keeping, food safety knowledge, structure, temperature control and food handling (Hodge and Balaba, 2012; see Figure 1).

In addition, food safety and sampling officers are rotated in a timely manner so risks to the community can be prevented (Figure 2).

FOOD PREMISES SCORE CHART

	Unacceptable (1)	Poor (2)	Average (3)	Good (4)	Excellent (5)
Temp (Food Temp)	• High risk foods not being maintained at correct temperatures and are dangerously out of temperature control • Foods to be seized.	• High risk foods not at correct temperature. • Too long in danger zone during cooling process. • Highly likely that if samples taken, several would fail.	• Display units over stocked. • Cold storage units overstocked. • These foods not maintained within temperature control.	• Majority of high risk foods maintained at correct storage temperatures. • 2hr/4hr rule adhered to, demonstrated in FSP (records 4 and 8).	• All high risk/ready to eat foods maintained at correct storage temperatures.
FSS (Food Safety Supervisor) Staff Awareness of Food safety	• FSS not qualified.	• FSS qualified but not imparting knowledge to staff.	• Staff have some understanding of good food safety practices but not implementing all of their knowledge.	• FSS has a good understanding of their role. • Most staff aware of good food safety practices. • Demonstrate good understanding of requirements. • More than one person trained in food safety.	• All staff trained and fully aware of good food safety practices. • Demonstrate excellent understanding of requirements. • Additional training, certification or quality control standards implemented.
Structure	• Many damaged surfaces. • Many holes throughout premises to allow pests to enter. • Equipment damaged that may cause contamination to food. • HWB not operational or not accessible. • Entire premises requires repainting.	• Potential for pests to enter premises. • Equipment worn/some damage. • Internal surfaces may require some structural work • HWB operational with cold water only. • Premises cluttered with excess equipment.	• HWB operational with warm running water. • Premises may require some painting or repairing of small gaps. • Worn flooring surfaces may require attention. • Sholving and the like may require resurfacing in places.	• Equipment generally in good working order, slight damage to utensils or section of equipment. • Small amount of gaps/holes require sealing. • Premises internal structures generally in good working order.	• All equipment in good working order. • All wall, floor and ceiling surfaces structurally sound. • No gaps or holes for pests to enter.
TOTAL score= /30					
					Date: 07/08/2006

* HWB = Hand Wash Basin Environmental Health Unit: 92381479

FIGURE 1 Scoring Matrix by City of Greater Dandenong.

Food Sampling Schedule 2012 – 2013.

Month	Sample Type	Responsible Officer	Area
July 2012	Ready to eat sandwiches	Chris + Eva	Mixed
August	Imported & Local Dried fruits & exposed Nuts	Terry + Stephen	All Areas
September	Swab of surfaces @ Takeaway Premises	Ray & Terry	Mixture
October	Labelling Follow up	Ray & Jim	All Areas
November	Kebab Meats and meals made with Kebab meats	Terry + Eva	Mixture
December	Pool water Samples	Ruth + Chris	Mixed
January 2013	Fruit Blend Drinks, Juice bars Cut salads	Stephen + Chris	Mixture
February	Class 1 Premises (15 each)	All In	Mixture
March	Nuts from Area 5 for Alfa toxins	Jim + Eva	Mixture
April	Supermarket Delis + Meats (SO2)	Chris + Ray	Mixture
May	Dips & afghan dips	Jim + Stephen	Mixture
June	Sushi & Asian Deserts	Ray + Terry	Mixture

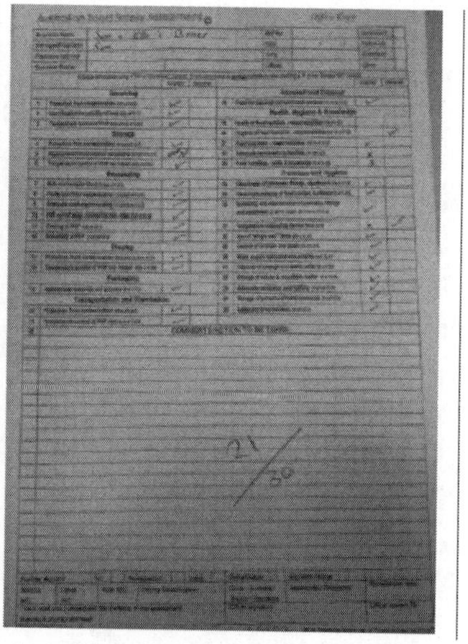

FIGURE 2 Food safety and sampling group and an example of a inspection report.

Food surveillance staff monitors and annually reviews the compliance of all food premises and assesses food safety programs as required for adequacy; they also investigate complaints related to food borne illness and collect food samples for analysis. However, posting results of the inspections are not required (http://www.greaterdandenong.com).

20.2 CANADA

Restaurants and households were the most common places associated with foodborne illness in Canada. Canada's inspection frequency is based on food safety risk. Food safety includes the proper supply, storage, preparation and distribution of food.

TABLE 1 Risk Levels for Food Establishments.

Risk level	Risk Assessment Criteria	Minimum Inspections
High	Any eating or drinking establishments that prepares hazardous food and meets a least one of the following criteria: Serve a high risk population Use processes involving many preparation steps and food frequently implicate as the cause of foodborne illness Implicated or confirmed as a source of foodborne illness/outbreak	Three times per year
Moderate	Any eating or drinking establishments that meets one or more of the following criteria: Prepare hazardous food without meeting the criteria for high risk Prepare nonhazardous foods with extensive handling or high volume	Two times per year
Low	Any eating or drinking establishment that does not prepare hazardous food and meets one or more of the following criteria: Serve prepackaged hazardous foods Prepare and/or service nonhazardous foods without meeting the criteria for moderate risk Are used as a food storage facility for non-hazardous foods only Public health concerns related primarily to sanitation and maintenance	One time per year

Source: From http://www.toronto.ca/health/dinesafe/system.htm.

Food safety inspections for foodservice establishments in Canada are done at three levels of government: provincial governments, municipalities and regional health authorities. Local food safety programs can be found at Canadian Food Inspection Agency (http://www.inspection.gc.ca/).

Ontario was the first site-implemented surveillance program for food and water-borne illness (Holley, 2010). Local health authorities carry out restaurant inspections. Food establishments receive one to three inspections each year based on the type of establishment. Table 1 shows three risk levels for food establishments in the city of Toronto. The failure of a food service operator to meet the minimum requirements or standards is referred to as an infraction. There are three types of infractions (see Table 2). The City of Toronto's Dinesafe program requires food establishments to be inspected and notices to be displayed near the entrance (Figure 3).

TABLE 2 Infraction types and examples.

Infraction Type	Examples
Minor infractions: Infractions that present a minimal health risk.	Walls, floors or other nonfood contract surfaces or equipment need cleaning or repair
Significant Infractions: Infractions that present a potential health hazard.	Food contact surfaces or equipment require cleaning or repair
Crucial Infractions: Infractions that present an immediate health hazard.	Not hot and cold running water under pressure in food preparation area or where utensils are washed.

Source: From http://www.toronto.ca/health/dinesafe/system.htm.

A Pass notice will be issued when only minor or no infractions are observed during an inspection.	A conditional pass indicating significant infractions such as inadequate cleaning of food contact surfaces or equipment, or refrigeration or mechanical dishwashing equipment requiring repair. A public health inspector will re-inspect the premises within 24 to 48 hours..	Closed, issued where crucial infractions are observed such as evidence of rodent or insect infestations or contaminated food.

FIGURE 3 Reporting and Disclosure.

20.3 DENMARK

In 1997, the Danish Veterinary and Food Administration (DVFA) unified all food safety matters, including inspections of food business operators. Regular inspections occur without advanced notice, except for certain cases when authorities may give prior warning before the inspection. Inspectors check for: 1) display of the inspection report with the smiley sign, 2) hygienic condition of the equipment, 3) record of self-inspections, and 4) other (IFC, 2010). The inspections are conducted on a need-oriented basis that may vary from inspection to inspection and from shop to shop (IFC, 2010; www.findsmiley.dk/en-US/Forside.htm). Inspections emphasize HACCP procedures, standard sanitation procedures and employee training.

In 2001, Denmark established a national smiley scheme and in 2008, an additional elite-smiley was introduced (Figure 4). Food establishments are inspected without advance notice, based on the risk-sensitiveness of the business. For example, hospital kitchens are inspected three times per year and retail packaged food product operations are inspected once every two years. All retail businesses are required to post the inspection report on the premises in a visible location to consumers. In addition, the reports must be placed on the home pages of their Web sites. The last four reports are available on the Danish Veterinary and Food Administration (DVFA) Web site at http://www.findsmiley.dk (DVFA, 2011) (Figure 5).

The Danish Food Act establishes, from farm to table, food and veterinary inspection authority and processes in restaurants and shops.

Elite-smiley: a rating of "1" was awarded four consecutive times
Score 1: inspector had no remarks
Score 2: inspector emphasized that certain rules must be obeyed
Score 3: inspector issued an injunction order or a prohibition
Score 4: inspector issued an administrative fine, reported the enterprise to the police, or withdrew an approval

FIGURE 4 Smiley Grading Inspection System in Denmark.

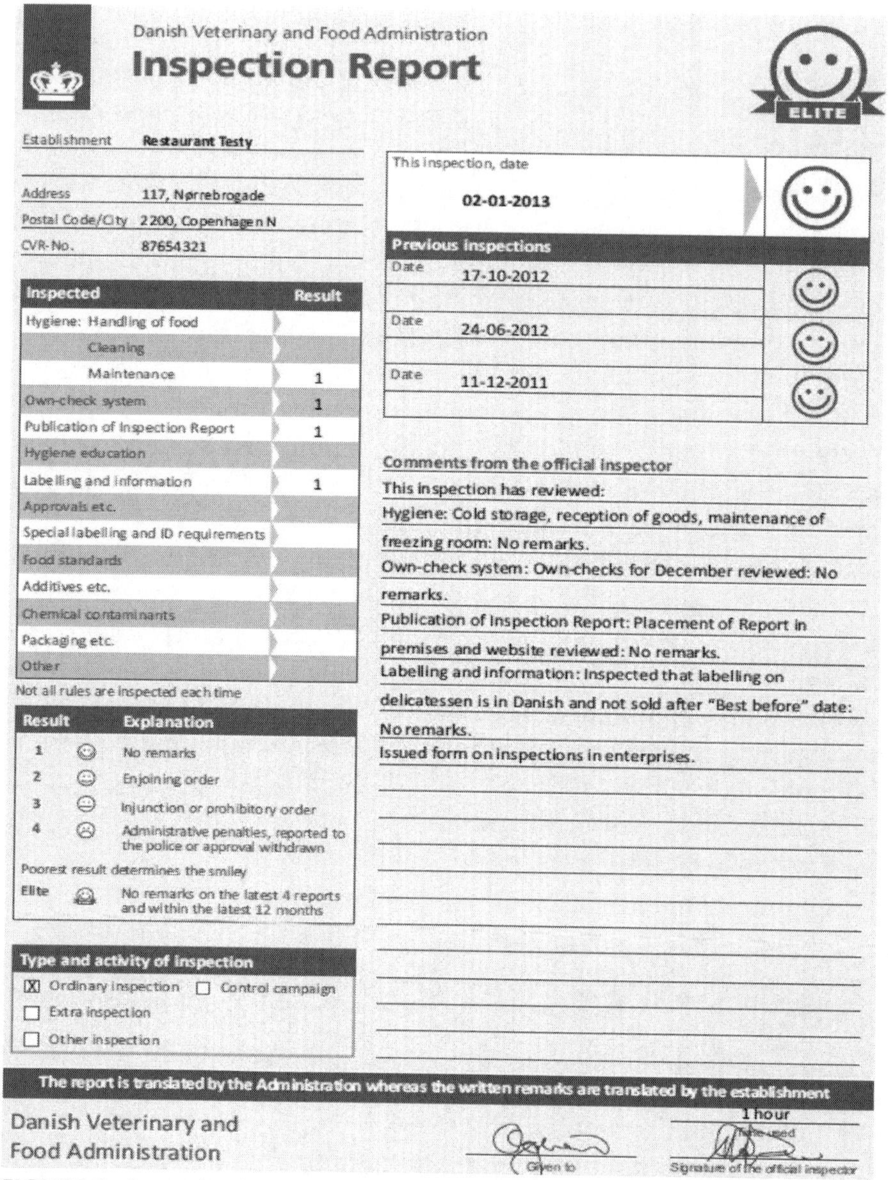

Danish Veterinary and Food Administration

Inspection Report

ELITE

Establishment	Restaurant Testy
Address	117, Nørrebrogade
Postal Code/City	2200, Copenhagen N
CVR-No.	87654321

Inspected	Result
Hygiene: Handling of food	
Cleaning	
Maintenance	1
Own-check system	1
Publication of Inspection Report	1
Hygiene education	
Labelling and information	1
Approvals etc.	
Special labelling and ID requirements	
Food standards	
Additives etc.	
Chemical contaminants	
Packaging etc.	
Other	

Not all rules are inspected each time

Result	Explanation
1	No remarks
2	Enjoining order
3	Injunction or prohibitory order
4	Administrative penalties, reported to the police or approval withdrawn

Poorest result determines the smiley

Elite: No remarks on the latest 4 reports and within the latest 12 months

Type and activity of inspection

- [X] Ordinary inspection
- [ ] Control campaign
- [ ] Extra inspection
- [ ] Other inspection

This inspection, date: 02-01-2013

Previous inspections

Date
17-10-2012
24-06-2012
11-12-2011

Comments from the official inspector

This inspection has reviewed:

Hygiene: Cold storage, reception of goods, maintenance of freezing room: No remarks.

Own-check system: Own-checks for December reviewed: No remarks.

Publication of Inspection Report: Placement of Report in premises and website reviewed: No remarks.

Labelling and information: Inspected that labelling on delicatessen is in Danish and not sold after "Best before" date: No remarks.

Issued form on inspections in enterprises.

The report is translated by the Administration whereas the written remarks are translated by the establishment

Danish Veterinary and Food Administration

Given to

1 hour

Time used

Signature of the official inspector

FIGURE 5 Inspection Report by Danish Veterinary and Food Administration.

20.4 GERMANY

Cities in Germany adopted and then abandoned the display of the inspection report. In 2007, the city of Zwickau, the capital city of the district of Zwickau in the eastern part of Germany, adopted Hygiene Passport Zwickau, of which 300 of about 600 eligible businesses participated. However, the scheme was abandoned in 2010 due to budget issues (Bavorova and Hirschauer, 2012; ZWICKAU, 2009). It was a voluntary program and only businesses with good inspection grades were announced and published on the Internet (http://www.hygiene-pass.de).

The grade was assigned based on a range of points from one (without health risk) to five (far above average). One point was considered a good grade awarded by the program. In 2007, North Rhine-Westphalia (NRW) started the smiley scheme for restaurants (Figure 6); the program expanded in 2011 to include food service businesses throughout 14 administrative districts of the state. The inspection is conducted against a list of 18 criteria and rated on a scale from one (very good) to five (insufficient). A smiley is awarded to businesses with an average good score and the results are posted on the premises and the authority's Web site (http://www.munlv.nrw.de/verbraucherschutz/lebensmittel/smiley/index.php).

FIGURE 6 NRW-Smiley.

20.5 REPUBLIC OF KOREA

In Korea, 510 foodborne disease outbreaks with 9,686 patients were reported in 2007; of these about 76 percent were linked to food service establishments (Park et al., 2010). With the occurrence of foodborne illness outbreaks, authorities tried to compel foodservice businesses to maintain proper food safety practices. Inspection scores are publicized in order to notify consumers about the food safety practices. In 2013 Seoul, the capital of South Korea, had two alphabetical rating programs (A, B or C) and (AAA, AA or A). The city of Seoul had started several programs that informed the public of food safety practices and communicated to managers the proper practices expected at their establishments. While the city tried to make the public aware of food safety practices in each establishment, the different programs may have confused consumers.

TABLE 3 Restaurant Hygiene Rating Programs in Seoul, Korea (Republic of).

Score	Plaque awarded as "Good Restaurant"	Restaurant Hygiene Ratings	
90 and above	Satisfy categories and approved by authorities	A	AAA
More than 80 less than 90	Satisfy categories and approved by authorities	B	AA
More than 70 less than 80	Satisfy categories and approved by authorities	C	A

Moreover, since the disclosure of inspection reports is not mandatory, some businesses that received ratings other than 'A' tended not to post their inspection ratings. In response, the City of Seoul is trying to move on to the "AAA" scheme so that all restaurants post their inspection ratings without hesitation. Table 3 shows the different types of programs that inform food safety practices in foodservice establishments.

In order to receive a plaque a "good restaurant," an establishment should meet the requirements of the Korean Food Hygiene Code. After receiving a 'good restaurant' designation from the authorities, hygiene inspection of the business can be waived for 2–3 years. Figure 7 shows three different programs run by the city of Seoul. Consumers can find both a list of 'good restaurants' and the list of restaurant ratings (either A, B or C or AAA, AA or A).

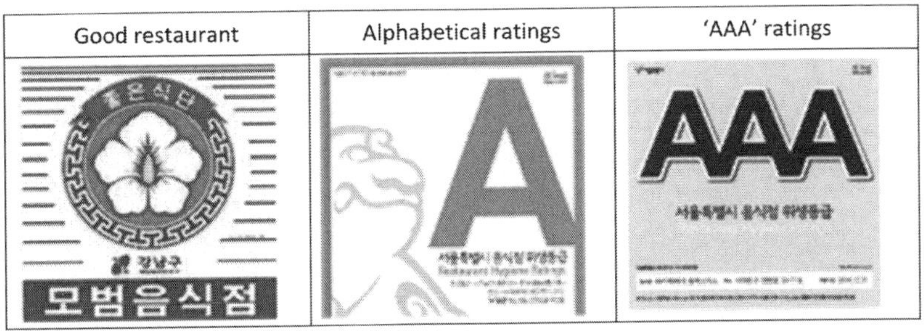

FIGURE 7 Examples of Postings of Restaurant Hygiene Rating in Seoul.

20.6 UK

Local authorities in partnership with the respective food standards agencies in England, Northern Ireland, Wales and Scotland run the rating scheme in the United Kingdom. In addition, United Kingdom food safety agencies maintain Web pages with food hygiene ratings (http://ratings.food.gov.uk/). The restaurant hygiene ratings are given to each foodservice business by local authorities. Figure 8 shows a food hygiene sticker scheme for England, Northern Ireland and Wales. This scheme has six standard hygiene levels; each business is given a rating that ranges from 0 (urgent improvement necessary) to 5 (very good).

FIGURE 8 Food Hygiene Sticker Scheme in England, Northern Ireland and Wales.

Restaurants can receive either a sticker or a certificate to display, showing consumers the inspection result for their businesses (Figure 9). 'Pass' shows that the establishment has met the food hygiene requirement. 'Improvement required' shows that the establishment has failed to meet requirement of food hygiene law. (http://www.food.gov.uk/multimedia/hygiene-rating-schemes/ratings-find-out-moreen/#anchor_2). The authorities also have an 'Eat Safe Award' program that is not part of the food hygiene-rating scheme. This award shows that the establishments have hygiene standards which are better than those required by law.

'Pass' certificate	'Improvement required' certificate	'Eat safe award'

FIGURE 9 Food hygiene sticker scheme in Scotland.

20.7 UNITED STATES

The variation in food codes from one state to another creates differences between the inspections. For example, although many states use a system that monitors and reports the number of critical and noncritical violations, others do not. Restaurants in Tennessee are subjected to inspections using a list of 44 checked items that are based on the FDA inspection criteria, which has a total possible score of 100 for no violations at all. The state of Mississippi uses a letter grade system, while New Hampshire uses a system of colors to indicate the inspection results.

A previous study investigated the availability and formats of the inspection reports on 111 county websites for more than 10,000 foodservice establishments (Choi and Almanza, 2012). Table 4 shows that violation descriptions were the most used format (54%) followed by a summary in the form of letter grades or numeric scores (41%). Also, Choi and Almanza (2012) investigated health inspection reports that were divided into two groups: (1) full report and (2) summary. In this study 85% of the counties posted a summary of the health inspection report and only 15% posted a full report on their Web site. Besides the different formats for inspection reports, the study found that 47% provided a single format while 36 counties (53%) provided more than one scoring format. The multiple formats included a full report, a letter grade or numeric score, violation descriptions, or a color-coded system.

TABLE 4 Regional Distribution of Different Forms of Posted Inspection Scores (Choi and Almanza, 2012).

Region	Number of Counties	Full† Report	Summary††	No inspection report information
West[a]	28 (25.5%)	3 (2.7%)	21 (19.1%)	4 (3.6%)
Midwest[b]	18 (16.4%)	4 (3.6%)	7 (6.4%)	7 (6.4%)
North-east[c]	34 (30.9%)	3 (2.7%)	9 (8.2%)	22 (20%)
South[d]	30 (27.3%)	1 (.9%)	20 (18.2%)	9 (8.2%)
Total	110 (100%)	11 (10%)	57 (51.8%)	42 (38.2%)

Notes: [a]Arizona, Colorado, Idaho, New Mexico, Montana, Utah, Nevada, Wyoming, Alaska, California, Hawaii, Oregon, and Washington.

[b]Indiana, Illinois, Michigan, Ohio, Wisconsin, Iowa, Kansas, Minnesota, Missouri, Nebraska, North Dakota, and South Dakota.

[c]Connecticut, Maine, Massachusetts, New Hampshire, Rhode Island, Vermont, New Jersey, New York, and Pennsylvania.

[d]Delaware, District of Columbia, Florida, Georgia, Maryland, North Carolina, South Carolina, Virginia, West Virginia, Alabama, Kentucky, Mississippi, Tennessee, Arkansas, Louisiana, Oklahoma, Texas.

† the websites offered all information from the health inspection report.

†† the websites provided only a summary of the report.

20.8 SUMMARY

Countries differ in their inspection formats and procedures. The most commonly used formats are numbered and alphabetical letter grading, whereas many countries have adopted their own symbolized grading formats such as smiley, colors, stickers, certificates etc. Announcing concise results of inspections to the public is important. One study suggested that a health hazard is likely to be found when authorities use enforcement measures in foodservice establishments (Luden, 2013). However, the first control needs to be an educational approach where the enforcement measures are justified from a human health perspective (Luden, 2013). Besides countries mentioned above, inspections in Slovenia are done based on risk level, where low risk establishments are audited once every 24 months, medium risk levels are audited once every 15 months and high risk food business operators once every nine months (IFC, 2010). Also, in Poland, Frequency of sanitary inspection is determined by specific instructions and ordinances. The sanitary inspector approves food businesses' self-inspection reports and provides implementation guidance for HACCP programs. Results of inspections are not publically accessible however they are shared between inspectors (IFC, 2010). Each country has adopted the most suitable system for food safety in foodservice establishments. These systems should be revised and corrected in timely manner in order to offer the most accurate information to consumers.

KEYWORDS

- **food safety programs**
- **foodborne illness**
- **HACCP procedures**
- **good restaurant**

REFERENCES

Anonymous (2005). Trends and sources of zoonoses, zoonotic agents and antimicrobial resistance in the European Union in 2004. The EFSA journal 310 ISBN 92–9199–016–7. accessed 8.27. 2007 www.efsa.europa.eu/en/science/monitoring_zoonoses/reports/1277.html In Much, P., Pichler, J., Kasper, S. S., and Allerberger, F. (2009). Foodborne outbreaks, Austria 2007. The Middle European Journal of Medicine. 121, 77–85.

Anonymous (2007). The community summary report on trends and sources of zoonoses, zoonotic agents, antimicrobial resistance and foodborne outbreaks in the European Union in 2006. The EFSA Journal 130 accessed 8.27. 2007

www.efsa.europa.eu/cs/Blobserver/DocumentSet/Zoon_report_2006_en.pdf?ssbinary=true

In Much, P., Pichler, J., Kasper, S. S., and Allerberger, F. (2009). Foodborne outbreaks, Austria 2007. *The Middle European Journal of Medicine*. 121, 77–85.

Barvorová, M., and Hirschauer, N. (2012). Producing compliant business behavior: disclosure of food inspection results in Denmark and Germany, *Journal of Consumer Protection and Food Safety*, 7, 45–53.

Brewer M. S., and Prestat, C. J. Consumer attitudes toward food safety issues. *Journal of Public Affairs*, 55, 10.

Buchholz, U., Run, G., Kool, J. L., Fielding, J., and Mascola, L. (2002). A risk-based restaurant inspection system in Los Angeles County. *Journal of Food Protection*, 65, 367–372.

Choi, J. and Almanza, B. (2012). Health department websites as a source of restaurant food safety information. *Journal of Culinary Science and Technology*, 10, 40–52.

DVFA (Danish Veterinary and Food Administration) (2011). Smiles keep food safety high in Denmark. Retrieved 04, 02, 2013. from http://www.findsmiley.dk/en-US/Forside.htm

EU Commission (2010). Communication from the Commission. Smart Regulation in the European Union. COM/2010/0543 In Barvorová, Miroslava and Hirschauer, Norbert. (2012). Producing compliant business behavior: disclosure of food inspection results in Denmark and Germany, *Journal of Consumer Protection and Food Safety*, 7, 45–53.

Greig, J. D., and Ravel, A. Analysis of foodborne outbreak data reported internationally for source attribution. *International Journal of Food Microbiology*. 1331, 210–212.

Hensen, S., Majowicz, S., Masakure, O., Sockett, P., Jones, A., Hart, R., Carr, D., and Knowles, L. (2006). Consumer assessment of the safety of restaurants: the role of inspection notices and other information cues. Journal of Food Safety. 26, 275–301.

Hodge, S. and Balaban, L. (2012). Comparative study of inspection scores on the microbiological quality of food and the processing environment in food premises. Retrieved May, 03, 2013 from http://www.ifeh2012.org/get.php?f.321

Holley, R. A. (2010). Smarter inspection will improve food safety in Canada. *Canadian Medical Association Journal*. 182(5), 471–473.

International Finance Corporation (2010). Food safety inspections: lessons learned from other countries, December, 2010.

Jones, T. F., Boris, I. P., LaFleur, B. J., Ingram, A., and Schaffer, W. (2004). Restaurant inspections scores and foodborne diseases. *Emergency Infectious Diseases*. 10. 688–692.

Kimura, A. H. (2011). Food education as food literacy: privatized and gendered food knowledge in contemporary Japan. *Agricultural Human Values*, 28, 465–482.

Luden, J. (2013). Reasons for using enforcement measures in food premises in Finland. *Food Control*, 31, 84–89.

Majowicz, S. E., McNab, W. B., Socket, P., Henson, T. S., Dore, K., Edge, V. L., Buffet, M. C., Fazil, A., Read, S., McEwen, S. A., Stacey, D., and Wilson, J. B. (2006), Burden and cost of gastroenteritis in a Canadian community. *Journal of Food* Protection. 69, 651–659.

Maurin, J. (2009), Verbraucher woolen smiley-system. Taz 29.04.2009. https://www.taz.de/1/zukunft/konsum/artikel/1/verbraucher-wollen-smiley-system/, retrieved 23.01.2011. In Barvorová, Miroslava and Hirschauer, Norbert. (2012). Producing compliant business behavior: disclosure of food inspection results in Denmark and Germany, *Journal of Consumer Protection and Food Safety*, 7, 45–53.

Much, P., Pichler, J., Kasper, S. S., and Allerberger, F. (2009). Foodborne outbreaks, Austria 2007. *The Middle European Journal of Medicine*. 121, 77–85.

OzFoodNet (2007). OzFoodNEt quarterly report, Communicable Diseases Intelligence, 33(1). Retrieved March 07, 2013 from
http://www.health.gov.au/internet/main/publishing.nsf/Content/cda-cdi3301l.htm

Park, S. H., Kwak, T. K., and Chang, H. J. (2010). Evaluation of the food safety training for food handlers in restaurant operations. *Nutrition Research and Practices*, 4, 58–68.

Scallan, E., Hoekstra, R. M., Angulo, F. J., Tauxe, R. V., Widdowson, M. A., Roy, S. L., Jones, J. L., and Griffin, P. M. (2011). Foodborne illness acquired in the United States – major pathogens. *Emergency Infectious Diseases*. 17. 1–12.

The OzFoodNet Working Group (2012). Monitoring the incidence and causes of diseases potentially transmitted by food in Australia: Annual report of the OzFoodNet network, 2010. Communicable Diseases Intelligence, 36(3), September 2012. Retrieved July, 2013 from http://www.health.gov.au/internet/main/publishing.nsf/Content/cda-cdi3603a.htm

Torp, K. (2007). Vergabe von Smileys durch die staatlichen Lebensmittelüberwachungsbehörden in Dänemark. Retrieved, April 20.2011. from http://www.lebensmittelkontrolle-sh.de/files/smiley_in_daenemark1.pdf In Barvorová, Miroslava and Hirschauer, Norbert. (2012). Producing compliant business behavior: disclosure of food inspection results in Denmark and Germany, *Journal of Consumer Protection and Food Safety*, 7, 45–53.

Vandevijvere, S. Lacjat, C., Kolsteren, P., and Oyen H. V. (2009). Eating out of home in Belgium: current situation and policy implications. *British Journal of Nutrition*. 102–921–928.

Wheeler, J. G., et al., (1999). Study of infectious intestinal disease in England: rates in the community, presenting to general practice, and reported to national surveillance. The infections Intestinal Disease Study Executive. *British Medical Journal*, 318, 1046–1050.

ZWICKAU (2009). Alle Teilnehmer von A bis Z. Stadt Zwickau http://www.zwickau.de/de/hygienepass/Hygiene_Pass_Teilnehmer.pdf retrieved 30.05.2011. In Barvorová, Miroslava and Hirschauer, Norbert. (2012). Producing compliant business behavior: disclosure of food inspection results in Denmark and Germany, *Journal of Consumer Protection and Food Safety*, 7, 45–53.

CHAPTER 21

TRAVEL HEALTH CONCERNS

KELLY WAY, PhD

Associate Professor, University of Arkansas

CONTENTS

21.1 INTRODUCTION

Almost any place on earth can be reached within 36 h, less than the incubation period for most infectious diseases (Global Health Education Consortium, 2011). Each year more than 500 million people cross international borders by aircraft alone and, with that travel, diseases are exported and imported like any commodity (GHEC, 2011). Familiar diseases, such as cholera, malaria, tuberculosis and yellow fever, once under control, are again on the increase, spreading to previously unaffected regions, including the United States. Avian flu is spreading on the wings of migrating birds and the HIV/AIDS pandemic continues to wreak havoc, primarily in the poorest countries of the world. But, beyond the obvious area maladies, there are other health concerns for the traveler that include: new drug resistant infectious diseases; natural and humanitarian crises; malnutrition and contaminated foods; toxic substances; biological and chemical terrorism; rising social inequalities and highly vulnerable populations; and inefficient or dysfunctional health systems. Realities like these are affecting the health of people all over the world and can become a major issue for travel (GHEC, 2011).

The fact remains, no matter where we live, no matter our economic status nor our political views, our well-being depends on how health issues are managed around the world and how the traveler protects against them while traveling.

21.2 TRAVEL CONCERNS: GLOBAL HEALTH

Globalization is not homogenization—for example, in the areas of ideas, wealth or health there remain a huge range of opinions and perspectives. Regarding health, there are tremendous disparities amidst signs of convergence. While many economically advanced countries find new methods to combat obesity and diabetes, in urban India, for example, the same cannot be said for the survival of Indian newborns and their mothers (Kruk, 2010). Absent public policies and social protection mechanisms along with inequities in access to health care and to the social determinants of health (some of which can be plausibly attributed to globalization) are contributing to asymmetries among generations and countries (Kruk, 2010).

These imbalances can be intensified by global trends: aging, urbanization, epidemiologic transition, climate change and the negative effects brought about by populations without the resources to mitigate these threats (Kruk, 2010). These may sound like local issues, but in today's globalized world they are a problem for everyone, especially those who travel.

The risk of becoming ill or possibly injured during travel depends on many factors including: region of the world visited, traveler's age, health status, length of trip and the diversity of planned activities (Whatley, and Kozarsky, 2012). The Centers for Disease Control and Prevention (CDC) provides international travel health information to address the range of health risks travelers might face at their destination. The goal is to assist travelers and clinicians in understanding the measures necessary to prevent illness and injury during international travel. This is completed through "The Yellow Book." Since first being published in 1967, it has long been a trusted travel companion to clinicians. The purpose of these measures is to ensure maximum security against the international spread of diseases, with minimum interference to world travel and commerce (CDC, 2012). In addition to reporting health events, the United States must also inform the public about health requirements for entering other countries, including, but not limited to vaccinations.

While traveling to different climates and environments abroad can expose one to disease and health risks, one should be aware of these dangers and understand how to stay healthy. Travelers must first ask themselves what risks they might encounter, for example:

- What type of accommodations will you stay in?
- Where will you eat your meals and get water to drink?
- What types of activities will you participate in on your trip?
- Are insects a problem where you are going?

Next, a traveler must anticipate medical and emergency care when traveling:

- Where would you get health care in case of an emergency during your trip?
- What first aid supplies should you bring with you?
- Do you have prescription medications you need to take with you?
- Are those medications allowed into the country where you are traveling?

Finally, what vaccine requirements are required for your travel?

- Which vaccines are required or recommended?
- What vaccines have you already had?

21.3 MODE OF TRAVEL-HEALTH CONCERNS

The expansion of international tourism has had a large impact on the discipline of transport geography. As of 2010, 877 million international tourists were accounted for, representing more than 10% of the global population (Rodrigue, 2013). Traveling has always been an important feature of society. -Initially-explorers traveled the world to learn more about geographical regions, potential markets and to exploit resources. As time moved on and as transportation became more reliable, traveling became a mundane activity taking place in an organized environment that became known as tourism. In the modern world, traveling ~~is~~ has become more centered on the annual holidays and, because of this, can be fairly well predicted.

Transport is the cause and the effect of the growth of tourism. To start with, the improved facilities have stimulated tourism, and the expansion of tourism has stimulated transport. Accessibility is the main function behind the basics of tourism transport. In order to access the areas that are the most popular destination spots, tourists will use any transportation mode available. However, air travel has become the main mode of travel for long distances especially when it comes to international tourism. Therefore, growth rates of international air traffic can be pegged with growth rates of international tourism (Rodrigue, 2013).

Independent means of travel and transports were popular when travel was meant mostly for the elite. However, mass tourism changed transportation and how all of society travels thereby giving birth to today's mass transportation systems. The two most common forms of mass transportation are air travel and sea travel. With mass transportation by air and sea come new risks that can expose~~s~~ passengers to any number of factors that may have an impact on health.

21.3.1 AIR TRAVEL

Air travel can lead to the rapid dissemination of infectious diseases faster than the incubation period of almost all infections (Abubakar, et al., 2012).

The efficient spread of infection and drug-resistant organisms is expatiated by this rapid mass movement and the mixing of infectious and susceptible populations. One aspect of that is the immense amount and varieties of disease spread by insects through international travel – thus, requiring the practice of aircraft disinsection. Aircraft disinsection is a public health measure that is mandated by the International Health Regulations (Annex 2) (WHO, 2013). Many countries require disinsection (to kill insects) of aircraft arriving from countries where diseases are spread by insects, such as malaria and yellow fever. Passengers are sometimes concerned about their exposure to insecticide sprays during air travel, and some have reported feeling unwell after the spraying of aircraft for disinsection. However, there is currently no evidence that the specified insecticide sprays are harmful to human health when used as recommended (WHO, 2013).

Research has shown that there is very little risk of any communicable disease being transmitted on board an aircraft. However, illnesses associated with air travel are very real and can be visible in the forms of motion sickness, stress, fear of flying, air rage and other psychological aspects associated with air travel. To minimize the risk of passing on infections, travelers who are unwell should delay their journey until they have recovered. Individuals with a known active communicable disease should not travel by air. Airlines may deny boarding to passengers who appear to be infected with a communicable disease.

Another concern of healthy air travel is air pressure on the plane. The aircraft cabin air quality is carefully controlled. Ventilation rates provide a total change of air 20–30 times per hour. Most modern aircraft have recirculation systems, which recycle up to 50% of the cabin air. The recirculated air is usually passed through HEPA (high-efficiency particulate air) filters, of the type used in hospital operating theaters and intensive care units, which trap dust particles, bacteria, fungi and viruses (WHO, 2013). Although aircraft cabins are pressurized, cabin air pressure at cruising altitude is lower than air pressure at sea level. At typical cruising altitudes in the range 11,000–12,200 m (36,000–40,000 feet), air pressure in the cabin is equivalent to the outside air pressure at 1800–2400 m (6000–8000 feet) above sea level. As a consequence, less oxygen is taken up by the blood (hypoxia) and gases within the body expand. The effects of reduced cabin air pressure are usually well tolerated by healthy passengers. Cabin air contains ample oxygen for healthy passengers and crew. However, because cabin air pressure is relatively low, the amount of oxygen carried in the blood is reduced compared with that at sea level. Passengers with certain medical conditions, particularly heart and lung diseases, and blood disorders such as anemia (in particular sickle-cell anemia), may not tolerate this reduced oxygen level (hypoxia) very well (WHO, 2013). The humidity in aircraft cabins is low, usually less than 20% (humidity in the home is normally over 30%). Low humidity may cause skin dryness and discomfort to the eyes, mouth and nose but presents no risk to health. The available evidence has not shown low humidity to cause internal dehydration and there is no need to drink more than usual.

Contraction of muscles is an important factor in helping to keep blood flowing through the veins, particularly in the legs. Prolonged immobility, especially when seated, can lead to pooling of blood in the legs, which in turn may cause swelling, stiffness and discomfort. It is known that immobility is one of the factors that may lead to the development of a blood clot in a deep vein – so-called "deep vein thrombosis" or DVT. The findings of the epidemiological studies indicate that the risk of venous thromboembolism is increased 2- to 3-fold after long-haul flights (more than 4 h) and also with other forms of travel involving prolonged seated immobility. The risk increases with the duration of travel and with multiple flights within a short period. In absolute terms, an average of 1 passenger in 6,000 will suffer from venous thromboembolism after a long-haul flight (WHO, 2013).

Jet lag is the term used for the symptoms caused by the disruption of the body's "internal clock" and the approximate 24-hour (circadian) rhythms it controls. Disruption occurs when crossing multiple time zones, i.e., when flying east to west or west to east. Jet lag may lead to indigestion and disturbance of bowel function, general malaise, daytime sleepiness, difficulty in sleeping at night, and reduced physical and mental performance. Its effects are often combined with tiredness caused by the journey itself. Jet lag symptoms gradually wear off as the body adapts to the new time zone.

21.3.2 SEA TRAVEL

The passenger shipping industry (cruise ships and ferries) has expanded considerably in recent decades. In 2012, 20 million passengers worldwide travelled on cruise ships (Global Travel Industry News, 2012). Cruise itineraries cover all continents, including areas that are not easily accessible by other means of travel. The average duration of a cruise is about 7 days, but cruise voyages can last from several hours to several months. A typical cruise ship now carries up to 3000 passengers and 1000 crew (Cruise Ship Industry Statistics, 2012). With these floating communities come greater health concerns for today's traveler.

The rapid movement of cruise ships from one port to another, with the likelihood of wide variations in sanitation standards and infectious disease exposure risks, often results in the introduction of communicable diseases by embarking passengers and crew members. In the relatively closed and crowded environment of a ship, disease may spread to other passengers and crew members; diseases may also be disseminated to the home communities of disembarking passengers and crew members. More than 100 disease outbreaks associated with ships have been identified in the past 30 years. This is probably an underestimate because many outbreaks are not reported and some may go undetected. Outbreaks of measles, rubella, varicella, meningococcal meningitis, hepatitis A, legionellosis, and respiratory and gastrointestinal illnesses among ship passengers have been reported. Such outbreaks are of concern because of their potentially serious health consequences and high costs to the industry. In recent years, influenza and Norovirus outbreaks have been public health challenges for the cruise industry (WHO, 2013).

Norovirus is the most common pathogen implicated in outbreaks. Symptoms often start with the sudden onset of vomiting and/or diarrhea. There may be fever, abdominal cramps and malaise. The virus can spread in food or water or from person to person; it is highly infectious and in an outbreak on a cruise ship, more than 80% of the passengers can be affected. To prevent or reduce outbreaks of gastroenteritis caused by Norovirus, ships are enhancing food and water sanitation measures and disinfection of surfaces; more ships are providing hand gel dispensers at strategic locations throughout the ship and passengers and crew are urged to use them. Some cruise companies ask that those who present with gastrointestinal symptoms at on-board medical centers be put into isolation until at least 24 h after their last symptoms, and some ships also isolate asymptomatic contacts for 24 h. In addition, respiratory tract infections are frequent health complaints among cruise ship passengers. Travelers from areas of the world where influenza viruses are in seasonal circulation may introduce such viruses to regions of the world where influenza is not in seasonal circulation (WHO, 2013).

Because of temperature and weather variations, changes in diet and physical activities, cruise ship passengers – particularly the elderly – may experience worsening of existing chronic health conditions. Cardiovascular events are the most common cause of mortality on cruise ships. Motion sickness can occur as well as injuries and dental emergencies.

Global standards regarding ship and port sanitation and disease surveillance, as well as response to infectious diseases, are in place and are sanctioned under Article 8 of the International Labor Organization Convention. Article 8 ensures that vessels carrying more than 100 crew members on an international voyage of three days or longer must provide a physician for the care of the crew. Article 8 also sets the standards for safe water, food, rodent control and waste disposal. These regulations do not apply to passenger vessels and ferries sailing for less than three days, even though the number of crew and passengers may exceed 1,000. Ferries often do not have an emergency room but a ship's officer or a nurse is designated to provide medical help.

21.4 HISTORY OF FOOD AS RELATED TO TOURISM

Throughout history there has always been an association between food and tourism. The earliest official food service was related to inns and monasteries with the first documented restaurant opening in Paris in 1765, serving traditional breads and soups. English taverns began serving food and drink along with offering various forms of lodging. Food service and lodging were eventually brought forth in the New England Colonial States and quickly made their way down the East Coast. From there this new service progressed to the West via stagecoaches, train service and eventually with the Model T automobile.

This leads to the question of "Why is food tourism so important?" Food tourism strengthens a region's or country's identity, assists in sustaining the cultural heritage, contributes to contesting fears of global food homogenization, and facilitates the regeneration of an area's sociocultural fabric. Food tourism also has a significant role in securing the "triple bottom line" of economic, social and environmental sustainability by increasing tourist expenditures, extending the season and encouraging sustainable development (Everett and Aitchison, 2008).

Food and tourism have long been a source of marketing images and experiences for the tourist. Food is considered a reflection of the culture of a country or region and its people. Therefore, it is the ideal product to offer as an attraction at a destination (Rand and Heath, 2006). But, with this offering can come gastrointestinal maladies or foodborne illness or disease where food safety is a concern.

21.5 GEOGRAPHIC BOUNDARIES AND FOOD SAFETY

Geographic boundaries frame a country. While that has been true in the past, nothing avoids or disrespects boundaries more than food safety. With globalization, boundaries were broken and goods and services began to flow more freely between countries, regions and even continents, thereby making food safety a global issue.

Food products now come to the United States from over 250,000 foreign establishments in 200 countries accounting for 15% of fruits, 20% of vegetables and 80% of seafood imported into the U.S. Accompanying and linked to these imports are increased recalls and foodborne illness outbreaks (Marler, 2012). Imported food into the U.S. is not the only problem: U.S. exports have raised concerns abroad as well. The United States Department of Agriculture Economic Research Service (USDA ERS) states that in 2011 the U.S. exported over $136 billion in agricultural products (up from $53 billion in 2001) (Marler, 2012). These exports also pose a global food safety risk.

The world today faces an uncertain and rapidly shifting picture of global food security. While the food price crisis of 2007–2008 shed some light on the global security picture, the picture itself is far from clear. Climate change, population growth, changing consumer tastes and biofuels demands will continue to have an impact on global food security; but, it remains to be seen exactly what will be that impact (Yang, 2012). Food safety issues originate in various parts of the world and are not any one country,' regions' or continents' problem. While it is imperative to maintain vigilance and confidence in food safety to protect the world's population it is equally imperative that travelers maintain the same vigilance whether traveling locally or abroad as these same safety issues can impact them as well.

21.6 GLOBALIZATION AND FOOD

Globalization is a defining economic and social trend of the past several decades that directly affects health and creates economic and health disparities between its beneficiaries and losers. Recently, there has been pressure put on the global community to address the issues of inequities in health and other determinants of human capability across countries. New stakeholders' initiatives, such as private foundations and multistakeholders, have contributed toward increased funding for global health, concentrating on identifying and addressing threats to the health of vulnerable populations worldwide (Mak, Lumbers, and Eves, 2012).

Globalization has influenced many aspects of human activity, including food production and consumption. While not being a new phenomenon, the world is seeing acceleration in the speed and scope facilitated by advances in transportation, information and communication technology (Mak, Lumbers, and Eves, 2012). There are multiple definitions of Globalization derived from various perspectives; however, the most accurate definition emphasizes that globalization brings about an intensified workplace interdependence and integration as well as an increased global consciousness (Mak, et al., 2012).

In the general context, food consumption is recognized as a collection of contextual and evolving social practices, whereby food no longer merely serves as sustenance but also initiates a way to relate to other people in social, cultural and political terms (Oosterveer, 2006). Food consumption is acknowledged to bear "symbolic" significance (social distinction) and as a means for encountering and experiencing other foodways and cultures (Chang, Kivela, and Mak, 2010) while providing energy and essential nutrients needed for body functions; thus, eating is often regarded as an "obligatory" tourist activity (Richards, 2002). Eating is an activity that embraces all five senses: sight (ophthalmoception), hearing (audioception), taste (gustaoception), smell (olfacoception or olfacception) and touch (tactioception). Therefore, embracing the five senses while eating contributes to a tourist's ultimate travel experience.

To understand tourist food behaviors, it is important to be aware of Schuetz's 1971study which generated two general categories of worldly food: novelty and familiarity. While the majority of tourist travel is motivated by novelty or unfamiliarity, there is a significant sector of tourists who will only travel to locales providing a certain degree of familiarity or awareness. This form of travel is often affiliated with tourism and food consumption.

Travel has long been associated with experiencing "otherness" (otherness denotes the sense of the strange and unfamiliar created by specific subject positions, which provide clear boundaries that divide individuals, cultures and races). For many tourists, this consists of consuming local delicacies and participating in local foodways that are essential for the tourist experience (Mak, et al., 2012). The risk posed by food consumption experiences is higher than other forms of a tourist experience due to the level of bodily involvement (Cohen and Avieli, 2004). For instance, viewing the cooking process of deep fried bugs by street vendors in Bangkok or watching a colleague consume chicken feet in China despite knowing where the chicken's feet have been can elicit feelings of repulsion among some tourists; even though the process of viewing involves less risk than consuming the fried insects or chicken feet (Elsrud, 2001.

21.7 FOOD SAFETY AND SECURITY

Foodborne illnesses are evident in all parts of the world. The toll on human life and suffering is enormous. Contaminated food contributes to 1.5 billion cases of diarrhea in children (Center for Science in the Public Interest, 2005) and 48 billion occurrences of foodborne illness or poisoning (Center for Disease Control and Prevention, 2013). These deaths and illnesses are shared by both developed and developing nations.

The symptoms of foodborne illness range from mild to life-threatening. Nausea, diarrhea, kidney and liver failure, brain and neural disorders, and death are all results of food borne illnesses. All humans are prone to foodborne illness: children, pregnant woman, the elderly and those whose immune systems have been compromised are especially susceptible.

Food safety challenges differ by region, due to differences in income level, diets, local conditions and government infrastructures. In developing countries, the food producer and the consumer often have a close connection. In these areas, there are fewer processed and packaged foods, most fresh food is traded in traditional markets and street vendors supply much of the food consumed outside the home. Perishable food is often prepared and consumed immediately and there is minimal storage of prepared foods (Center for Science in the Public Interest, 2005). Food safety concerns in these countries typically include:

- The inappropriate use of agricultural chemicals
- The use of untreated or partially treated wastewater
- The use of sewage or animal manure on crops
- The absence of food inspection, including meat inspection
- A lack of infrastructure, such as adequate refrigeration
- Poor hygiene, including a lack of clean water supplies

Food safety is an umbrella term that envelopes many facets of handling, preparation and storage of food to prevent illness and injury. Chemical, microphysical and microbiological aspects of food safety all fall under this umbrella. **Chemical** properties of food include vitamin and mineral content that affect the overall quality of the food but are not as significant in terms of food safety. **Microphysical** particles such as glass, bone, stones and metal can be hazardous and cause serious injuries. **Pathogenic** bacteria, viruses and toxins produced by microorganisms are all possible contaminants of food and impact food safety (Hanning, O'Bryan, Crandall, and Ricke, 2012).

The Center for Disease Control and Prevention (2013) estimates that each year 1 in 6 Americans (48 million people) get sick due to food borne illnesses. Of those 48 million, 128,000 are hospitalized and 3,000 will die. The 2011 estimates provide the most accurate picture yet of which foodborne bacteria, viruses, microbes ("pathogens") are causing the most illnesses in the United States, as well as estimating the number of foodborne illnesses without a known cause. The estimates show that there is still much work to be done—specifically in focusing efforts on the top known pathogens (31 of them) and identifying the causes of foodborne illness and death without a known cause.

The most common pathogens causing food borne illnesses in the United States are: Norovirus, Salmonella, Clostridium perfringens, Campylobacter, and Staphylococcus aureus (CDC, 2013). Many food animals can carry bacteria that are pathogenic to humans. For this reason, foods of primarily animal origin were thought to carry the majority of foodborne illnesses. However, there is an increasing rise in foodborne illness outbreaks due to contaminated fruits and vegetables (Hanning, et al., 2012). To help prevent foodborne illnesses there are a few tips: boil it, peel it or don't eat it; eat only steaming-hot foods; drink bottled water; avoid dairy products and take acidophilus tablets.

Foodborne illnesses and contaminated food make an appearance in many settings. The most reported cases of foodborne illnesses (48%) were a result of food consumed in a restaurant, deli, catering or from street vendors as compared to 21% that was consumed in a private home (CDC, 2012). Food can also be contaminated or compromised in production, processing, and distribution.

All foods have the potential to be hazardous to one's health; however, there are some foods that are more prone to contamination and increasing the risk of foodborne illnesses. Microorganisms generally grow rapidly in moist, high protein foods that have not been acidified or otherwise further processed to prevent such growth: (1) Animal foods that are raw or heat treated such as: milk or milk products including cheese, sour cream and whipped butter; meats including raw or partially cooked bacon; shell eggs; poultry and poultry products; shellfish and fish. (2) Food derived from plants that are heat-treated including: onions (cooked and rehydrated), cooked rice, soy protein products (example: tofu), potatoes (baked or broiled). (3) Food derived from plants that consist of: cut melons or raw seed sprouts. (4) Garlic-in-oil and other vegetable-in-oil mixtures that are not treated to prevent the growth and toxin production of *C. botulinum*. And, (5) certain sauces, breads, and pastries containing potentially hazardous food (examples: meat, cheese, cooked vegetables or cream).

Alcohol provides another risk for contamination, while developed countries have health permits and laws that ensure the processes are in place to guarantee the preservation of alcohol, many undeveloped countries do not. Contaminated alcohol can lead to outbreaks of alcohol poisoning. Most deaths associated with alcohol are a result of drinking and driving. Alcohol related to injuries and deaths include: diseases of the circulatory and respiratory system, fire and flames, accidental drowning, suicides, falls and homicides. In addition, excessive drinking of alcohol can lead to alcohol poisoning resulting in seizures, low blood pressure, hypothermia, irregular heart rhythms and ultimately death.

21.8 GLOBAL HEALTH THREATS

COMMUNICABLE AND INFECTIOUS DISEASES (HUMAN AND VETERINARY)

Travel medicine is based on the concept of the reduction of risk. Risk refers to the possibility of harm during the course of a planned trip. While some risks might be avoidable, others may not. Vaccine-preventable diseases may be mostly avoidable, depending on the risk of the disease and the protective efficacy of the vaccine. Non-disease risks, such a motor vehicle accidents or drowning, account for a much higher percentage of deaths among travelers than infectious diseases (Shlim, 2012).

BEFORE YOU GO! HEALTH CONCERNS

Research Health Concerns

- ○ Check Center for Disease Control and Prevention and the World Health Organization
- ○ Get Immunizations
- ○ Some immunizations need to be administered months in advance
- ○ Some medications need to be taken before, during, and after your trip
- ○ Get an International Health Certificate from your doctor documenting your immunizations
- ○ Review First –Aid Skills
- ○ Take a class
- ○ See Your Doctor
- ○ Get checkup, visit dentist and optometrists
- ○ Easier to address minor problems at home

✓ Choose Travel Insurance

✓ Depends on one's personal insurance policies

- ✓ Pack First-Aid Kit
- ✓ Include basic supplies
    - Prescription medicine (in original containers)
    - Extra glasses and contacts (if applicable)
    - Water treatment
    - Antibiotics (topic and oral)

Although the movement of pathogens through travel is not a new phenomenon, today's increasing pace and scale of global movement has enhanced the opportunities for disease spread. Some symptoms may be delayed for years; others have very short incubation periods – allowing for rapid spread illustrating the role travel plays in the translocation of infectious disease (Ostroff, 2012).

By far the most common form of illness when travelling is diarrhea or turista. This is usually caused by the difference in water purity in the form of water consumption and food prepared with water. However, there are many common communicable and infectious diseases that can be contracted while traveling. Below are the most common:

Dengue Fever: With more than one-third of the world's population living in areas at risk for transmission, dengue infection is a leading cause of illness and death in the tropics and subtropics infecting as many as 100 million people yearly. Dengue has emerged as a worldwide problem only since the 1950s. Although dengue rarely occurs in the continental United States, it is endemic in Puerto Rico, and in many popular tourist destinations in Latin America and South-east Asia; periodic outbreaks occur in Samoa and Guam. Dengue is caused by any one of four related viruses transmitted by mosquitoes in these tropical tourist locales.

Influenza: Two new forms of influenza have recently emerged. Avian influenza A (H5N1), while primarily affecting poultry, it does have an alarming fatality ratio in humans. Influenza A (H1N1) is often aided by infectious travelers-during its incubation stage. The H1N1 pandemic demonstrates the potential for global dissemination of pathogens in a highly interconnected world.

Mad Cow Disease: BSE (bovine spongiform encephalopathy) is a progressive neurological disorder of cattle that results from infection by an unusual transmissible agent called a prion. The nature of the transmissible agent is not known. For reasons that are not yet understood, the normal prion protein changes into a pathogenic (harmful) form that then damages the central nervous system of cattle.

Malaria: Malaria is a mosquito-borne disease caused by a parasite. People with malaria often experience fever, chills, and flu-like illness. Left untreated, they may develop severe complications and die. In 2010, an estimated 219 million cases of malaria occurred worldwide and 660,000 people died, most (91%) in the African Region.

Meningococcal Meningitis: *N. meningitidis* is found worldwide. At any time, 5%–10% of the population may be carriers of *N. meningitidis*. Young children have the highest risk for meningococcal disease, but 60% of cases occur in adolescents and adults. Risk is highest in travelers who have prolonged contact with local populations in the meningitis belt during an epidemic.

SARS: Severe Acute Respiratory Syndrome was an epidemic if 2003 and is another example of the role travel played in the spread of infectious diseases in the Twenty-first Century resulting in 8,098 cases and 774 deaths in 29 countries. SARS had a major influence on the revisions to the International Health Regulations in 2005.

Schistosomiasis: Also known as bilharzia, is a disease caused by parasitic worms. Although the worms that cause schistosomiasis are not found in the United States, more than 200 million people are infected worldwide. In terms of impact this disease, it is second only to malaria as the most devastating parasitic disease. One can become infected when the skin comes in contact with contaminated freshwater.

Sexually Transmitted Diseases: Sexually transmitted diseases (STDs) are the infections and resulting clinical syndromes caused by more than 25 infectious organisms. Sexual activity is the predominant mode of transmission, through genital, anal or oral mucosal contact. Most commonly, sexually transmitted diseases are HIV/AIDS, syphilis, herpes, genital warts, chlamydia and gonorrhea.

Tuberculosis: TB is a disease caused by a bacterium called *Mycobacterium tuberculosis*. The bacteria usually attack the lungs, but TB bacteria can attack any part of the body such as the kidney, spine and brain. Travelers should avoid close contact or prolonged time with known TB patients in crowded, enclosed environments (for example, clinics, hospitals, prisons, or homeless shelters).

Yellow Fever: Yellow fever virus is found in tropical and subtropical areas in South America and Africa. The virus is transmitted to humans by the bite of an infected mosquito. Yellow fever disease is diagnosed based on symptoms, physical findings, laboratory testing and travel history, including the possibility of exposure to infected mosquitoes.

CHILDHOOD DISEASES

Vaccines have reduced, and, in some cases, eliminated many diseases that killed or severely disabled people in past generations. If travelers keep vaccinating now, we can trust that diseases won't continue to cripple, infect or kill in the future. However, many of these diseases still exist and are thriving in low income and vaccination-challenged countries where health care is limited. The risk of contamination and the spread of these diseases are evident in these locales. Some of these diseases to be aware of when traveling are:

Diphtheria: Direct person-to-person transmission by contact with respiratory secretions and cutaneous lesions. Circulation of this disease appears to continue in some settings even in populations with >80% childhood immunization rates. Respiratory diphtheria presents as a sore throat with low-grade fever and an adherent pseudo membrane of the tonsils, pharynx or nose. Neck swelling is usually present in severe cases of the disease.

Measles: Measles is a highly contagious respiratory disease caused by a virus. The disease of measles and the virus that causes it share the same name. The disease is also called rubeola. For every 1,000 children who get measles, one or two will die.

Mumps: Mumps is a contagious disease that is caused by the mumps virus. Mumps typically starts with a few days of fever, headache, muscle aches, tiredness loss of appetite, and is followed by swelling of salivary glands. Anyone who is not immune either through a previous mumps infection or from vaccination can get mumps.

Pertussis: Also known as whooping cough is a highly contagious respiratory disease. Pertussis is known for uncontrollable, violent coughing which often makes it hard to breathe. It most commonly affects infants and young children and can be fatal, especially in babies less than 1 year of age.

Polio: Polio is a crippling and potentially deadly infectious disease caused by a virus that spreads from person to person invading the brain and spinal cord and causing paralysis. Because polio has no cure, vaccination is the best protection and the only way to stop the disease from spreading. The spread of polio has never stopped in Afghanistan, Nigeria and Pakistan. Poliovirus has been reintroduced and continues to spread in Chad and Democratic Republic of the Congo after the spread of the virus was previously stopped.

Tetanus: Tetanus is an infection caused by bacteria called *Clostridium tetani*. When the bacteria invade the body, they produce a poison (toxin) that causes painful muscle contractions. Another name for tetanus is "lockjaw" because it often causes a person's neck and jaw muscles to lock, making it hard to open the mouth or swallow.

TOP 10 TRAVEL HEALTH PROBLEMS

1. Diarrhea:
   a. Most common problem that can affect a trip.
   b. Avoid tap water, ice and vegetables washed in tap water.
   c. Stay hydrated
2. Blisters:
   a. Common complaint
   b. Carry a common antibiotic when you travel
   c. Make sure shoes fit well
   d. Caused by friction and heat
   e. Use a skin lubricants and moleskin
3. Respiratory Infections:
   a. Common cold
   b. Use nasal decongestants and acetaminophen
   c. Stay hydrated
4. Bladder Infections:
   a. Most common in woman travelers
   b. Infections can spread to kidneys
   c. Common antibiotic medications used to Urinary Tract Infections (UTI) may be needed

5. Tooth Injuries:
   a. Clove oil can be placed on a cotton ball to relieve localized mouth pain
   b. Temporary filling material can be used to fill holes
   c. Candle Wax can work in a tough spot
   d. Warm salt water rinses can keep areas clean
6. Superficial Skin Infections:
   a. Cuts and scrapes
   b. Local infections are marked by redness or swelling
   c. Make sure tetanus vaccine is current
   d. Use topical antibiotic medications
7. Mosquito Bites:
   a. Avoid being bitten
   b. Wear long sleeves and long pants
   c. Use permethrin spray on your clothing, tents, and curtains
   d. Use a personal protection containing a minimum of 35% DEET
8. Muscle Aches and Pains:
   a. Uncomfortable sleeping positions on planes, hotels, trains etc.
   b. Use Ibuprofen for muscle soreness
   c. Acetaminophen is another great option, although it does not reduce inflammation
9. Jet Lag:
   a. No immediate cure for jet lag
   b. Bright light stimulates your body to be awake, so if you need to be awake, open the curtains or get outside
   c. Caffeine can help give a boost, allowing you to stay up a bit more or feel less fatigued
10. Sexually Transmitted Diseases:
    a. Nobody wants to come home with an unwanted "souvenir" from his or her trip, like a sexually transmitted infection
    b. All contact with a new partner should be made using a latex barrier, like a condom
    c. Common infections include syphilis, gonorrhea and herpes. Not to mention HIV

Source: *McLaughlin, Erik (2009).

CROWDING AND LACK OF SANITATION

Mass Gatherings

Mass gatherings have been found to be associated with the occurrence of clusters of infectious diseases, particularly respiratory infections and gastrointestinal illness (Severi et al., 2012). The lack of infrastructure at many mass gatherings contributes to public health emergencies or disasters including: public health, health care or emergency services. The concentration of people temporally and spatially and the unique socioeconomic characteristics of participants at mass gatherings (i.e., specific events that might attract participants in particular risk groups, either increasing their chance of being a source of or becoming susceptible to infection) compound routine disease factors, such as susceptibility and effectiveness of transmission, leading to the emergence of infectious diseases and creating challenges of prevention and control of these diseases (Abubakar et al., 2012).

Communicable diseases are typically seen in the form of fecal-oral transmissions (despite advances in food and water hygiene, modern methods for food preparation and distribution) have the potential to cause outbreaks of gastrointestinal diseases and food-borne illnesses. Also present are blood borne diseases such as sexually transmitted diseases and infections from attraction activities including those using potentially unsanitary tools and practices (i.e., tattooing or injecting drugs). Other areas to take into consideration are: air pollution, gathering related injuries, weather related injuries (heat wave) and alcohol-associated morbidity.

Respiratory transmission requires living in close proximity; and, therefore, can be rampant at mass gatherings with overcrowded accommodations. The duration of contact and the amount of shared air are key determinants for the spread of an infection. Examples of infections transmitted through the respiratory route include: influenza, tuberculosis, measles, mumps and meningococcal meningitis (Abubaker et al., 2012).

Vector-borne diseases (transferred from one host to another host:, that is, a mosquito) can cause outbreaks even in countries where they are not an endemic if a traveler is infected and an appropriate vector is present. Such vector-borne diseases are malaria, dengue, Nile encephalitis and yellow fever. Lastly, infectious diseases caused by known human pathogens or occasionally emerging infections can be transmitted from animals to people. Zoonotic infections can be spread through direct contact or contaminated food and water including Escherichia coli 0157, which is transmitted through cattle feces. Escherichia coli 0157 was responsible for the contaminated mud at the Glastonbury Festival in the United Kingdom in 1997 (Abubaker, et al., 2012).

All travelers to large events should visit a health care provider 4–6 weeks before traveling in order to assess any risk the traveler might encounter (Abubaker, et al., 2012) and take actions to manage the risk. Management of risk might include vaccines, drugs and travel advice. In addition, knowledge of the country or region being visited is essential to reduce the potential to increase the risk of introduction of new and emerging infections.

21.9 TRAVEL MEDICINES

Ideally, travelers would seek medical advice before traveling internationally to ensure they are current on all vaccines required to enter their destination. However, Hamer and Conner (2004) discovered that the overall level of knowledge of risk and the practice concerning preventive travel health measures, especially the use of itinerary-specific immunizations, was low: only 36% of travelers sought travel health advice, despite the fact that more than half prepared for their trip at least a month in advance. Awareness of the risk of vaccine-preventable diseases was also woefully inadequate. In addition, actual knowledge of previous or recent vaccination was low. It is imperative for travelers to remember that while they may have been previously vaccinated, those vaccines may not be up-to-date.

Vaccines are recommended to protect travelers from illnesses present in other parts of the world and to prevent the importation of infectious diseases across international borders. The required vaccination depends on a number of factors including: destination, the time spent in rural areas, the season of the year, age, health status and previous immunizations.

The only vaccine required by International Health Regulations is yellow fever vaccination for travel to certain countries in sub-Saharan Africa and tropical South America. Meningococcal vaccination is required by the government of Saudi Arabia for annual travel during the Hajj. All other vaccinations needed for travel are considered "recommended" by the Center for Disease Control and Prevention. Travel adds additional variables: additional medications; environmental changes such as altitude, heat and cold exposure and, sometimes travel-related diseases such as gastroenteritis, for example. Travelers should review their medications with a medical expert in conjunction with their travel plans prior to departure.

29.9.1 THE ENVIRONMENT

Companies spend billions of dollars creating products that are environmentally friendly. The environment and protecting it is now one of the most discussed topics globally. Through the years the biggest debated tourism/environmental issues have been mass tourism and deforestation. "Going green" now has so many definitions and meanings it is hard to pinpoint what being "eco" really means.

One of the areas affecting the global environment is loss of biological diversity. Biological diversity is the term given to the variety of life on earth and the natural patterns it forms. The effects of loss of biodiversity include:

- Threat to food supplies, opportunities for recreation and tourism and sources of wood, medicines and energy.
- Interference with essential ecological functions such as species balance, soil formation and greenhouse gas absorption.
- Reduction of the productivity of ecosystems – thereby shrinking nature's basket of goods and services, from which we constantly draw.
- Destabilization of ecosystems and weakening of their ability to deal with natural disasters such as floods, droughts and hurricanes, as well as with human-caused stresses, such as pollution and climate change.

Tourism, especially nature tourism, is closely linked to biodiversity and the attractions created by a rich and varied environment. It can also cause loss of biodiversity when land and resources are strained by excessive use, and when impacts on vegetation, wildlife, mountain, marine and coastal environments and water resources exceed the carrying capacity. This loss of biodiversity in fact means loss of tourism potential. Tourists and suppliers–often unwittingly– can bring in species (insects, wild and cultivated plants and diseases) that are not native to the local environment, which can cause enormous disruption and even destruction of ecosystems (United Nations Environment Program, 2001).

The ozone layer, which is situated in the upper atmosphere (or stratosphere) at an altitude of 12–50 kilometers (Tourism Concern, 2013), protects life on earth by absorbing the harmful wavelengths of the sun's ultraviolet (UV) radiation, which in high doses is dangerous to humans and animals. This increased exposure to UV radiation is one of the reasons scientists propose for the global decrease of amphibian populations.

Ozone depleting substances (ODSs), such as CFCs (chlorofluorocarbon) and halons, have contributed to the destruction of this layer. The tourism industry may be part of the problem; direct impacts start with the construction of new developments and continue during daily management and operations. Refrigerators, air conditioners and propellants in aerosol spray cans, among others, contain ODSs and are widely used in the hotel and tourism industry. Emissions from jet aircraft are also a significant source of ODSs. According to Tourism Concern (2013), scientists predict that by 2015 half of the annual destruction of the ozone layer will be caused by air travel.

Climate scientists now generally agree that the Earth's surface temperatures have risen steadily in recent years because of an increase in the so-called greenhouse gases in the atmosphere, which trap heat from the sun. One of the most significant of these gases is carbon dioxide (CO_2), which is generated when fossil fuels, such as coal, oil and natural gas are burned (e.g., in industry, electricity generation and automobiles) and when there are changes in land use, such as deforestation. In the long run, the accumulation of CO_2 and other greenhouse gases in the atmosphere can cause global climate change–a process that may already be occurring. Tourism not only contributes to climate change, but is affected by it as well. Climate change is likely to increase the severity and frequency of storms and severe weather events, which can have disastrous effects on tourism in the affected regions. Some of the other impacts that the world risks as a result of global warming are drought, diseases and heat waves.

Global tourism is closely linked to climate change. Tourism involves the movement of people from their homes to other destinations and accounts for about 50% of traffic movements; rapidly expanding air traffic contributes to about 2.5% of the production of CO_2. Tourism is thus a significant contributor to the increasing concentrations of greenhouse gases in the atmosphere (Mountain Forum, 2013).

Air travel itself is a major contributor to the greenhouse effect. Passenger jets are the fastest growing source of greenhouse gas emissions. The number of international travelers is expected to increase from 594 million in 1996 to 1.6 billion by 2020, adding greatly to the problem unless steps are taken to reduce emissions (Tourism Concern, 2013).

Catastrophes like floods, earthquakes, wildfires, volcanoes, typhoons, avalanches, tornados, drought and diseases can have a serious effect on inbound and domestic tourism and therefore also affect local tourism industries. The outbreak of epidemics has severely affected the inbound tourism market over the past few years. However, catastrophes have spawned a new segment of tourism: disaster tourism. At first glance, natural disasters and tourism don't seem to mix. But recent world disasters, like Hurricane Katrina, suggest otherwise. In New Orleans, Louisiana, several tour companies ran "post-Katrina" tours, taking tourists to see the destruction of the lower ninth ward. Some of these tours can be viewed as a double edged sword: while travelers must be sensitive to the communities where buses and tours are coming through, it also can be perceived as insensitive. But, on the other hand, it does provide education to people and encourages recovery donations and volunteer work replacing fascination with awareness.

The environment and tourism have a long-standing relationship; however, with the deterioration of the environment, we may see many changes affecting tourism such as:

- Less snowfall at ski resorts–meaning shorter skiing seasons.
- In already hot areas like Asia and the Mediterranean, tourists will stay away because of the heat and for fear of diseases and/or water shortages.
- Harm to vulnerable ecosystems such as rainforests and coral reefs because of rising temperatures and less rainfall. A major risk to coral reefs is bleaching. This occurs when coral is stressed by temperature increases, high or low levels of salinity, lower water quality and an increase in suspended sediments. These conditions cause the zooxanthallae (the single-celled algae which forms the colors within the coral) to leave the coral. Without the algae, the coral appears white, or "bleached" and rapidly dies. The Great Barrier Reef, which supports

a $640 (US) million tourism industry, has been experiencing coral bleaching events for the last 20 years.

- Rising sea levels, the result of melting glaciers and polar ice, will threaten coastal and marine areas with widespread floods in low-lying countries and island states, increasing the loss of coastal land. Beaches and islands that are major tourism attractions may be the first areas to be affected.
- Increased events of extreme weather, such as tornadoes, hurricanes and typhoons. These are already becoming more prevalent in tourist areas in the Caribbean and South East Asia. Wind damage, storm waves, heavy rains and flooding caused major losses in the local tourism sector. (Environmental Impacts of Tourism, 2013).

21.9.2 EMERGING HEALTH THREATS

According to Rodin, Nunn, Miribel, and Brillant (2013) the world has made significant strides in tackling major public health challenges over the last several decades. However, the medical community has only been successful in the eradication of one disease, smallpox, but is close to doing so with polio and guinea worm. The medical community continues to make great strides and progress on other debilitating illnesses, including malaria, tuberculosis and HIV/AIDS. Yet, even as technology and practices improve, new threats arise.

Rodin, et al. (2013) elaborate that in the last two decades, the global community has seen some 30 new zoonotic diseases emerge, from SARS to hantavirus to Ebola and more. Population pressures and economic growth push humans into ever closer contact with animals while disturbing ecosystems and creating ripe conditions for new pathogens to jump from animals to humans. Added to this is the incredible growth in global travel and trade. Mass gatherings and the risk of new diseases is quickly spreading worldwide—this risk has never been greater. Globally, our ability to respond continually improves, but the challenges we face increase as well.

Early detection and rapid identification of novel infections help to make up a successful fight against emerging infectious diseases. If new pathogens can be detected early enough resulting in isolation in the emerging region, the medical community can contain the pathogen and reduce the transition to new regions, countries and continents. While this procedure may be costly and difficult, it contributes to eliminating a potential epidemic with the probable outcome of jeopardizing global health. The continuing critical gap in global public health will soon be filled with disaster management, biosecurity, terrorism and violence, psychological health and morbidity.

21.9.3 RECOMMENDATIONS

Running a quick internet search grants travelers an overwhelming amount of travel-related information and services. Internet users search for destination information such as points of interest, historical data, weather conditions, and for products such as flight and hotel packages. These recommendations can strongly influence a traveler's destination choice and activities. With over 940 million international journeys (WHO, 2013) occurring each year, the exposure to health related risks is astronomical. If precautions are taken before, during and after travel, the majority of these risks can be minimized. Travelers must be educated on vaccinations, treatments, personal protection again insects and other disease vectors, and overall safety in different environmental settings. Travelers need resources to identify principal risks to their health relevant to infectious diseases (including causative agents and modes of transmission), clinical features and geographical distribution as well as preventive measures.

The responsibility to seek the appropriate information from the medical profession and the travel industry in order to understand the risks involved with travel and ensure they are taking the proper precautions to protect their health while traveling lies with the traveler. Some risks are minuscule but when there are changes in altitude, humidity, temperature, and exposure to various infectious diseases, illness can result. Other risks are more compelling and may arise in areas where accommodations are of poor quality, hygiene and sanitation are inadequate, medical services are lacking and clean water is unavailable. Unfortunately, mortality risks do exist; however, it should be noted that the most common cause of morbidity and mortality in travel is related to accidents.

All individuals planning to travel should seek advice on the potential hazards-at the chosen destinations by researching all travel warnings and notices. Information related to risks can be found on various websites on the internet; however, the most trusted and up-to-date information can be found at:

- *United States Department of State* – International travel is discussed in detail and recommendations by country are made related to travel notices and warnings. Information concerning traveling aboard, living aboard and passports and visas can be found. www.travel.state.gov
- *World Health Organization* – Global world health and diseases are showcased on this website. Disease information can be searched by country and is accompanied by disease specific maps. Vaccinations are detailed by destination and general precautions are explored. Mode of travel consideration and general health risks are scrutinized. www.who.int/en/
- *Centers for Disease Control and Prevention* – The CDC offers different types of notices for international travelers. Travel notices are posted and determined by the watch level: watch, alert and warning. One can also access the infamous "Yellow Book" published by the CDC. The Yellow Book is a reference book published every two years concerning health risks that may be encountered by international travelers. www.cdc.gov

All individuals planning travel should seek advice on the potential hazards at their chosen destinations and understand how best to protect their health and minimize the risk of acquiring disease. Another method of finding relevant information is to check the destinations' or countries' website for health risks information. Forward planning, appropriate preventive measures and careful precautions can protect traveler's health and minimize the risks of accident and of acquiring disease.

21.10 SUMMARY

Every year more and more people are traveling internationally—for vacation, business, and volunteerism, and to visit friends and family. It is imperative that all travelers be prepared, protected and proactive when it comes to health risks during travel.

Travel alerts are issued to disseminate information about short-term conditions that pose significant risks to the security of all travelers. In today's society, travel alerts are not only focused on health epidemics but also natural disasters, terrorist attacks, coups, election-related demonstrations or violence and high-profile events such as international conferences or regional sports events. Travelers should familiarize themselves with their destinations, both to get the most enjoyment out of the visit and to avoid known dangers.

KEYWORDS

- **aircraft disinsection**
- **globalization**
- **Immunizations**
- **The Yellow Book**

REFERENCES

Abubakar, I., Gautret, P., Brunette, G., Blumberg, L., Johnson, D., Poumerol, G., Memish, Z., Barbeschi, M. and Khan, A. (2012). Mass Gatherings Health 2: Global perspectives for prevention of infectious diseases associated with mass gatherings. *Lancet Infect,* 12, p. 66–74.

Centers for Disease Control and Prevention. (2012). Retrieved from: http://www.cdc.gov/foodborneburden/2011-foodborne-estimates.html

Center for Disease Control and Prevention. (2013). Retrieved from: http://www.cdc.gov/foodborneburden/2011-foodborne-estimates.html

Center for Science in the Public Interest. (2005). Global and Local Food Safety Around the World.

Chang, R., Kivela, J. and Mak, A. (2010). Food preferences of Chinese tourists. *Annals of Tourism Research, 37*(4), p. 989–1011.

Cohen, E. and Avieli, N. (2004). Food in tourism: attraction and impediment. *Annals of Tourism Research, 31*(4), p. 755–778.

Cruise Ship Industry Statistics (2012). Retrieved from: http://www.statisticbrain.com/cruise-ship-industry-statistics/

Elsrud, T. (2001). Risk creation in traveling: backpacker adventure narration. *Annals of Tourism Research, 28*(3), p. 597–617.

Environmental Impacts of Tourism. (2013). Environmental Industry Impacts at the global level. Retrieved from: http://www.gdrc.org/uem/ecotour/envi/two.html

Everett, S. and Aitchison, C. (2008). The role of food tourism in sustaining regional identity: a case study of Cornwall, South West England. *Journal of Sustainable Tourism, 16*(2), p.150–167.

Global Health Education Consortium (2011). Retrieved from: http://globalhealtheducation.org/SitePages/Home.aspx

Global Travel Industry News, (2012). 20 million people took a cruise in 2012. Retrieved from: http://www.eturbonews.com/31296/20-million-people-took-cruise-2012

Hamar, D. and Conner, B. (2004). Travel health knowledge, attitudes and practices among United States travelers. *Journal of Travel Medicine, 11*, p.23–26.

Hanning, I., O'Bryan, C., Crandall, P., and Ricke, S. (2012). Food safety and food security. *Nature Education Knowledge, 3*(10): 9.

Kruk, M. (2010). Globalization and global health governance: implications for public health.

Mak, A.H.N., Lumbers, M. and Eves, A. (2012). Globalization and Food Consumption in Tourism, *Annals of Tourism Research, 39*(1), pp.171–196.

Marler, B. (2012, May 27). Publisher's Platform: Food Safety—It's a Global Issue. *Food Safety News.* Retrieved: January, 2013 from: http://www.foodsfetynews.com/2012/050publishers-platform-food-safety-its-a-global-issue/

McLaughlin, E. (2009). Top 10 travel health problems and how to handle them. *Matador Network.* Retrieved from: http://matadornetwork.com/notebook/top-ten-travel-health-problems-how-to-handle-them/

Mountain Forum. (2013). Retrieved from: http://www.mtnforum.org/

Oosterveer, P. (2006). Globalization and sustainable consumption of shrimp: consumers and governance in the global space of flows. *International Journal of Consumer Studies, 30*(5), p. 465–476.

Ostroff, S. (2012). Perspectives: the role of the traveler in translocation of disease. *Centers for Disease Control and Prevention.* Yellowbook. Retrieved from: http://wwwnc.cdc.gov/travel/yellowbook/2012/chapter-1-introduction/perspectives-the-role

Rand, G.E. and Heath, E. (2006). Towards a framework for food tourism as an element of destination marketing. *Current Issues in Tourism, 9*(3), p.206–234.

Richards, G. (2002). Gastronomy: an essential ingredient in tourism production and consumption? In A.M. Hjalager and G. Richards (Eds.), *Tourism and Gastronomy*, p.3–20. London: Routledge.

Rodin, J., Nunn, S., Miribel, B., and Brillant, L. (2013). Letter. *Emerging Health Threats Journal, Supplement 1,* p. 5.

Rodrigue, J.P. (2013). International tourism and transport. *The Geography of Transport Systems,* 3rd ed. Routledge, New York, NY.

Schuetz, A. (1971). *Collected Papers, Vol. 1, The Problem of Social Reality.* The Hague: Martinus Nijoff.

Severi, E., Heinsbroek, E., Watson, C., Catchpole, M., HPA Olympics Surveillance Work Group (2012). Infectious disease surveillance for the London 2012 Olympic and Paralympic Games. Euro Surveill. 2012:17(31):pii=20232.

Shlim, D. (2012). Perspectives: risks travelers face. *Centers for Disease Control and Prevention.* Yellowbook. Retrieved from: http://wwwnc.cdc.gov/travel/yellowbook/2012/chapter-1-introduction/perspectives-risks-travelers

Tourism Concern. (2013). Retrieved from: http://www.tourismconcern.org.uk/

United Nations Environment Program. (2001). Climate Change. Retrieved from: http://www.unep.org/climatechange/

Whatley, A., and Kozarsky, P. (2012). Introduction to travel health and the yellowbook. *Centers for Disease Control and Prevention.* Yellowbook. Retrieved from: http://wwwnc.cdc.gov/travel/yellowbook/2012/chapter-1-introduction/introduction-to-travel

World Health Organization. (2013). International Travel and Health, 2012. WHO Press, Geneva, Switzerland.

World Health Organization (2013). Water Sanitation Health. Retrieved from: http://www.who.int/water_sanitation_health/diseases/diarrhea/en/

Yang, A. (2012, Winter). Food Matters: U.S. Food Policy for the 21st Century.

GLOSSARY

The definitions in this glossary have been taken from the following resources:

Food and Drug Administration. (April 2006). Managing Food Safety: A Manual for the Voluntary Use of HACCP Principles for Operators of Food Service and Retail Establishments. OMB Control No. 0910–0578

National Restaurant Association Educational Foundation. (2008). Servsafe Coursebook (5th ed). Chicago: National Restaurant Association Solutions.

APPROVED SOURCE. An acceptable supplier to the regulatory authority based on a determination of conformity with principles, practices, and generally recognized standards that protect public health.

BACTERIA. Single-cell microorganisms that are widely distributed in nature and reproduce by fission or by forming spores.

CCP. Critical Control Point.

CLEANING. The process of removing food and other types of soil from a surface.

CONTAMINATION. The unintended presence in food of potentially harmful substances, including microorganisms, chemicals, and physical objects.

CORRECTIVE ACTION. An activity that is taken by a person whenever a critical limit is not met.

CRITICAL CONTROL POINT (CCP). An operational step in a food preparation process at which control can be applied and is essential to prevent or eliminate a hazard or reduce it to an acceptable level.

CRITICAL LIMIT. One or more prescribed parameters that must be met to ensure that a CCP effectively controls a hazard.

CROSS-CONTAMINATION. The transfer of harmful substances or disease-causing microorganisms to food by hands, food-contact surfaces, sponges, cloth towels and utensils that touch raw food, are not cleaned, and then touch ready-to-eat foods. Cross-contamination can also occur when raw food touches or drips onto cooked or ready-to-eat foods.

DANGER ZONE. The temperature range between 5°C (41°F) and 57°C (135°F) that favors the growth of pathogenic microorganisms.

FAIRS. An outdoor public event generally with a variety of activities and food.

FARMER'S MARKET. Outdoor markets originally intended as a place where farmers could sell their products. Farmer's markets today tend to sell much more than just produce.

FDA. United States Food and Drug Administration.

FESTIVALS. A short, celebratory period usually involving activities, cultural events, entertainment and food.

FOOD CODE. A model that provides governmental units with scientifically grounded rules and regulations for the retail and foodservice industries (FDA, 2009b).

FOODBORNE ILLNESS. Sickness resulting from the consumption of foods or beverages contaminated with disease-causing microorganisms, chemicals, or other harmful substances.

FOODBORNE OUTBREAK. The occurrence of two or more cases of a similar illness resulting from the ingestion of a common food.

FUNGI. A group of unicellular, multicellular, or syncytial organisms, such as molds, yeasts, and mushrooms that produce spores and feed on organic matter.

HACCP. Hazard Analysis and Critical Control Point.

HAZARD. A biological, physical, or chemical property that may cause a food to be unsafe for human consumption.

HAZARD ANALYSIS AND CRITICAL CONTROL POINT (HACCP). A prevention-based food safety system that identifies and monitors specific food safety hazards that can adversely affect the safety of food products.

HOME-BASED VENDOR (HBV). A vendor who prepares their products in an uninspected home kitchen for sale in a market or roadside stand

HOME MANUFACTURED FOODS. Specific foods, generally nonpotentially hazardous, prepared in uninspected home kitchens

MARKET MANAGER. The supervisor who assigns market space, collects fees, and maintains records in accordance with established market rules.

MICROORGANISM. A form of life that can be seen only with a microscope; including bacteria, viruses, yeast, and single-celled animals.

PARASITE. An organism that lives on or in another, usually larger, host organism in a way that harms or is of no advantage to the host.

PATHOGEN. A microorganism (bacteria, parasites, viruses, or fungi) that causes disease in humans.

pH. A measure of the acidity/alkalinity of a product. The pH scale ranges from 0 to 14.0. A pH between 7.1 and 14 is alkaline, while a pH between 0.0 and 6.9 is acidic. Foodborne microorganisms grow well in food that has a neutral to slightly acidic pH (7.5 to 4.6).

POTABLE WATER. Water that is safe to drink.

POTENTIALLY HAZARDOUS FOOD (PHF). PHF/TCS food requires T/T control for safety to limit pathogenic microorganism growth or toxin formation.

READY-TO-EAT (RTE). Those foods that are intended for immediate consumption, including foods that are in a form that is edible without any subsequent preparation.

RISK. An estimate of the likely occurrence of a hazard.

RISK FACTOR. One of the broad categories of contributing factors to foodborne illness outbreaks, as identified in the Centers for Disease Control and Prevention (CDC) Surveillance Report for 1993–1997, that directly relates to foodborne safety concerns within retail and food service establishments. The factors are Food from Unsafe Sources, Inadequate Cooking Temperatures, Improper Holding Temperatures, Contaminated Equipment, and Poor Personal Hygiene.

SANITIZING. The process of reducing the number of microorganisms on that surface to safe levels.

SPORE. A very tough, dormant form of certain bacterial cells that is very resistant to desiccation, heat, and a variety of chemical and radiation treatments that are otherwise lethal to vegetative cells. Spores may also be a form of reproduction in molds.

STANDARD OPERATING PROCEDURE (SOP). A written method of controlling a practice in accordance with predetermined specifications to obtain a desired outcome.

TCS FOOD. Food that contains moisture and protein and has a neutral or slightly acidic pH. Such food requires time/temperature control for safety to limit pathogen growth or toxin formation.

TEMPORARY FOODSERVICE ESTABLISHMENT (TFE). Food production business that operate for no more than 14 consecutive days in conjunction with a single event.

TIME AND TEMPERATURE ABUSE. Food that has been held in the danger zone beyond allowable time limits.

USDA. United States Department of Agriculture.

VIRUS. A submicroscopic parasite consisting of nucleic acid (DNA or RNA) surrounded by a protein coat, and sometimes also encased in a lipid and glycoprotein envelope. Viruses are completely dependent on a living host cell to survive and multiply, and therefore cannot multiply in or on food.

WATER ACTIVITY (A_w). Generally speaking, it is the amount of water available in the product to allow bacteria to live and grow. The quotient of the water vapor pressure of the substance, divided by the vapor pressure of pure water at the same temperature.

YEAST. Type of fungus that can cause food spoilage. Yeasts are considered beneficial in fermentation of certain foods and beverages and leavened baked products.

INDEX

D

G

H

I

Q

R

S

T

U

V

W